Advances in Molecular Microbiology, Genetics, and Bioinformatics of Multiple-Drug-Resistant Bacteria in Public Health

Advances in Molecular Microbiology, Genetics, and Bioinformatics of Multiple-Drug-Resistant Bacteria in Public Health

Guest Editors

Pedro Panzenhagen
Dália Dos Prazeres Rodrigues
Carlos Adam Conte Junior

MDPI

Basel • Beijing • Wuhan • Barcelona • Belgrade • Novi Sad • Cluj • Manchester

Guest Editors
Pedro Panzenhagen
Department of Biochemistry
Federal University of Rio de Janeiro
Rio de Janeiro
Brazil

Dália Dos Prazeres Rodrigues
Oswaldo Cruz Institute
Oswaldo Cruz Foundation - FIOCRUZ
Rio de Janeiro
Brazil

Carlos Adam Conte Junior
Department of Biochemistry
Federal University of Rio de Janeiro
Rio de Janeiro
Brazil

Editorial Office
MDPI AG
Grosspeteranlage 5
4052 Basel, Switzerland

This is a reprint of the Special Issue, published open access by the journal *Genes* (ISSN 2073-4425), freely accessible at: https://www.mdpi.com/journal/genes/special_issues/A18L2WXX3P.

For citation purposes, cite each article independently as indicated on the article page online and as indicated below:

Lastname, A.A.; Lastname, B.B. Article Title. *Journal Name* **Year**, *Volume Number*, Page Range.

ISBN 978-3-7258-8067-6 (Hbk)
ISBN 978-3-7258-8068-3 (PDF)
https://doi.org/10.3390/books978-3-7258-8068-3

Contents

About the Editors

Pedro Panzenhagen

Pedro Panzenhagen is a Visiting Professor at the Graduate Program in Biochemistry at the Institute of Chemistry, Federal University of Rio de Janeiro, Brazil. He holds degrees in Veterinary Medicine, with a PhD in Veterinary Medicine from Fluminense Federal University and a PhD in Food Science from the Federal University of Rio de Janeiro, including a research period at Washington State University, USA. His research focuses on food microbiology, molecular epidemiology, and the characterization of antimicrobial resistance and virulence in bacterial pathogens using advanced genomic and bioinformatic approaches. He has extensive experience in microbiological quality control and food safety, applying molecular techniques such as PCR, whole-genome sequencing, and comparative genomics. He has received two highly competitive postdoctoral fellowships from FAPERJ and has authored more than 50 peer-reviewed articles and book chapters, accumulating over 1000 citations. He actively collaborates with international institutions, including the FDA and research centers in the United States, and contributes to the training of graduate students in microbiology and genomics. He also serves as a Reviewer and Guest Editor for several international scientific journals.

Dália Dos Prazeres Rodrigues

Dália Dos Prazeres Rodrigues is a Public Health Researcher at the Oswaldo Cruz Foundation (Fiocruz), Brazil, and Head of the Laboratory of Enterobacteria at the Oswaldo Cruz Institute. She holds a degree in Biological Sciences, a Master's degree in Veterinary Medicine from Fluminense Federal University, and a PhD in Microbiology from the Federal University of Rio de Janeiro. She also coordinates the National Reference Laboratory for Cholera and Bacterial Enteric Infections and serves as the national coordinator of the PulseNet surveillance network. Her research focuses on molecular epidemiology, pathogenicity mechanisms, and the antimicrobial resistance of enteric bacterial pathogens circulating in the food chain, as well as their relevance to public health and animal health. She has extensive experience in bacterial subtyping, laboratory diagnosis, and genomic surveillance, integrating classical microbiological methods with molecular and whole-genome sequencing approaches. Throughout her career, she has led and participated in numerous national and international research initiatives in collaboration with organizations such as PAHO, WHO, and CDC. She is actively involved in training programs, laboratory capacity building, and public health surveillance networks, contributing significantly to the monitoring and control of foodborne diseases and antimicrobial resistance within a One Health framework.

Carlos Adam Conte Junior

Carlos Adam Conte Junior is an Associate Professor at the Department of Biochemistry, Institute of Chemistry, Federal University of Rio de Janeiro, Brazil, and a leading researcher in food science, molecular microbiology, and genomic analysis. He holds degrees in Veterinary Medicine, Food Science, and Veterinary Sciences, including a PhD with distinction from the Universidad Complutense de Madrid and postdoctoral training in international institutions such as the Karolinska Institutet and the University of California. His research integrates food safety, antimicrobial resistance, nanotechnology, and whole-genome sequencing in collaboration with global partners, including the FDA. He is a CNPq Research Fellow (level 1A) and a FAPERJ's Scientist of the State of Rio de Janeiro, and has published more than 500 scientific articles, book chapters, and patents, with over 18,000 citations and a high h-index across major databases. He has supervised dozens of postgraduate researchers and coordinated major national and international research networks and funding projects. He currently serves in leadership roles at UFRJ and national funding agencies and has been recognized among the most influential scientists in food science worldwide.

genes

MDPI

Editorial

Editorial for the Special Issue "Advances in Molecular Microbiology, Genetics, and Bioinformatics of Multiple-Drug-Resistant Bacteria in Public Health"

Pedro Panzenhagen [1,*] and Carlos Adam Conte-Junior [1,2,3,4,5,6,7]

1 Graduate Program in Biochemistry (PPGBq), Institute of Chemistry (IQ), Federal University of Rio de Janeiro (UFRJ), Cidade Universitária, Rio de Janeiro 21941-909, RJ, Brazil; conte@iq.ufrj.br
2 Center for Food Analysis (NAL), Technological Development Support Laboratory (LADETEC), Federal University of Rio de Janeiro (UFRJ), Cidade Universitária, Rio de Janeiro 21941-598, RJ, Brazil
3 Laboratory of Advanced Analysis in Biochemistry and Molecular Biology (LAABBM), Department of Biochemistry, Federal University of Rio de Janeiro (UFRJ), Cidade Universitária, Rio de Janeiro 21941-909, RJ, Brazil
4 Graduate Program in Veterinary Hygiene (PGHIGVET), Faculty of Veterinary Medicine, Fluminense Federal University (UFF), Niterói 24230-340, RJ, Brazil
5 Graduate Program in Food Science, Institute of Chemistry, Federal University of Rio de Janeiro, Rio de Janeiro 21941-909, RJ, Brazil
6 Graduate Program in Microbiology (PPGMicro), Department of General Microbiology, Institute of Microbiology Paulo de Góes (IMPG), Federal University of Rio de Janeiro, Rio de Janeiro 21941-971, RJ, Brazil
7 Analytical and Molecular Laboratory Center (CLAn), Institute of Chemistry (IQ), Federal University of Rio de Janeiro (UFRJ), Cidade Universitária, Rio de Janeiro 21941-598, RJ, Brazil
* Correspondence: pedropanzen@iq.ufrj.br

Antimicrobial resistance (AMR) has emerged as one of the most significant global threats to public health, food safety, and environmental sustainability. The rapid emergence and dissemination of multidrug-resistant (MDR) bacteria compromise the effectiveness of available antimicrobial therapies and complicate the management of infectious diseases worldwide. Advances in molecular microbiology, genomics, and bioinformatics have significantly improved the ability to investigate bacterial evolution, resistance mechanisms, and pathogen dissemination at high resolution. These approaches enable the identification of resistance determinants, virulence factors, and mobile genetic elements that contribute to the adaptation and persistence of bacterial pathogens in diverse ecological niches. This Special Issue brings together nine contributions that explore bacterial evolution, genomic diversity, resistance and virulence determinants, and potential strategies for combating MDR pathogens across clinical, environmental, and food-related contexts.

Several articles focus on the genomic characterization of bacterial pathogens associated with food production systems. Zhang et al. conducted whole-genome sequencing and comparative genomic analysis of goat-derived *Klebsiella oxytoca*, providing insights into the genomic features, resistance determinants, and potential virulence factors of this opportunistic pathogen. Their findings highlight the relevance of genomic surveillance in understanding zoonotic risks and the potential dissemination of antimicrobial resistance through livestock reservoirs [1]. Panzenhagen et al. investigated the global population structure of *Salmonella* Saintpaul using comparative genomic analyses. By examining genomic diversity and epidemiological patterns, the authors described the evolutionary dynamics and outbreak potential of this important foodborne pathogen. Their study provides valuable insights into the global dissemination of this serovar and highlights the role of genomic epidemiology in foodborne disease surveillance [2]. Another contribution explored the genomic characteristics of *Salmonella*. Using whole-genome sequencing and comparative genomic analysis, Nunes et al. identified distinct virulence gene profiles, antimicrobial resistance determinants, and plasmid content among strains of *Salmonella* Cerro and *Salmonella* Schwarzengrund isolated from food products in Brazil. These findings demonstrate the value of genomic approaches in characterizing foodborne pathogens and assessing their potential impact on public health [3].

Comparative genomic approaches are also applied to investigate the emergence of globally disseminated resistant lineages. Ahmed et al. analyzed the metabolic and genomic adaptations of fluoroquinolone-resistant *Salmonella* Kentucky ST198, a lineage associated with human infections worldwide. Their results revealed lineage-specific metabolic traits and enhanced respiratory activity that may contribute to the ecological success and global spread of this pathogen [4].

Clinical pathogens are also represented in this Special Issue. Cavka et al. investigated multidrug-resistant *Pseudomonas aeruginosa* isolates related to ventilator-associated pneumonia and ventilator-associated tracheobronchitis in an intensive care unit. Whole-genome sequencing revealed the presence of multiple resistance determinants, virulence genes, and high-risk sequence types, highlighting the complexity of antimicrobial resistance and virulence in hospital-associated infections [5].

In addition to pathogenic bacteria, some studies address the role of environmental and commensal microorganisms as reservoirs of antimicrobial resistance genes. Mikołajczuk-Szczyrba et al. examined lactic acid bacteria isolated from livestock environments and identified multiple resistance determinants through genomic analysis. The presence of plasmid-encoded resistance genes suggests that commensal bacteria may contribute to the horizontal dissemination of antimicrobial resistance across microbial communities [6]. Similarly, Zawiasa and Olejnik-Schmidt investigated antimicrobial resistance in *Listeria innocua* isolated from food products and processing environments. Although this species is generally considered non-pathogenic, the study revealed heterogeneous resistance profiles and the presence of resistance-associated genes, emphasizing the importance of monitoring non-pathogenic species as potential reservoirs of antimicrobial resistance [7].

Alternative antimicrobial strategies are also explored in this collection. Andrzejczuk et al. evaluated the antimicrobial and antibiofilm activity of *Arnica montana* extracts against *Acinetobacter baumannii*, an opportunistic pathogen known for its high levels of antibiotic resistance and ability to form persistent biofilms. The study demonstrated that plant-derived extracts can inhibit biofilm formation and downregulate genes associated with bacterial adhesion, suggesting potential applications for natural compounds in antimicrobial therapy [8].

Finally, Sánchez-Corona et al. explored the ecological relationship between heavy metal contamination and antimicrobial resistance. Their work highlights how heavy metal resistance genes frequently co-occur with antibiotic resistance determinants within mobile genetic elements, promoting co-selection and persistence of multidrug-resistant bacteria in contaminated environments. These findings reinforce the importance of considering environmental reservoirs in the global epidemiology of antimicrobial resistance [9].

Collectively, the articles presented in this collection demonstrate the importance of integrating molecular microbiology, genomic technologies, and bioinformatic tools to advance our understanding of antimicrobial resistance. These approaches provide essential insights into pathogen evolution, resistance dissemination, and potential strategies for combating MDR bacteria. We thank all authors, reviewers, editors and the editorial staff who contributed to this Special Issue and hope that this collection will serve as a valuable reference for researchers working in the fields of microbial genomics, antimicrobial resistance, and public health microbiology.

Conflicts of Interest: The authors declare no conflicts of interests.

https://doi.org/10.3390/genes17040366

References

1. Zhang, Y.; Zhang, Z.; Wang, Z.; Chen, Y.; Liao, L.; Du, L.; Gao, H.; Chen, Q.; Man, C.; Chen, S.; et al. Whole Genome Sequencing and Comparative Genomics Analysis of Goat-Derived *Klebsiella oxytoca*. *Genes* **2024**, *16*, 13. [CrossRef] [PubMed]
2. Panzenhagen, P.; Shah, D.H.; Rodrigues, D.D.P.; Conte Junior, C.A. Worldwide Population Dynamics of Salmonella Saintpaul: Outbreaks, Epidemiology, and Genome Structure. *Genes* **2025**, *16*, 254. [CrossRef] [PubMed]
3. Nunes, N.B.; Castro, V.S.; Da Cunha-Neto, A.; Carvalho, F.T.; Carvalho, R.C.T.; Figueiredo, E.E.D.S. Integrated Whole-Genome Sequencing and In Silico Characterization of *Salmonella cerro* and Schwarzengrund from Brazil. *Genes* **2025**, *16*, 880. [CrossRef] [PubMed]
4. Ahmed, J.; Soltys, R.C.; Shringi, S.; Guard, J.; Haley, B.J.; Shah, D.H. Comparative Genomics and Functional Profiling Reveal Lineage-Specific Metabolic Adaptations in Globally Emerging Fluoroquinolone-Resistant *Salmonella* Kentucky ST198. *Genes* **2025**, *16*, 1051. [CrossRef] [PubMed]
5. Cavka, M.; Kvesic Ivankovic, M.; Maravic, A.; Dzelalija, M.; Marinovic, J.; Goic-Barisic, I.; Tonkic, M.; Novak, A. Phenotypic and Whole-Genome Sequencing-Based Profiling of Antimicrobial Resistance and Virulence in *Pseudomonas aeruginosa* Isolated from Patients with Ventilator-Associated Pneumonia and Ventilator-Associated Tracheobronchitis in a Croatian Intensive Care Unit. *Genes* **2026**, *17*, 130. [CrossRef]
6. Mikołajczuk-Szczyrba, A.; Wnęk-Auguścik, K.; Średnicka, P.; Shymialevich, D.; Jaroszewska, E.; Wojtczak, A.; Zapaśnik, A.; Bucka-Kolendo, J.; Cieślak, H.; Nasiłowska, J. Genomic and Phenotypic Landscape of Antibiotic Resistance in Gut Lactic Acid Bacteria from Livestock Environments. *Genes* **2025**, *16*, 1518. [CrossRef] [PubMed]
7. Zawiasa, A.; Olejnik-Schmidt, A. Antibiotic Resistance Profiles and Genetic Determinants of *Listeria innocua* Isolated from Food Sources in Poland. *Genes* **2025**, *16*, 1455. [CrossRef] [PubMed]
8. Andrzejczuk, S.; Sozoniuk, M.; Sugier, D. Preliminary Assessment of *Arnica montana* L. Extract: Antimicrobial Activity Against *Acinetobacter baumannii* and Biofilm-Related Gene Expression Profiling. *Genes* **2025**, *16*, 1473. [CrossRef] [PubMed]
9. Sánchez-Corona, C.G.; Gonzalez-Avila, L.U.; Hernández-Cortez, C.; Rojas-Vargas, J.; Castro-Escarpulli, G.; Castelán-Sánchez, H.G. Impact of Heavy Metal and Resistance Genes on Antimicrobial Resistance: Ecological and Public Health Implications. *Genes* **2025**, *16*, 625. [CrossRef] [PubMed]

https://doi.org/10.3390/genes17040366

genes

MDPI

Article

Whole Genome Sequencing and Comparative Genomics Analysis of Goat-Derived *Klebsiella oxytoca*

Yu Zhang [1], Zhenxing Zhang [1], Ziying Wang [1], Yimei Chen [1], Lianjie Liao [1], Li Du [1], Hongyan Gao [1], Qiaoling Chen [1], Churiga Man [1], Si Chen [2,*] and Fengyang Wang [1,*]

[1] Hainan Key Lab of Tropical Animal Reproduction, Breeding and Epidemic Disease Research, Animal Genetic Engineering Key Lab of Haikou, Hainan University, Haikou 570228, China

[2] School of Tropical Agriculture and Forestry, Hainan University, Haikou 570228, China

* Correspondence: chensi@hainanu.edu.cn (S.C.); fywang68@163.com (F.W.)

Abstract: Background: This research aims to enhance the genomic database of *Klebsiella oxytoca* by identifying virulence genes through the whole genome sequencing and comparative analysis of a goat-derived *K. oxytoca* (KOHN1) strain, while clarifying the relationship between its genetic evolution and virulence, ultimately providing a theoretical foundation for clinical prevention and diagnosis. **Methods:** Third-generation Oxford Nanopore Technologies (ONT) sequencing and second-generation Illumina sequencing were used to sequence the strain and analyze the database annotations. Screening for 10 virulence genes was conducted using PCR. Comparative genomic analyses of the strain KOHN1 with four human-derived *K. oxytoca* model strains were performed using collinearity analysis, taxonomy classification through ANI analysis, and gene function family analysis. **Results:** The genome size of the KOHN1 strain was 5,817,806 bp, and the GC content was 55.14%. It contained 5227 predicted coding genes, including 25 rRNA genes, 85 tRNA genes, and 53 sRNA genes. A total of 14 type VI secretion system effector proteins and 146 virulence factor-related genes were annotated. Additionally, eight virulence genes—*fimA*, *fimH*, *entB*, *mrkD*, *clpV*, *rmpA*, *vgrG*, and *hcp*—were detected through PCR identification. The strain has 448 drug resistance genes, mainly against β-lactams and fosfomycins. Comparative genomic analysis indicated that its closest relation is the human isolate ASM338647. **Conclusions:** In this study, the whole genome sequence of a goat-derived *K. oxytoca* (KOHN1) strain was obtained, revealing its evolutionary relationship with domestic and foreign isolates and providing a reference for future studies on the mechanisms of antimicrobial resistance and the pathogenicity of *K. oxytoca*.

Keywords: *Klebsiella oxytoca*; genome sequencing; virulence-related genes; comparative genomics analysis

1. Introduction

K. oxytoca belongs to the genus *Klebsiella* of Enterobacteriaceae. It is a Gram-negative opportunistic pathogen that can cause zoonotic infectious diseases and is the second most important clinical species of *Klebsiella* after *K. pneumoniae* [1]. *K. oxytoca*, a zoonotic pathogen, is capable of causing multiple types of infections, such as antibiotic-associated hemorrhagic colitis (AAHC), urinary tract infections, and bacteremia [2]. It mainly infects the intestines by producing enterotoxins during the infant period of animals, resulting in complications such as bacteremia [3]. These complications may ultimately lead to significant morbidity and mortality in both humans and animals [4,5]. At present, the potential transmission of this pathogen and its associated drug resistance genes from animals to humans is a pressing public health concern [6].

In recent years, the misuse of antibiotics has led to an increase in antibiotic resistance (AMR), particularly to β-lactams [7], fosfomycin, and fluoroquinolones [8]. This has resulted in treatment failures in many cases and the emergence of multi-drug-resistant strains, which have serious economic implications for goat farms. As a multi-drug-resistant bacterium, studies indicate that *K. oxytoca* may exhibit greater resistance than the more commonly encountered *K. pneumoniae*. Furthermore, infections caused by *K. oxytoca* isolates demonstrate higher invasiveness compared to those caused by *K. pneumoniae* isolates [9]. Therefore, although the prevalence of *K. oxytoca* is low, it causes more serious invasive infections and is more harmful to humans and animals. This finding may be related to the increased frequency of specific virulence factors.

At present, the main virulence genes related to *K. oxytoca* include *wabG*, *fimA*, *fimH*, *mrkD*, *ureA*, and others [10]. Lipopolysaccharide-related genes, such as *wabG*, are crucial virulence factors in *K. pneumoniae* that contribute to the development of invasive infections [11]. Adhesin and fimbrial genes of type 1 (*fim*) and type 3 (*mrk*) are also important factors in the pathogenicity of *Klebsiella*. Currently, research into the infection mechanisms of *K. oxytoca* remains in the exploratory phase. It is believed that the early immune response in the respiratory tract to *K. oxytoca* exhibits specific characteristics and corresponding immune reactions, which may be related to a microbial imbalance [12]. *K. oxytoca* produces adhesins that enhance interactions between the bacteria and host cells, as well as with other microbial species [13]. In addition, the study found that *K. oxytoca* was positively correlated with the *in vitro* growth of *Clostridium* difficile and negatively correlated with *Staphylococcus aureus*. The interaction of multiple pathogens during infection may be the cause of the complex pathogenesis of *K. oxytoca* [14]. Therefore, the infection process of *K. oxytoca* is influenced by multiple mechanisms, and the specific pathogenic mechanisms still require further study.

Genome research is one of the most important directions in the field of modern microbiology, significantly contributing to our understanding of bacterial characteristics and the management of related diseases. Whole genome sequencing (WGS) technology, which integrates third-generation ONT [15] and second-generation Illumina methods, allows for the concurrent detection of multiple gene mutations with a high throughput and reduced time consumption [16]. This approach leverages the advantages of third-generation sequencing, characterized by long read lengths that effectively address assembly challenges such as repetitive sequences and structural variations, alongside the high accuracy of second-generation sequencing, which produces high-quality genomes through assembly without relying on a reference genome. These WGS methods offer comprehensive biological insights across entire genomes and can concurrently identify numerous pathogens in clinical samples [17]. The data generated through WGS can be processed with various bioinformatics tools [18], allowing for the assessment of the genome quality and the identification of the species and genotypes of infecting strains, as well as predictions regarding drug sensitivity, resistance, and epidemiological studies [19]. This information is of great value for early diagnosis, effective treatment, and epidemiological investigation. At present, there have been reports on the whole genome study of *K. oxytoca*, such as Campos [20], using Illumina and nanopore platforms for whole genome sequencing and species identification. Dickman [21] used Illumina Mi Seq for sequencing and compared this with the classification results obtained based on phenotypic techniques to determine the type of strain as *K. oxytoca*. Lo [22] used the Illumina Next Seq platform for sequencing and used the genomic results to perform an epidemiological analysis on the *K. oxytoca* collected in the experiment to evaluate its benefits for monitoring and understanding the outbreak of Enterobacteriaceae in the community. At present, the second-generation and third-generation mixed sequencing methods have not been reported in research. The related sequencing technology is mostly used for other common bacterial research areas such as research on *Mycoplasma* [23], *Salmonella* [24], and *Escherichia coli* [25] and tumor studies such as those on breast cancer [26], rectal cancer [27], etc.

The purpose of this study was to use ONT and Illumina sequencing techniques to assemble the genomic sequence of *K. oxytoca* and perform bioinformatics analyses to reveal its genomic structure and functional annotation. This study is of great significance for understanding the metabolic pathways of *K. oxytoca* and its mechanism of adapting to the environment and for analyzing its virulence genes and drug resistance genes, laying a foundation for future research. Furthermore, comparing the genome of KOHN1 with those of other distinct strains can aid in identifying significant drug targets and offer valuable insights for preventing disease outbreaks.

2. Materials and Methods

2.1. Strains

The KOHN1 strain of *K. oxytoca* was successfully isolated from the lung disease material of a Hainan black goat in this experiment. It was identified through colony-staining microscopy, a biochemical test, and 16S rRNA sequencing. A volume of 800 μL of goat-derived *K. oxytoca* solution was stored in 200 μL of 75% glycerol and stored in our laboratory's ultra-low-temperature refrigerator.

2.2. Whole Genome Sequencing

Genomic DNA was extracted using the CTAB method [28]. After a quantitative quality inspection through NanoDrop microspectrophotometer detection (Nanodrop 2000, Thermo Fisher Scientific, Waltham, MA, USA), agarose gel electrophoresis (DYY-6C, Beijing Liuyi Instrument Factory, Beijing, China) and the use of a Qubit® 2.0 Fluorometer (Life Technologies, Carlsbad, CA, USA), the ONT system and Illumina sequencing system were used for whole genome sequencing.

2.3. ONT and Illumina Sequencing

Following the standard protocol from ONT, the genomic DNA of the KOHN1 strain was fragmented using G-tubes (Covaris, Woburn, MA, USA) and subsequently end-repaired with the NEBNext FFPE DNA Repair kit (M6630, NEB, Ipswich, MA, USA) along with the NEBNext End Repair/dA-tailing Module reagents (E7546, NEB, Ipswich, MA, USA). Then, the fragments were ligated to adaptors to construct the sequencing library using NEBNext Quick Ligation Module reagents (E6056, NEB, Ipswich, MA, USA). After purification using the AMPure XP system (Beckman Coulter, Brea, CA, USA), library sequencing was performed on the long-read sequencing platform Oxford Nanopore PromethION (Oxford Nanopore Technologies, Oxford, UK) according to standard protocols. After that, the Illumina platform was used for calibration. Qualified genomic DNA was initially subjected to random sonication, followed by end repair, A-tailing, and adaptor ligation using the NEBNext® MLtra™ DNA Library Prep Kit for Illumina (NEB, Ipswich, MA, USA) in accordance with the preparation protocol. DNA fragments ranging from 300 to 400 bp were then enriched through PCR. Finally, the PCR products were purified with the AMPure XP system (Beckman Coulter, Brea, CA, USA), and the libraries were assessed for size distribution using the 2100 Bioanalyzer (Agilent, Santa Clara, CA, USA) and quantified via real-time PCR. Genome sequencing was carried out on the Illumina NovaSeq 6000 sequencer employing paired-end technology (PE 150).

2.4. Genome Assembly

The sequence data obtained from ONT and Illumina sequencing were influenced by the characteristics of ONT sequencing, particularly its longer read lengths, which enhance the overall quality of genome assembly. The assembly process employs third-generation sequencing data, followed by the application of second-generation data through Fastp (version 0.20.0, parameter: -q 20 -u 50 -n 15 -l 50) to refine the assembly outcomes. After filtering the initial ONT and Illumina sequencing data, we acquired clean data. We first used Flye (version 2.9, parameter: -g 6 m -min_overlap 2000 -iterations 3) to assemble the filtered readings from scratch to produce a gapless overlap group. Pilon software (version 1.23, parameter: -mindepth 2 -mingap 10) was employed to refine the data, enhancing the accuracy and reliability of the assembly results. The initial ONT and Illumina sequencing data underwent filtering to generate clean data.

2.5. Genome Annotation and Bioinformatics Analysis

The bioinformatics analysis of the whole genome sequence included the identification of clustered regularly interspaced short palindromic repeats (CRISPRs), transposons, non-coding RNAs, repetitive sequences, coding sequences, genomic islands (GIs), and precursors. The identification of CRISPRs and transposons was carried out using CRISPRfinder (version 4.2.17) and TransposonPSI (version 20100822). The tRNA gene and rRNA gene were predicted using tRNAscan (version 1.3.1, parameter: -B) and RNAmmer (version 1.2, parameter: -S bac -m tsu, lsu, ssu), respectively. Cmscan (version 1.1.2, with the parameters --cpu 4 -E 1×10^{-5}) was employed to compare the predicted sSRNA against the Rfam database. The dispersed repeat sequences were analyzed using RepeatMasker (version 4.0.5, parameter: -nolow -no_is -norna), and the coding sequences were identified utilizing the NCBI Prokaryotic Genome Annotation pipeline. TRF (version 4.09, parameter: 2 7 7 80 10 50 500 -m -d) was used to analyze tandem repeats. Gene islands were identified using IslandPath-DIMOB (version 1.0), while Phage_finder (version 2.0) was utilized to locate prophages.

2.6. Gene Function Annotation

Gene function analysis was performed using the databases of the Cluster of Orthologous Groups of proteins (COG) [29], Gene Ontology (GO) [30], the Kyoto Encyclopedia of Genes and Genomes (KEGG) [31], the Non-Redundant Protein Database (NR) [32], the Carbohydrate-Active enZYmes Database (CAZy), and TNSSs (Type N secretion systems, Types I–VII) [33]. The CARD (Comprehensive Antibiotic Research Database) [34] and the VFDB (Virulence Factors of Pathogenic Bacteria) [35] databases were used to predict and analyze the drug resistance genes and virulence genes in the genome. Details are shown in Table S1 and Figure S1.

2.7. Genome Visualization Analysis

For the assembled genome sequence, we integrated the predicted coding gene results and utilized Circos software (version 0.69-9) to create a circular map of the entire genome for the KOHN1 strain, so as to fully display the characteristics of the genome, such as the COG, KEGG, and GO functional classifications of the gene, the ncRNA prediction results, the distribution of the gene on the sense and antisense chains, the GC content, and so on. Finally, through the comprehensive display of the circle diagram, we could easily understand some basic characteristics of the KOHN1 strain genome.

2.8. Colinearity Analysis and SNP, InDels, and SV Statistics

The genomes of four human-derived model strains of *K. oxytoca* were selected (ATCC 13182, ASM381292, ASM2573217, and ASM3386047; see Table S2) as the target genomes. The KOHN1 genome served as the reference for conducting a collinearity analysis with the target genome. Firstly, Mummer (version 3.1, parameter: nucmer --maxmatch -c 100 -b 200 -l 50, 'delta-filter -m -i 80 -l 100) was used to compare ATTC 13182, ASM381292, ASM2573217, ASM3386047, and KOHN1 to determine the large-scale collinearity between the genomes and to detect and analyze single nucleotide polymorphisms (SNPs). Subsequently, Syri (version 1.4, parameters: delta-filter -m -i 80 -l 100) was employed to assess the local positional arrangement of the samples and to check and analyze structural variations (SVs).

2.9. ANI, Gene Family Analysis, and Phylogenomic Tree Construction

Following the collinearity analysis, we employed PYANI (version 0.6.7, parameter: -m ANIb -g --workers 4) to compute the Average Nucleotide Identity (ANI) between the KOHN1 genome and the target genome. Subsequently, gene family analysis was performed on the target and reference genomes using the bidirectional best hit (BBH) criterion. This criterion stipulates that 80% of the shortest protein sequences exhibit a minimum of 40% amino acid similarity. Diamond (version 2.0.7, parameter: blastp -k 0 -f 6 -p 8 -e 1×10^{-7} --salltitles --more-sensitive) and Ortho MCL (version 1.4, parameter: mcl -I

1.5) were employed to compare and cluster the amino acid sequences, thereby obtaining homologous gene clusters and species distribution. Finally, single-copy gene families from both the KOHN1 and reference genome were selected to construct a phylogenetic tree using IQ-TREE (version 1.6.3, parameter: -nt 8 -m GTR+I+G -bb 1000 -st DNA) to study the evolutionary relationships among species, employing the maximum likelihood method to explore the phylogenetic relationships among strains.

2.10. Virulence Gene Detection

According to the reference for identifying the virulence gene primers of *K. oxytoca* (Table 1 and Table S3), related primers were synthesized. Using KOHN1 genomic DNA as a template, 10 virulence genes were detected through polymerase chain reaction (PCR) amplification. After the PCR (ETC811, Dongsheng Xingye Scientific Instrument Co., Ltd., Suzhou, China) amplification reaction, 1% agarose gel electrophoresis (Biowest Agarose; supplier: Beijing Mengyimei Business Center; brand: Biowest, Nuaillé, France) was used to determine the results.

Table 1. Virulence gene primer information.

Virulence Gene	Primer Sequences (5′-3′)	Annealing Temperature (°C)	Reference
vgrG	F: GCATCTTCCAACTCAACAC R: GTACACCAGCCCTTCTTC	55	Zhang A et al., 2021 [36]
fimA	F: GCACCGCGATTGACAGC R: CGAAGGTTGCGCCATCCAG	61	Omer FH et al., 2022 [37]
fimH	F: GCTCTGGCCGATACCACGACGG R: GCGAAGTAACGTGCCTGGAACGG	55	Brisse S et al., 2009 [16]
mrkD	F: TAT TGGCTTAATGGCGCTGG R: TAATCGTACGTCAGGTTAAAGACC	60	Brisse S et al., 2009 [16]
hcp	F: TGCTGAACGTGTTGAACATT R: TAACAGCACCTTTAGCAGTA	58	Hu YY et al., 2023 [38]
clpV	F: CGAGGTCATTAAAGTTGCCCAATC R: CGCCGAAACCGACATACCCT	60	Luo G et al., 2019 [39]
cf29a	F: GACTCTGATTGCACTGGCTGTG R: GTTATAAGTTACTGCCACGTTC	51	Brisse S et al., 2009 [16]
wabG	F: CGGACTGGCAGATCCATATC R: ACCATCGGCCATTTGATAGA	58	Lin WH et al., 2010 [40]
entB	F: ATTTCCTCAACTTCTGGGGC R: AGCATCGGTGGCGGTGGTCA	56	Candan ED et al., 2015 [41]
rmpA	F: CGTATGAAGGCTCGATGGAT R: TGTGCACCATTTTTCATCAG	60	Turton GF et al., 2010 [42]

3. Results

3.1. General Characteristics of KOHN1 Genome

As shown in Table 2, the whole genome sequence of KOHN1 was 5,817,806 bp and the GC content was 55.14%. A total of 5227 genes were encoded by all genes. The total length of all genes was 5,087,988 pb, with an average length of 32,936.12 bp, and the total length of the coding region accounted for 87.46% of the whole genome. The number of copies of tandem repeats was 106, and the total length was 26,896 bp, accounting for 0.46% of the total length of the genome. There were 85 tRNA genes, 25 rRNA genes (including 8 23S rRNA genes, 8 16S rRNA genes, 9 5S rRNA genes), and 53 sRNA genes. There were eight gene islands and one prophage with a total length of 59,576 bp. The circular map of the KOHN1 genome was generated using Circos software, as illustrated in Figure 1 below.

Table 2. General characteristics of the KOHN1 genome.

Item	Number	Item	Number
Genome size (bp)	5,817,806	Number of rRNA genes	25
Total length of gene (pb)	5,087,988	Number of tRNA genes	85
Genome GC content (3%)	55.14	Number of sRNA genes	53
Average length of gene (bp)	32,936.12	Number of DNA elements	13
Number of genes	5227	Number of short interspersed elements	33
Number of gene islands	8	Number of long interspersed repeated sequences	27
Total length of coding region (%)	87.46	Total potential CRISPRs	7
Total length of gene island (bp)	263,489	Number of prophages	1
Number of tandem repeats	106	Total length of prophages (pb)	59,576

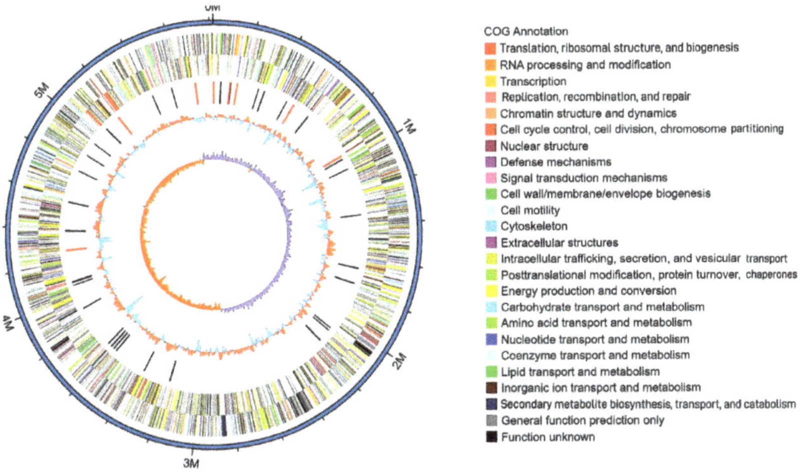

Figure 1. Whole genome map of KOHN1.The outermost circle illustrates the positional coordinates of the genome sequences, while the subsequent inner circles display various annotations and features, including the coding genes, COG database annotations, KEGG database annotations, GO database annotations, GC content, GC skew, and ncRNA. The GC skew was used to identify the starting and ending points on circular chromosomes, with both the window and step sizes configured to 1000. The outermost circle is the position coordinate of the genome sequence. From the outer circle to the inner circle, the order is the following: the positive chain gene, negative chain gene, ncRNA (black represents tRNA; red represents rRNA), GC content (red represents greater than the mean; blue represents less than the mean), and the GC skew (GC offset, used to measure the relative content of G and C, used to mark the starting point and end point in the circular chromosome; gC skew = $(G - C)/(G + C)$; purple means greater than 0; orange means less than 0).

3.2. Gene Function Annotation Results

In the GO functional analysis (Figure 2A), the cellular processes and single-organism processes enriched 3969 and 3658 genes, respectively. In the COG analysis (Figure 2B), 4316 protein-coding genes were annotated into 25 categories from J to S, among which the pathway for general function prediction was the most enriched with 670 genes. Carbohydrate transport and metabolism enriched 638 genes, while amino acid transport and metabolism enriched 636 genes. In the NR analysis (Figure 2C), 5210 genes were annotated through a comparison with the NR database, with the highest number of genes annotated to *K. oxytoca*, totaling 4546. In the KEGG analysis (Figure 2D), the pathway with the highest enrichment was the Global and Overview maps at the metabolic level, which enriched 1163 genes. A total of 4102 genes were annotated in the four gene libraries (Figure S1).

The analysis indicated that the pathways with the highest gene enrichment in the COG, GO, and KEGG databases were predominantly metabolic processes. This underscores the significance of metabolism in the development of bacterial life, as it is essential for meeting the nutritional requirements necessary for growth and replication.

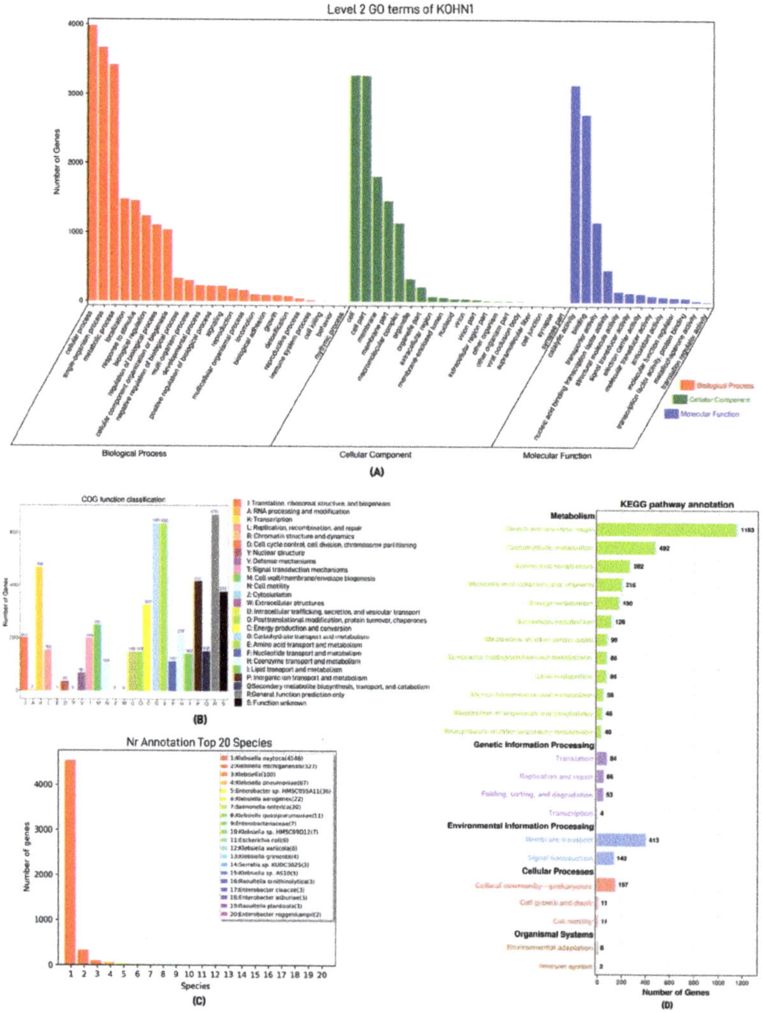

Figure 2. Analysis results of four different databases. (**A**) GO gene library annotation results. The x-axis illustrates the GO functional classification from the sample annotation, while the left y-axis shows the count of genes in the annotation. (**B**) NR gene library annotation results. The x-axis denotes the species ID, while the y-axis indicates the quantity of genes in the annotation. (**C**) COG gene library annotation results. The x-axis illustrates the functional classifications, shown in various colors, while the y-axis indicates the count of predicted genes associated with each classification. (**D**) KEGG gene library annotation results. The x-axis indicates the gene count, while the y-axis reflects the functional classifications from the KEGG database. The level 1 classifications are displayed in black font, and the level 2 classifications are shown in colored font. The figures on the bar graph represent the quantity of genes included in the annotation.

3.3. Results of CAZy Database Annotation

A total of 859 genes in the KOHN1 strain were annotated to the CAZy database (Figure 3), including 89 genes related to Carbohydrate-Binding Modules (CBMs), 47 genes related to Carbohydrate Esterases (CEs), 380 genes related to Glycoside Hydrolases (GHs), 326 genes related to Glycosyl Transferases (GTs), 11 genes related to Polysaccharide Lyases (PLs), and 6 genes related to Auxiliary Activities (AAs). Among the carbohydrate-active enzymes, Glycoside Hydrolases (GHs) represented the largest proportion, followed by GTs.

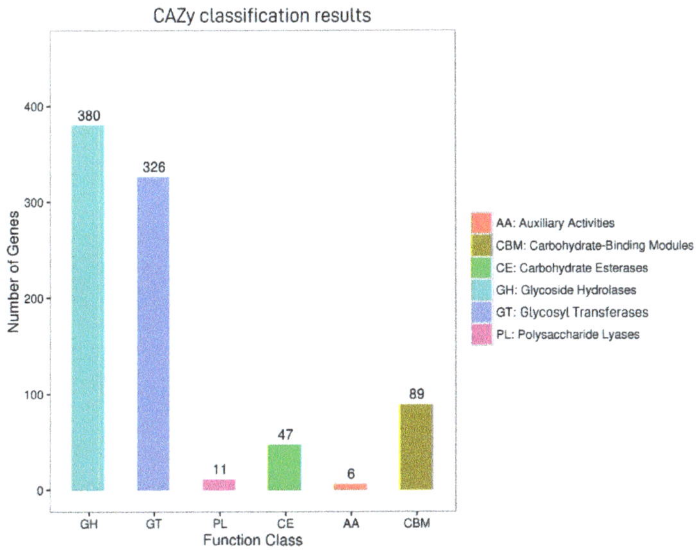

Figure 3. CAZy annotation results of KOHN1 gene. The x-axis denotes the classification of enzymes, while the y-axis indicates the count of genes that fall under each classification.

3.4. Results of Drug Resistance Gene Annotation

The CARD database was used to annotate all the genes of the KOHN1 strain. It was found that 448 genes in the KOHN1 strain were annotated to the CARD database. Seven genes were deleted under the condition of an 80% matching degree (Table S4). These drug resistance genes were mainly produced by encoding various enzymes and enhancing efflux. The analysis found that the strain contained one or more drug resistance genes related to β-lactams, tetracyclines, fluoroquinolones, cephalosporins, glycylcyclines, and fosfomycins corresponding to the multi-drug resistance phenotype of the KOHN1 strain.

3.5. The Results of the TNSS and Virulence Gene Analysis

In the TNSS analysis (Table S5), the KOHN1 strain harbored 14 T6SS effector proteins. Among these, *vgrG* and *hcp* serve as markers for T6SS activation, while *clpV* is a candidate key gene involved in the T6SS infection process. As a significant virulence gene in KOHN1, *clpV* also tested positive in subsequent PCR assays. In the comparative analysis with the VFDB database, a total of 146 genes were annotated, with the majority of annotations pertaining to virulence factor secretion systems and iron uptake (Table S6). Ten virulence genes were screened using VFDB analysis and detected through PCR. The results are shown below (Figure 4A). Eight virulence genes were detected, *fimA*, *fimH*, *entB*, *mrkD*, *rmpA*, *vgrG*, *clpV*, and *hcp*, while the PCR results for the *cf29a* and *wabG* genes were negative. Figure 4B shows the VFDB database annotation results of related virulence genes in five different *K. oxytoca* strains. All five strains carried the aforementioned eight genes; however, the annotation results for the *cps I* gene showed variations, with only the reference strain ATCC 13182 testing positive.

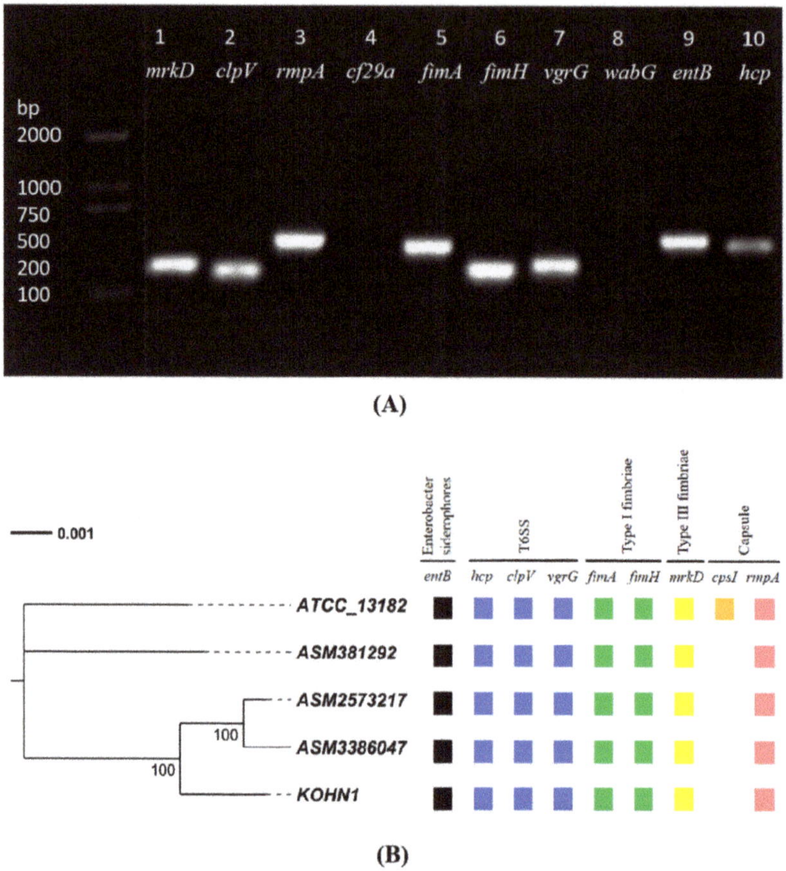

Figure 4. Virulence gene results. (**A**) PCR results of virulence genes of KOHN1 strain. M: DL2000 DNA Marker; 1: *mrkD*; 2: *clpV*; 3: *rmpA*; 4: *cf29a*; 5: *fimA*; 6: *fimH*; 7: *vgrG*; 8: *wabG*; 9: *entB*; 10: *hcp*. (**B**) Distribution of virulence factors of five strains.

3.6. The Collinearity Analysis Results of the Five Strains

The collinearity results were statistically analyzed. On the basis of these results, a graph of parallel collinear results was drawn (Figure 5). According to the diagram, the alignment sequence region of the four target genes of KOHN1 accounts for more than 90% of the whole genome, indicating that the evolutionary distance between the five strains is relatively close. The strain with the closest evolutionary distance to KOHN1 is ASM3386047, and the strain with the farthest evolutionary distance from KOHN1 is ATCC 13182.

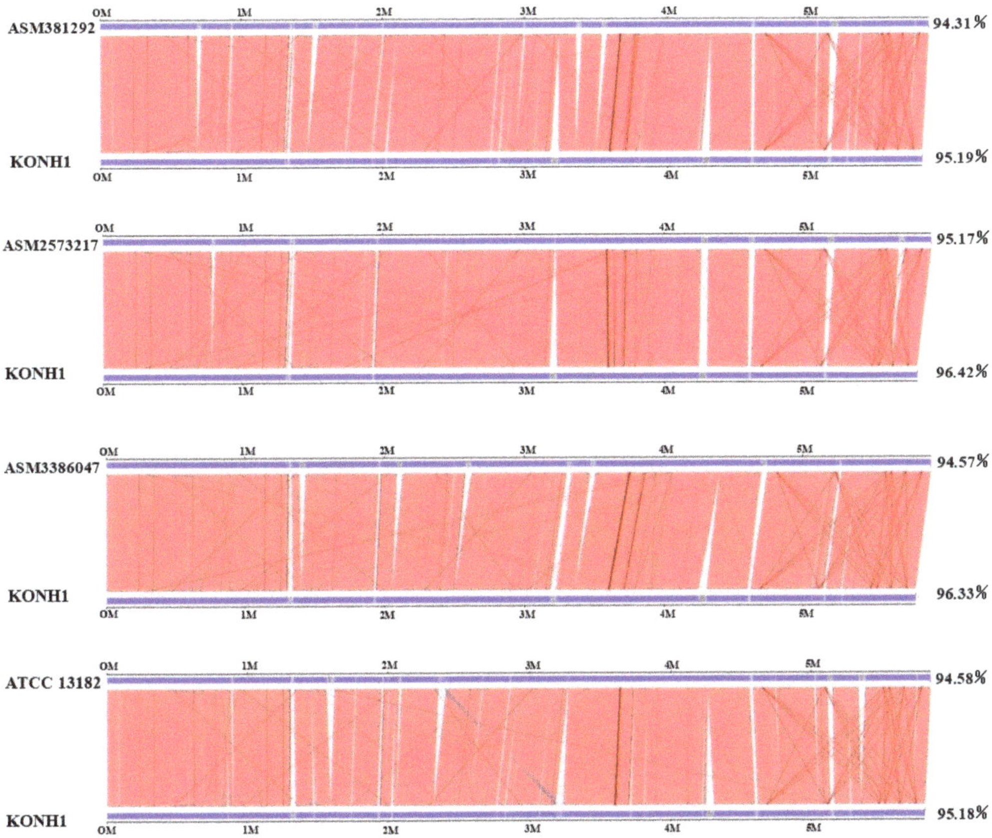

Figure 5. Parallel collinear result plot. The top axis illustrates the genome of the target species, while the bottom axis depicts the KOHN1 genome. The red line indicates the positive alignment of the corresponding regions, whereas the blue line represents the reverse alignment of those regions.

3.7. ANI Analysis and Phylogenetic Tree Analysis

Based on the collinear alignment, the ANI value was determined to create a heat map (Figure 6A). The results showed that the ANI value of ASM2573217 was the largest compared with KOHN1. This indicates that KOHN1 is closely related to ASM2573217. At the same time, the ANI value between ASM3386047 and ASM2573217 was the highest, indicating that there is a close relationship between them. Through the above diagram, we know that ASM2573217 has the closest relationship with KOHN1. The genetic distance between ASM3386047 and ASM2573217 was the closest. Due to the high similarity among the genomes of these five *K. oxytoca* strains, the genetic relationships between them cannot be accurately assessed through the analysis mentioned above.

Therefore, we further evaluated the genetic evolutionary distances between the strains using a phylogenetic tree. According to the results from the phylogenetic tree (Figure 6B), the genetic distance between KOHN1 and ASM3386047 was the smallest, followed closely by ASM2573217.

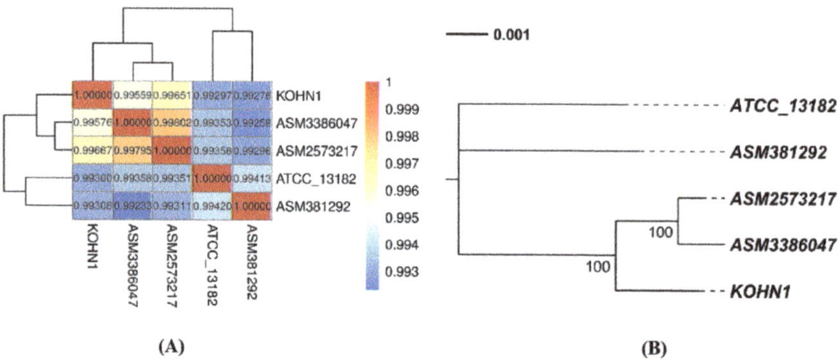

Figure 6. (**A**) ANI analysis heat map. (**B**) Phylogenetic tree based on characteristic genes. The phylogenetic tree depicts the evolutionary links and structural affiliations between different species. Each species is located at the tree's tips, while the intersection points signify common ancestors. The length of the branches reflects the relative genetic distance among the species, and the numbers along the branches indicate their reliability.

3.8. Gene Family Analysis and Data Results of SNPs, InDels, and SV

The whole genomes of five strains of *K. oxytoca* comprised a total of 26,570 genes, with the core genome consisting of 4822 genes. These 4822 genes are responsible for encoding the essential biological functions and key phenotypic traits of *K. oxytoca* (Table 3 and Figure 7). Among them, KOHN1 contains 113 specific genes, such as *PGB18_19815*, *PGB18_19945*, *PGB18_02255*, and so on.

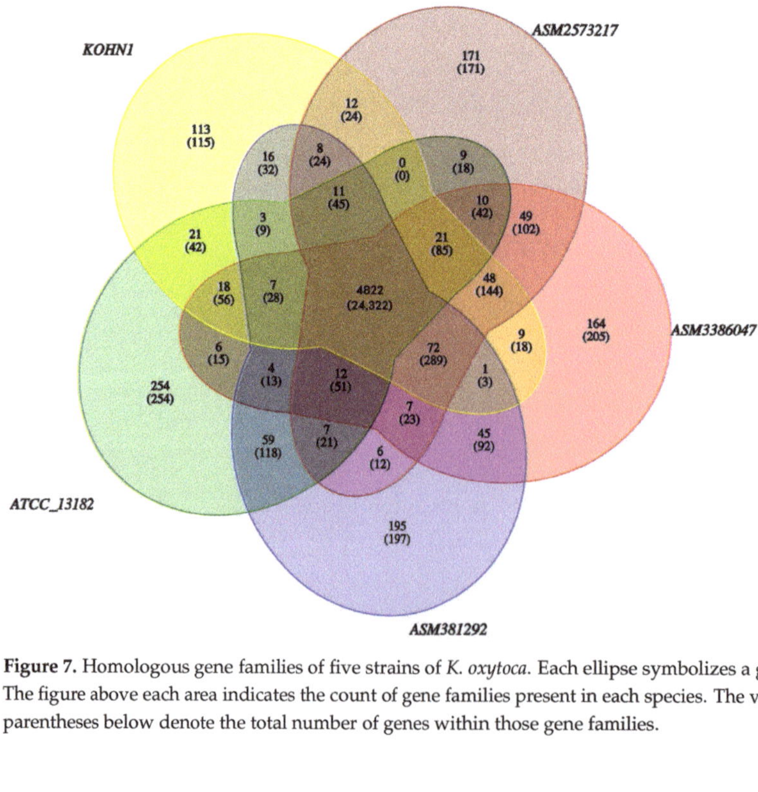

Figure 7. Homologous gene families of five strains of *K. oxytoca*. Each ellipse symbolizes a genome. The figure above each area indicates the count of gene families present in each species. The values in parentheses below denote the total number of genes within those gene families.

Table 3. Gene family analysis of five strains of *K. oxytoca*.

Strain Name	Total Genes Number	Number of Genes in the Family	Number of Specific Families	Unclustered Genes Number	Families
ATTC 13182	5311	5057	254	254	5010
ASM381292	5332	5161	171	171	5094
ASM2573217	5388	5255	164	133	5162
ASM3386047	5312	5118	195	194	5081
KOHN1	5227	5116	113	111	5071

Finally, we analyzed the quantitative relationship between SNPs, Insertions, Deletions, and SVs between KOHN1 and the target strain.

As shown in Table S7 and Figure S2, the genomes of KOHN1 and ASM2573217 exhibit the fewest SNPs, Insertions, and Deletions. Additionally, the number of SVs between KOHN1 and ASM3386047 is greater than that between KOHN1 and ASM2573217. Therefore, compared to KOHN1, ASM2573217 displays the least genomic variation, followed by ASM3386047.

4. Discussion

As of September 2024, the NCBI database included 1072 whole genome sequences of *K. oxytoca*, most of which were of low quality. Most of the early sequences are of low quality, at the level of Contingent or Scaffold. There are many species sources of *K. oxytoca* that have been included, but there is no goat source. In this study, the complete genome information of a goat-derived *K. oxytoca* was obtained for the first time, and the genome database information of *K. oxytoca* was expanded.

The whole genome size of KOHN1 was 5,817,806 bp, and the GC content was 55.14%. There were 2389 protein-coding genes in the whole genome of KOHN1 strain, and the total coding length of all genes was 5,087,988 bp. Simultaneously, seven potential CRISPRs were identified in *K. oxytoca* on the initial occasion. CRISPRs serve as a natural immune system in bacteria, allowing them to protect against phage sequences and foreign plasmids by effectively silencing these invading genetic elements [43]. In addition, compared with the average readings of the strains measured in Lo's study [22], the genome size of this strain is higher than the average, which means that the strain may have a more complex structure and function of the organism, and has great potential in subsequent research.

CARD is a database based on antibiotic resistance ontology (ARO), which correlates information such as antibiotic modules, drug resistance mechanisms, and gene mutations. It provides data on drug resistance-related gene names and types of resistant antibiotics. In a previously reported antimicrobial susceptibility monitoring project for Europe [44], it was observed that the proportion of AMR in *Klebsiella* isolated from acute mastitis was low, with the highest resistance observed against tetracycline antibiotics. In this experiment, through the analysis of the CARD database combined with the previous experimental results in the laboratory, it was found that the strains of *K. oxytoca* isolated from goats were resistant to β-lactams, tetracyclines, fluoroquinolones, cephalosporins, aminoglycosides, fosfomycins, and other related drugs. The most common was the resistance to β-lactams. Differently from the report, the resistance of the goat-derived strain to cephalosporins and fluoroquinolones was observed in this study. It is speculated that it may be a multi-drug-resistant bacterium with high drug resistance.

The results of CAZy database analysis showed that the main enzymes of KOHN1 in the metabolic process were GH and GT. GH can decompose the glycosidic bonds between carbohydrate molecules and plays an important role in the hydrolysis and synthesis of polysaccharides in organisms. GT enzymes facilitate the attachment of activated sugars to various receptor molecules, including proteins, nucleic acids, and oligosaccharides, resulting in glycosylation products that serve numerous biological functions. In the study conducted by Vuillemin [45], it was noted that the GH family has emerged as a significant area of research in recent years, demonstrating a widespread presence in symbiotic intestinal bacteria and potentially participating in the degradation of animal glycans. In humans, it catalyzes the degradation of lactic acid-n-tetrasaccharide (LNT), an important component of human milk oligosaccharides that supports the development of healthy infant intestinal microbiota. We reasonably infer that this may be related to the pathogenesis of intestinal diseases caused by *K. oxytoca*.

In most studies on the number of *Klebsiella*, the T6SS is usually annotated [46]. In this study, the annotation results of the TNSS showed that the most important secretion system in this strain was also the T6SS. Research has demonstrated that the T3SS, T4SS, and T6SS are crucial for the pathogenicity of Gram-negative bacteria [47]. For *K. oxytoca*, some studies have mentioned that the bacteria will be killed in the host intestinal tract in a T6SS-dependent manner [48]. Therefore, this study concentrated on the T6SS. Research has shown that the T6SS, a transmembrane structure primarily found in Gram-negative bacteria, facilitates the direct transport of effectors from the cytoplasm to target cells or the extracellular environment through transmembrane channels. Pathogens exploit the T6SS to secrete effectors into extracellular spaces or host cells, thereby influencing critical processes such as immune responses and cell death, which can ultimately result in pathological reactions. In addition to the pathogenicity of bacteria, the T6SS also enhances their ability to adapt to various environments and engage in both intraspecific and interspecific competition. This is because the T6SS secretes some toxic proteins that can inhibit the growth of competitive bacteria and even kill them [49]. In this study, the main secreted protein of the T6SS was also an important virulence gene of KOHN1.

The VFDB is a comprehensive resource for studying the pathogenic factors of bacteria, including the pathogenic strains of Chlamydia and Mycoplasma. It provides essential evidence for investigating the mechanisms by which these pathogens invade host organisms. Combined with comparative genomics and other methods, the composition and genomic distribution of the virulence factors of different pathogenic bacteria were analyzed. The virulence factors used in the experiment have an important influence on *K. oxytoca*, in which *fimA* and *fimH* mainly encode type 1 fimbriae, which primarily function to enhance adhesion to host cells and tissues. These fimbriae have the ability to mediate adhesion, which is inhibited by soluble mannose-containing compounds *in vitro*, so they are often referred to as mannose-sensitive fimbriae [50]. Like *fim*, *mrkD* mediates adhesion by encoding type 3 biofilms [51]. Studies have shown that *fimA* and *mrkD* may regulate the formation of bacterial biofilms, which may be one of the mechanisms affecting the pathogenicity of *K. oxytoca*. As mentioned in Yang's study [2], only a limited number of these genes have been experimentally investigated to assess their relationship with the virulence of *K. oxytoca*. In this study, PCR detection was used to determine the virulence relationship between these genes and *K. oxytoca*, which is an important virulence gene of the bacteria. *VgrG*, *hcp*, and *clpV*, as markers and key virulence genes of T6SS [52], also play an important role in the infection of *K. oxytoca*. According to Chen, these virulence genes, which are part of the T6SS, contribute to the development of bacterial biofilms, enhancing cell adhesion, promoting the survival and toxicity of bacteria, and enhancing the pathogenicity of *K. oxytoca*, which may be one of the pathogenic mechanisms of the bacteria. In addition, *hcp* is also regulated by iron ions, and *entB* acts as a siderophore gene that mediates iron uptake [53]. Since iron is involved in the biochemical metabolic reactions, electron transport, and other life activities of bacteria, it promotes the growth and reproduction of bacteria. Therefore, genes involved in iron uptake are also crucial to the pathogenicity of bacteria.

Through the comparative genomic analysis of KOHN1 and four other human-derived strains, we determined a high similarity between the goat-derived KOHN1 genome and the human-derived strains. The heat map and phylogenetic tree of the ANI analysis revealed the evolutionary and genetic relationship between the strains. Therefore, we infer that the infection route of the goat-derived isolate may be environmental infection. Based on the above results, we speculate that the pathogen may be transmitted between humans and animals. This provides a new perspective for the infection and spread of *K. oxytoca* and the consideration of biological health.

In this experiment, *K. oxytoca* was isolated from a Hainan black goat for the first time. Through genomic analysis, we could clarify its significant biological characteristics, including metabolic pathways, pathogenic genes, and drug resistance genes. This research is crucial for understanding the infection mechanisms of the bacteria, as well as for developing diagnostic methods and treatment strategies. Additionally, we obtained 113 unique genomes of *K. oxytoca* through comparative genomics, which included metabolism-related GT4, GT8, and GH3, potentially linked to its fermentation capabilities. Furthermore, studying its evolutionary relationship with human strains can enhance our understanding of the evolution and spread of *K. oxytoca*, as well as the differences among various strains. This knowledge is essential for monitoring and controlling the spread of pathogenic bacteria and guiding clinical treatment.

5. Conclusions

In this study, we obtained the whole genome sequence of a goat-derived *K. oxytoca* (KOHN1) and analyzed its fundamental biological functions. A total of 448 drug resistance genes were identified, and their multi-drug resistance to various antibiotics, including β-lactams and tetracyclines, was assessed in conjunction with previous experiments. A total of 146 virulence genes were obtained and virulence genes such as *VgrG*, *hcp*, *clpV*, and *fimA* were studied through PCR experiments. Through comparative genomics analysis, it was found that the strain contained 113 specific genes, including GH3, GT4, and GT8, related to carbohydrate metabolism. At the same time, this study also revealed the strain's evolutionary relationship with human isolates. It provides a reference for future research on the drug resistance mechanism, transmission route, and pathogenic mechanism of *K. oxytoca*.

Supplementary Materials: The following supporting information can be downloaded at https://www.mdpi.com/article/10.3390/genes16010013/s1: Table S1: Summary of genome function analysis of KOHN1 strain; Table S2: Reference genome information table; Table S3: Function of virulence gene; Table S4: CARD analysis results; Table S5: TNSS analysis results; Table S6: VFDB analysis results; Table S7: Statistical results of SNPs, InDels, and SVs between KONH1 and target strains; Figure S1: The Venn diagram of the annotation results of four databases; Figure S2: SNP matrix diagram.

Author Contributions: Y.Z. wrote and typeset the manuscript. Z.Z. carried out the experiments and analyzed the data. Z.W., Y.C. and L.L. contributed to the data collection. C.M. and H.G. managed the experiments. Q.C. and L.D. performed and supervised the research. S.C. reviewed the data and authored the manuscript. F.W. designed the study and revised the manuscript. All authors have read and agreed to the published version of the manuscript.

Funding: This work was financially supported by the Biological Breeding-Major Projects (2023ZD0405103) and the China Agriculture Research System of MOF and MARA (CARS-38).

Institutional Review Board Statement: Not applicable.

Informed Consent Statement: Not applicable.

Data Availability Statement: All data are presented in the article, and the raw sequence data reported in this paper have been deposited in the Genome Sequence Archive (Genomics, Proteomics & Bioinformatics 2021) in the National Genomics Data Center (Nucleic Acids Res 2022), China National Center for Bioinformation/Beijing Institute of Genomics, Chinese Academy of Sciences (GSA: CRA019326), which is publicly accessible at https://ngdc.cncb.ac.cn/gsa (accessed on 11 December 2024).

Conflicts of Interest: The authors declare no conflicts of interest.

Abbreviations

The following abbreviations are used in this manuscript:

ONT	Oxford Nanopore Technologies
AMR	Antibiotic resistance
WGS	Whole genome sequencing
PCR	Polymerase chain reaction
COG	Cluster of Orthologous Groups of proteins
GO	Gene Ontology
KEGG	Kyoto Encyclopedia of Genes and Genomes
NR	Non-Redundant Protein Database
CAZy	Carbohydrate-Active enZYmes Database
TNSS	Type N secretion systems, Type I-VII
VFDB	Virulence Factors of Pathogenic Bacteria
CARD	Comprehensive Antibiotic Research Database
SNP	Single nucleotide polymorphisms
SV	Structural variation
ANI	Average Nucleotide Identity
GH	Glycoside Hydrolases
GT	Glycosyl Transferases
PL	Polysaccharide Lyases
AA	Auxiliary Activities
CE	Carbohydrate Esterases
CBM	Carbohydrate-Binding Modules

References

1. Stewart, J.; Judd, L.M.; Jenney, A.; Holt, K.E.; Wyres, K.L.; Hawkey, J. Epidemiology and genomic analysis of *Klebsiella oxytoca* from a single hospital network in Australia. *BMC Infect. Dis.* **2022**, *22*, 704. [CrossRef] [PubMed]
2. Yang, J.; Long, H.; Hu, Y.; Feng, Y.; McNally, A.; Zong, Z. *Klebsiella oxytoca* Complex: Update on Taxonomy, Antimicrobial Resistance, and Virulence. *Clin. Microbiol. Rev.* **2022**, *35*, e0000621. [CrossRef] [PubMed]
3. Osbelt, L.; Almási, É.D.H.; Wende, M.; Kienesberger, S.; Voltz, A.; Lesker, T.R.; Muthukumarasamy, U.; Knischewski, N.; Nordmann, E.; Bielecka, A.A.; et al. *Klebsiella oxytoca* inhibits Salmonella infection through multiple microbiota-context-dependent mechanisms. *Nat. Microbiol.* **2024**, *9*, 1792–1811. [CrossRef] [PubMed]
4. Podschun, R.; Ullmann, U. *Klebsiella* spp. as nosocomial pathogens: Epidemiology, taxonomy, typing methods, and pathogenicity factors. *Clin. Microbiol. Rev.* **1998**, *11*, 589–603. [CrossRef]
5. Neog, N.; Phukan, U.; Puzari, M.; Sharma, M.; Chetia, P. *Klebsiella oxytoca* and Emerging Nosocomial Infections. *Curr. Microbiol.* **2021**, *78*, 1115–1123. [CrossRef]
6. Lee, D.; Oh, J.Y.; Sum, S.; Park, H.M. Prevalence and antimicrobial resistance of *Klebsiella* species isolated from clinically ill companion animals. *J. Vet. Sci.* **2021**, *22*, e17. [CrossRef]
7. Harada, K.; Shimizu, T.; Mukai, Y.; Kuwajima, K.; Sato, T.; Usui, M.; Tamura, Y.; Kimura, Y.; Miyamoto, T.; Tsuyuki, Y.; et al. Phenotypic and Molecular Characterization of Antimicrobial Resistance in *Klebsiella* spp. Isolates from Companion Animals in Japan: Clonal Dissemination of Multidrug-Resistant Extended-Spectrum β-Lactamase-Producing *Klebsiella pneumoniae*. *Front. Microbiol.* **2016**, *7*, 1021. [CrossRef]
8. Marques, C.; Menezes, J.; Belas, A.; Aboim, C.; Cavaco-Silva, P.; Trigueiro, G.; Telo Gama, L.; Pomba, C. *Klebsiella pneumoniae* causing urinary tract infections in companion animals and humans: Population structure, antimicrobial resistance and virulence genes. *J. Antimicrob. Chemother.* **2019**, *74*, 594–602. [CrossRef]
9. Cuénod, A.; Wüthrich, D.; Seth-Smith, H.M.B.; Ott, C.; Gehringer, C.; Foucault, F.; Mouchet, R.; Kassim, A.; Revathi, G.; Vogt, D.R.; et al. Whole-genome sequence-informed MALDI-TOF MS diagnostics reveal importance of *Klebsiella oxytoca* group in invasive infections: A retrospective clinical study. *Genome Med.* **2021**, *13*, 150. [CrossRef]
10. McNerney, R.; Zignol, M.; Clark, T.G. Use of whole genome sequencing in surveillance of drug resistant tuberculosis. *Expert Rev. Anti Infect. Ther.* **2018**, *16*, 433–442. [CrossRef]
11. Massé, J.; Dufour, S.; Archambault, M. Characterization of *Klebsiella* isolates obtained from clinical mastitis cases in dairy cattle. *J. Dairy Sci.* **2020**, *103*, 3392–3400. [CrossRef] [PubMed]
12. Hua, M.; Duan, A.; Li, Q.; Yue, J.; Liu, X.; Yuan, L.; Liu, J.; Chen, C. Alteration of microbiota and immune response of mice gavaged with *Klebsiella oxytoca*. *Microbes Infect.* **2022**, *24*, 104977. [CrossRef] [PubMed]
13. Ghasemian, A.; Mohabati Mobarez, A.; Najar Peerayeh, S.; Talebi Bezmin Abadi, A.; Khodaparast, S.; Mahmood, S.S. Expression of adhesin genes and biofilm formation among Klebsiella oxytoca clinical isolates from patients with antibiotic-associated haemorrhagic colitis. *J. Med Microbiol.* **2019**, *68*, 978–985. [CrossRef] [PubMed]
14. García-Pérez, A.N.; de Jong, A.; Junker, S.; Becher, D.; Chlebowicz, M.A.; Duipmans, J.C.; Jonkman, M.F.; van Dijl, J.M. From the wound to the bench: Exoproteome interplay between wound-colonizing Staphylococcus aureus strains and co-existing bacteria. *Virulence* **2018**, *9*, 363–378. [CrossRef]

15. Levkova, M.; Chervenkov, T.; Angelova, L.; Dzenkov, D. Oxford Nanopore Technology and its Application in Liquid Biopsies. *Curr. Genom.* **2023**, *24*, 337–344. [CrossRef]
16. Brisse, S.; Fevre, C.; Passet, V.; Issenhuth-Jeanjean, S.; Tournebize, R.; Diancourt, L.; Grimont, P. Virulent clones of *Klebsiella pneumoniae*: Identification and evolutionary scenario based on genomic and phenotypic characterization. *PLoS ONE* **2009**, *4*, e4982. [CrossRef]
17. Nasser, K.; Mustafa, A.S.; Khan, M.W.; Purohit, P.; Al-Obaid, I.; Dhar, R.; Al-Fouzan, W. Draft Genome Sequences of Six Multidrug-Resistant Clinical Strains of Acinetobacter baumannii, Isolated at Two Major Hospitals in Kuwait. *Genome Announc.* **2018**, *6*, e00264-18. [CrossRef]
18. Singh, R.; Kusalik, A.; Dillon, J.R. Bioinformatics tools used for whole-genome sequencing analysis of Neisseria gonorrhoeae: A literature review. *Brief. Funct. Genom.* **2022**, *21*, 78–89. [CrossRef]
19. Safar, H.A.; Alatar, F.; Nasser, K.; Al-Ajmi, R.; Alfouzan, W.; Mustafa, A.S. The impact of applying various de novo assembly and correction tools on the identification of genome characterization, drug resistance, and virulence factors of clinical isolates using ONT sequencing. *BMC Biotechnol.* **2023**, *23*, 26. [CrossRef]
20. Campos-Madueno, E.I.; Sigrist, T.; Flückiger, U.M.; Risch, L.; Bodmer, T.; Endimiani, A. First report of a blaVIM-1 metallo-β-lactamase-possessing *Klebsiella* michiganensis. *J. Glob. Antimicrob. Resist.* **2021**, *25*, 310–314. [CrossRef]
21. Dieckmann, R.; Hammerl, J.A.; Hahmann, H.; Wicke, A.; Kleta, S.; Dabrowski, P.W.; Nitsche, A.; Stämmler, M.; Al Dahouk, S.; Lasch, P. Rapid characterisation of *Klebsiella oxytoca* isolates from contaminated liquid hand soap using mass spectrometry, FTIR and Raman spectroscopy. *Faraday Discuss.* **2016**, *187*, 353–375. [CrossRef] [PubMed]
22. Lo, S.; Lolom, I.; Goldstein, V.; Petitjean, M.; Rondinaud, E.; Bunel-Gourdy, V.; Tran Dinh, A.; Wicky, P.H.; Ruppé, E.; d'Humières, C.; et al. Simultaneous Hospital Outbreaks of New Delhi Metallo-β-Lactamase-Producing Enterobacterales Unraveled Using Whole-Genome Sequencing. *Microbiol. Spectr.* **2022**, *10*, e0228721. [CrossRef] [PubMed]
23. Zhang, Z.; Jiang, J.; He, M.; Li, H.; Cheng, Y.; An, Q.; Chen, S.; Du, L.; Man, C.; Chen, Q.; et al. First Report and Comparative Genomic Analysis of a Mycoplasma mycoides Subspecies capri HN-A in Hainan Island. *Microorganisms* **2022**, *10*, 1908. [CrossRef] [PubMed]
24. Holt, K.E.; Parkhill, J.; Mazzoni, C.J.; Roumagnac, P.; Weill, F.X.; Goodhead, I.; Rance, R.; Baker, S.; Maskell, D.J.; Wain, J.; et al. High-throughput sequencing provides insights into genome variation and evolution in Salmonella Typhi. *Nat. Genet.* **2008**, *40*, 987–993. [CrossRef] [PubMed]
25. Clawson, M.L.; Keen, J.E.; Smith, T.P.; Durso, L.M.; McDaneld, T.G.; Mandrell, R.E.; Davis, M.A.; Bono, J.L. Phylogenetic classification of *Escherichia coli* O157:H7 strains of human and bovine origin using a novel set of nucleotide polymorphisms. *Genome Biol.* **2009**, *10*, R56. [CrossRef]
26. Rossing, M.; Sørensen, C.S.; Ejlertsen, B.; Nielsen, F.C. Whole genome sequencing of breast cancer. *APMIS* **2019**, *127*, 303–315. [CrossRef]
27. Wang, R.; Li, J.; Zhou, X.; Mao, Y.; Wang, W.; Gao, S.; Wang, W.; Gao, Y.; Chen, K.; Yu, S.; et al. Single-cell genomic and transcriptomic landscapes of primary and metastatic colorectal cancer tumors. *Genome Med.* **2022**, *14*, 93. [CrossRef]
28. Quiñones, K.J.O.; Gentallan, R.P., Jr.; Timog, E.B.S.; Cruz, J.R.A.V.; Macabecha, C.G.A.; Papa, I.A.; Coronado, N.B.; Bartolome, M.C.B.; Ceribo, D.B.; Madayag, R.E.; et al. Liquid-nitrogen-free CTAB DNA extraction method from silica-dried specimens for next-generation sequencing and assembly. *MethodsX* **2024**, *12*, 102758. [CrossRef]
29. Galperin, M.Y.; Makarova, K.S.; Wolf, Y.I.; Koonin, E.V. Expanded microbial genome coverage and improved protein family annotation in the COG database. *Nucleic Acids Res.* **2015**, *43*, D261–D269. [CrossRef]
30. Ashburner, M.; Ball, C.A.; Blake, J.A.; Botstein, D.; Butler, H.; Cherry, J.M.; Davis, A.P.; Dolinski, K.; Dwight, S.S.; Eppig, J.T.; et al. Gene ontology: Tool for the unification of biology. The Gene Ontology Consortium. *Nat. Genet.* **2000**, *25*, 25–29. [CrossRef]
31. Kanehisa, M.; Goto, S.; Hattori, M.; Aoki-Kinoshita, K.F.; Itoh, M.; Kawashima, S.; Katayama, T.; Araki, M.; Hirakawa, M. From genomics to chemical genomics: New developments in KEGG. *Nucleic Acids Res.* **2006**, *34*, D354–D357. [CrossRef] [PubMed]
32. Li, W.; Jaroszewski, L.; Godzik, A. Tolerating some redundancy significantly speeds up clustering of large protein databases. *Bioinformatics* **2002**, *18*, 77–82. [CrossRef] [PubMed]
33. Singh, R.P.; Kumari, K. Bacterial type VI secretion system (T6SS): An evolved molecular weapon with diverse functionality. *Biotechnol. Lett.* **2023**, *45*, 309–331. [CrossRef]
34. Jia, B.; Raphenya, A.R.; Alcock, B.; Waglechner, N.; Guo, P.; Tsang, K.K.; Lago, B.A.; Dave, B.M.; Pereira, S.; Sharma, A.N.; et al. CARD 2017: Expansion and model-centric curation of the comprehensive antibiotic resistance database. *Nucleic Acids Res.* **2017**, *45*, D566–D573. [CrossRef] [PubMed]
35. Chen, L.; Xiong, Z.; Sun, L.; Yang, J.; Jin, Q. VFDB 2012 update: Toward the genetic diversity and molecular evolution of bacterial virulence factors. *Nucleic Acids Res.* **2012**, *40*, D641–D645. [CrossRef] [PubMed]
36. Zhang, A.; Han, Y.; Huang, Y.; Hu, X.; Liu, P.; Liu, X.; Kan, B.; Liang, W. vgrG is separately transcribed from *hcp* in T6SS orphan clusters and is under the regulation of IHF and HapR. *Biochem. Biophys. Res. Commun.* **2021**, *559*, 15–20. [CrossRef]
37. Omer, F.H.; Al-Khafaji, N.S.K.; Al-Alaq, F.T.; Al-Dahmoshi, H.O.M.; Memariani, M.; Saki, M. Synergistic effects of silybin and curcumin on virulence and carbapenemase genes expression in multidrug resistant *Klebsiella oxytoca*. *BMC Res. Notes* **2022**, *15*, 330. [CrossRef]
38. Hu, Y.Y.; Chen, S.; Zhang, Y.D.; Lu, Q.W.; Wang, F.; Ren, A.; Liu, C.X. Value of T6SS Core Gene *hcp* in Acinetobacter baumannii Respiratory Tract Infection. *Indian J. Microbiol.* **2023**, *63*, 291–298. [CrossRef]

39. Luo, G.; Xu, X.; Zhao, L.; Qin, Y.; Huang, L.; Su, Y.; Yan, Q. clpV is a key virulence gene during in vivo Pseudomonas plecoglossicida infection. *J. Fish Dis.* **2019**, *42*, 991–1000. [CrossRef]
40. Lin, W.H.; Wang, M.C.; Tseng, C.C.; Ko, W.C.; Wu, A.B.; Zheng, P.X.; Wu, J.J. Clinical and microbiological characteristics of *Klebsiella pneumoniae* isolates causing community-acquired urinary tract infections. *Infection* **2010**, *38*, 459–464. [CrossRef]
41. Candan, E.D.; Aksöz, N. *Klebsiella pneumoniae*: Characteristics of carbapenem resistance and virulence factors. *Acta Biochim. Pol.* **2015**, *62*, 867–874. [CrossRef] [PubMed]
42. Turton, J.F.; Perry, C.; Elgohari, S.; Hampton, C.V. PCR characterization and typing of *Klebsiella pneumoniae* using capsular type-specific, variable number tandem repeat and virulence gene targets. *J. Med. Microbiol.* **2010**, *59 Pt 5*, 541–547. [CrossRef] [PubMed]
43. Dong, M.B.; Tang, K.; Zhou, X.; Zhou, J.J.; Chen, S. Tumor immunology CRISPR screening: Present, past, and future. *Trends Cancer* **2022**, *8*, 210–225. [CrossRef] [PubMed]
44. de Jong, A.; Garch, F.E.; Simjee, S.; Moyaert, H.; Rose, M.; Youala, M.; Siegwart, E.; VetPath Study Group. Monitoring of antimicrobial susceptibility of udder pathogens recovered from cases of clinical mastitis in dairy cows across Europe: VetPath results. *Vet. Microbiol.* **2018**, *213*, 73–81. [CrossRef]
45. Vuillemin, M.; Muschiol, J.; Zhang, Y.; Holck, J.; Barrett, K.; Preben Morth, J.; Meyer, A.S.; Zeuner, B. Discovery of Lacto-*N*-Biosidases and a Novel *N*-Acetyllactosaminidase Activity in the CAZy Family GH20: Functional Diversity and Structural Insights. *ChemBioChem* **2024**, e202400710. [CrossRef]
46. Storey, D.; McNally, A.; Åstrand, M.; Sa-Pessoa Graca Santos, J.; Rodriguez-Escudero, I.; Elmore, B.; Palacios, L.; Marshall, H.; Hobley, L.; Molina, M.; et al. *Klebsiella pneumoniae* type VI secretion system-mediated microbial competition is PhoPQ controlled and reactive oxygen species dependent. *PLoS Pathog.* **2020**, *16*, e1007969. [CrossRef]
47. Meir, A.; Macé, K.; Vegunta, Y.; Williams, S.M.; Waksman, G. Substrate recruitment mechanism by gram-negative type III, IV, and VI bacterial injectisomes. *Trends Microbiol.* **2023**, *31*, 916–932. [CrossRef]
48. Sana, T.G.; Flaugnatti, N.; Lugo, K.A.; Lam, L.H.; Jacobson, A.; Baylot, V.; Durand, E.; Journet, L.; Cascales, E.; Monack, D.M. Salmonella Typhimurium utilizes a T6SS-mediated antibacterial weapon to establish in the host gut. *Proc. Natl. Acad. Sci. USA* **2016**, *113*, E5044–E5051. [CrossRef]
49. Joshi, A.; Kostiuk, B.; Rogers, A.; Teschler, J.; Pukatzki, S.; Yildiz, F.H. Rules of Engagement: The Type VI Secretion System in Vibrio cholerae. *Trends Microbiol.* **2017**, *25*, 267–279. [CrossRef]
50. Liu, Q.; Zhu, J.; Liu, N.; Sun, W.; Yu, B.; Niu, H.; Liu, D.; Ouyang, P.; Ying, H.; Chen, Y.; et al. Type I fimbriae subunit fimA enhances *Escherichia coli* biofilm formation but affects L-threonine carbon distribution. *Front. Bioeng. Biotechnol.* **2022**, *10*, 904636. [CrossRef]
51. Stahlhut, S.G.; Chattopadhyay, S.; Kisiela, D.I.; Hvidtfeldt, K.; Clegg, S.; Struve, C.; Sokurenko, E.V.; Krogfelt, K.A. Structural and population characterization of *MrkD*, the adhesive subunit of type 3 fimbriae. *J. Bacteriol.* **2013**, *195*, 5602–5613. [CrossRef] [PubMed]
52. Douzi, B.; Brunet, Y.R.; Spinelli, S.; Lensi, V.; Legrand, P.; Blangy, S.; Kumar, A.; Journet, L.; Cascales, E.; Cambillau, C. Structure and specificity of the Type VI secretion system ClpV-TssC interaction in enteroaggregative *Escherichia coli*. *Sci. Rep.* **2016**, *6*, 34405. [CrossRef] [PubMed]
53. Chen, Z.; Mao, Y.; Song, Y.; Dou, M.; Shang, K.; Yu, Z.; Ding, K.; Chen, S. Refined egoist: The toxin-antitoxin immune system of T6SS. *Microb. Pathog.* **2024**, *196*, 106991. [CrossRef] [PubMed]

genes

MDPI

Article

Worldwide Population Dynamics of *Salmonella* Saintpaul: Outbreaks, Epidemiology, and Genome Structure

Pedro Panzenhagen [1,2,3,4,*], Devendra H. Shah [5], Dalia dos Prazeres Rodrigues [2] and Carlos Adam Conte Junior [1,3,4,6,7,8]

[1] Center for Food Analysis (NAL), Technological Development Support Laboratory (LADETEC), Federal University of Rio de Janeiro (UFRJ), Cidade Universitária, Rio de Janeiro 21941-598, RJ, Brazil; conte@iq.ufrj.br
[2] National Reference Laboratory for Diagnosis of Enteric Bacteria (LABENT/LRNEB), Oswaldo Cruz Foundation, Av. Brasil, 4365-Manguinhos, Rio de Janeiro 21040-900, RJ, Brazil; daliarodrigues@yahoo.com.br
[3] Analytical and Molecular Laboratorial Center (CLAn), Institute of Chemistry (IQ), Federal University of Rio de Janeiro (UFRJ), Cidade Universitária, Rio de Janeiro 21941-909, RJ, Brazil
[4] Laboratory of Advanced Analysis in Biochemistry and Molecular Biology (LAABBM), Department of Biochemistry, Federal University of Rio de Janeiro (UFRJ), Cidade Universitária, Rio de Janeiro 21941-909, RJ, Brazil
[5] School of Veterinary Medicine, Texas Tech University, Amarillo, TX 79106, USA; devendra.shah@ttu.edu
[6] Graduate Program in Food Science (PPGCAL), Institute of Chemistry (IQ), Federal University of Rio de Janeiro (UFRJ), Cidade Universitária, Rio de Janeiro 21941-909, RJ, Brazil
[7] Graduate Program in Chemistry (PGQu), Institute of Chemistry (IQ), Federal Universit of Rio de Janeiro (UFRJ), Cidade Universitária, Rio de Janeiro 21941-909, RJ, Brazil
[8] Graduate Program in Biochemistry (PPGBq), Institute of Chemistry (IQ), Federal University of Rio de Janeiro (UFRJ), Cidade Universitária, Rio de Janeiro 21941-909, RJ, Brazil
* Correspondence: panzenhagen@ufrj.br

Abstract: Background/Objectives: *Salmonella* Saintpaul (SSa) is increasingly linked to foodborne outbreaks in Brazil and globally. Despite its rising public health significance, its epidemiology, genomic diversity, and pathogenic potential remain underexplored. This study addresses these gaps through a comprehensive global analysis of SSa population dynamics, outbreak patterns, and genetic structures, along with an in-depth phenotypic and genomic characterization of strain PP_BR059, isolated from a hospitalized patient in Ceará, Brazil. Methods: We analyzed 1,953 publicly available SSa genomes using core-genome multi-locus sequence typing (cgMLST), antimicrobial resistance (AMR) profiling, pan-genome analysis, and phylogenetic inference. A genome-wide association study (GWAS) identified genetic determinants of virulence and AMR. The invasiveness and intracellular survival of PP_BR059 were assessed using in vitro macrophage infection assays, while whole-genome sequencing (WGS) provided genetic insights. Results: Phylogenetic analysis identified 49 sequence types (STs), with ST-50 (787 genomes) and ST-27 (634 genomes) being most prevalent. ST-50 included all clinical strains from South America, including PP_BR059. AMR analysis showed 60% of SSa genomes were pan-susceptible, while ST-27 had the highest proportion of AMR strains. GWAS revealed distinct evolutionary lineages within ST-50 and ST-27. PP_BR059 exhibited lower macrophage invasion (3.82%) but significantly higher intracellular survival at 2 h (68.72%) and 20 h (25.68%) post-infection. WGS confirmed a pan-susceptible AMR profile and plasmid absence. Conclusions: This study highlights SSa's global dissemination, evolutionary trends, and pathogenic variability, emphasizing the need for molecular surveillance to inform public health interventions.

Keywords: motility; intramacrophage survival; bacterial virulence; phenotypic characteristics; antimicrobial resistance; plasmid; MLST; WGS

1. Introduction

Non-typhoidal *Salmonella* (NTS) is a leading cause of bacterial foodborne illnesses worldwide [1]. NTS poses a significant public health threat and imposes a substantial economic burden for both developed and developing nations [2]. Among over 1400 known NTS serovars, *Salmonella* Saintpaul (SSa) ranks as one of the top ten most frequently detected serovars in humans and animals globally, accounting for approximately 1.6% of severe *Salmonella* infections [3]. In the United States, SSa has caused notable outbreaks, including those in 2008, 2009, and 2013, linked to contaminated peppers, alfalfa sprouts, and cucumbers, primarily sourced from Mexico [4–8]. For instance, the 2008 outbreak affected 1442 individuals across 43 states and D.C., leading to 286 hospitalizations and two fatalities. Initially traced to tomatoes, the outbreak strain was later linked to jalapeno peppers from various locations, including a farm in Mexico [9]. In the 2009 outbreak, alfalfa sprouts contaminated with SSa sickened 235 people in 14 states, with a hospitalization rate of 3% [4]; the Food and Drug Administration (FDA) traced the source to multiple sprout production facilities. The 2013 cucumber outbreak affected 84 people across 18 states, resulting in 17 hospitalizations [10], with the contamination traced back to farms in Mexico.

In Brazil, SSa is among the top 10 most prevalent *Salmonella* serovars, with 1222 cases (0.16% of 62,030 isolates) reported between 2011 and 2020 [11].These isolates originated from a variety of sources, including food (57%), the environment (16%), animals (14%), raw materials (8%), and humans (5%). While large-scale outbreaks of SSa have not been reported in Brazil, localized cases have been identified in specific neighborhoods or cities. SSa is also one of the top ten most frequently isolated *Salmonella* serovars from human sources in Brazil [11]. Recent studies have documented SSa isolation in the country [12–14]; however, despite extensive epidemiological data on its prevalence, there is a lack of comprehensive studies consolidating genetic data from sequenced SSa genomes associated with foodborne outbreaks. Furthermore, research on the invasiveness and virulence potential of clinical SSa strains is still limited.

This study aims to fill this gap by analyzing the global population dynamics of SSa, focusing on major outbreaks, epidemiological patterns, and genomic structures. Through this multidimensional approach, we seek to identify evolutionary trends, genetic determinants, and potential resistance and virulence factors that contribute to the pathogenicity and spread of SSa. Additionally, we provide a detailed phenotypic and genotypic characterization of the clinical SSa strain (PP_BR059) isolated from a hospitalized patient in the Ceará outbreak in northeast Brazil, specifically addressing its invasiveness and virulence potential in the context of human infection, which is similar to that of other NTS serovars.

2. Materials and Methods

2.1. Phenotypic Characterization of SSa Strain PP_BR059

Strain PP_BR059 is a clinical isolate of SSa obtained from the fecal sample of a patient hospitalized during a 2011 outbreak in Ceará, Brazil. The infection was characterized by severe symptoms, such as diarrhea and high fever, raising concerns about its potentially enhanced virulence and invasiveness compared to typical NTS strains. The strain was sent to the National Reference Laboratory for serotyping analysis and was included in the Enterobacteria Collection at the Oswaldo Cruz Foundation's National Reference Laboratory. Later, PP_BR059 served as a critical model for this research, studying the virulence and pathogenicity of SSa, contributing to our understanding of its clinical severity, and aiding in the development of public health strategies. The strain was preserved in 15% (v/v) glycerol at −80 °C and cultured in either Brain Heart Infusion (Difco, USA) or LB broth Lennox (LB) medium for 16 h at 37 °C. For the experiments, a single colony was inoculated into 5 mL of LB medium and incubated overnight (16 h) at 37 °C, with shaking at 200 rpm.

Antimicrobial susceptibility testing of the SSa strain PP_BR059 was performed using the disk diffusion method on Mueller–Hinton agar (Difco, Becton Dickinson, MD, USA), according to the guidelines provided by the Clinical and Laboratory Standards Institute [15]. The following antibiotic disks (BD, Becton Dickinson, MD, USA) were utilized: ampicillin—10 μg (Am), amoxicillin/clavulanic acid—20/10 μg (Amc), amikacin—30 μg (An), chloramphenicol—30 μg (C), ciprofloxacin—5 μg (Cip), sulfisoxazole—0.25 μg (G), gentamicin—10 μg (Gm), nalidixic acid—30 μg (Nal), streptomycin—10 μg (S), sulfamethoxazole–trimethoprim—23.75/1.25 μg (Sxt), tetracycline—30 μg (Te), and ceftiofur—30 μg (XNL). The control strains included *Escherichia coli* ATCC 25922, *Staphylococcus aureus* ATCC 25923, and *Pseudomonas aeruginosa* 10536.

The motility of SSa strain PP_BR059 was determined following a protocol described previously [16]. Briefly, the LB medium supplemented with 0.3% (*w*/*v*) motility agar (Difco, USA) was used. A 2 μL suspension from overnight cultures of the strain PP_BR059 and reference strains *S.* Typhimurium SL1344 (ST19) and LT2 (ST19) were spot-inoculated onto the surface of the semisolid medium. The plates were then incubated at 37 °C for 7 h, after which the diameter of the growth halo was measured using a ruler. Each strain was tested in triplicate and the experiment was repeated independently three times. To analyze the data, the percentage of motility for the strain PP_BR059 was calculated relative to the reference strains, which were set as 100%. Statistical differences in motility between the strains and the reference were determined using one-way analysis of variance (ANOVA) with Dunnett's post hoc test ($p < 0.05$).

The intra-macrophage survival of SSa strain PP_BR059 was determined following a previously described protocol [17]. Human macrophages (THP-1 cells) were seeded into a well plate with complete RPMI 1640 medium (HyClone Laboratories, Cytiva, Utah, USA) and activated by adding 10 mM phorbol 12-myristate 13-acetate (PMA) for 18 h at 37 °C with 5% CO_2. An aliquot from the overnight culture was then transferred to an LB medium and incubated at 37 °C for 3 h before cell infection. The activated THP-1 cells were infected with a bacteria-to-cell ratio of 50:1 (bacteria:cell). To facilitate contact between the bacteria and cells, the plates were centrifuged for 10 min at 400× *g* at room temperature. After a 30 min incubation at 37 °C in a 5% CO_2 incubator, the cells were washed thrice with phosphate-buffered saline (PBS). Subsequently, each well received an RPMI 1640 medium containing 200 μg/mL gentamicin. The cells were lysed at 30 min (for uptake/invasion), 2 h, and 20 h (for survival) of incubation using 200 μL of 0.5% Triton X-100 (Fisher Biotech, Waltham, MA, USA). Serial 10-fold dilutions were plated on LB agar to determine viable counts. Each isolate was tested in triplicate in three independent experiments. The percentage of intracellular counts was calculated as follows: the colony-forming unit (CFU) at t = 0 divided by the CFU at t = 30 min multiplied by 100 (for % uptake/invasion); the CFU at t = 30 min divided by the CFU at t = 2 h multiplied by 100 (for % intracellular survival at 2 h); and the CFU at t = 2 h divided by the CFU at t = 20 h multiplied by 100 (for % intracellular survival at 20 h). Statistical differences ($p < 0.05$) in intracellular survival between the strains were determined using one-way analysis of variance (ANOVA) with Dunnett's post hoc test, comparing the strains to the reference *S.* Typhimurium strains, LT2 and SL1344 (ST19).

2.2. WGS of SSa Strain PP_BR059

The genome of the SSa strain PP_BR059 was sequenced using the Illumina© MiSeq platform. Paired-end (2 × 250 bp) Nextera XT libraries (Illumina©, San Diego, CA, USA) were prepared following the manufacturer's suggested protocol. DNA samples were multiplexed using Illumina-supplied barcodes; DNA pools were size-selected to be in the range of 600–1000 bp (with an average peak at ~800 bp). The read quality was assessed using FastQC v0.11.5. Reads were processed with Trimmomatic 0.36 [18]; short reads were length-adjusted using FLASh v1.2.11 [19]. The final treated reads were assembled using SPAdes v3.11.1 [20]. Genome statistics were obtained using QUAST v4.6.0 [21]. The assembly of strain PP_BR059 generated 38 contigs, with 24 contigs greater than or equal to 300 bp. The assembly covered a total of 4687,206 bp, with an N50 value of 379,274 bp and an average read coverage of 188×. The average G + C content was 52.23%; the genomic features, including genes, CDS, and transcripts, numbered 4604. The whole-genome assembly of strain PP_BR059 has been deposited in the NCBI GenBank under the BioSample number SAMN05505497. The raw reads can be found in the Single Read Archive (SRA) run SRR4046393; the assembled contigs are available under the accession number GCA_009537095.1.

2.3. Comparative Genomic Analysis of *Salmonella* Saintpaul

A total of 1953 assembled genomes of SSa strains were obtained from the Pathogen Detection Web Browser on the National Center for Biotechnology Information (NCBI) website "https://www.ncbi.nlm.nih.gov/pathogens/ (accessed on 30 August 2020). These records were retrieved by using the search keyword "*Salmonella* Saintpaul" and were downloaded for all available assemblies as of 30 August 2020 (Supplementary Table S1). The genomes consist of isolates collected from 16 countries spanning North America, Latin America, Asia, Africa, Europe, and Oceania. These isolates were sourced from diverse origins, including the environment, food, animal waste, human feces, and blood. The isolation dates of the samples range over 30 years, from 1988 to 2020. The quality of the selected genomes was assessed using STARAMR [22] with the following three exclusion criteria: (1) a final genome size of less than 4 Mb or greater than 6 Mb; (2) an N50 value exceeding 10 kb; and (3) at least one contig equal to or greater than 300 bp. All follow-up in silico genomics analyses were conducted using the sequenced genome of SSa strain PP_BR059 and 1953 SSa genomes downloaded from a publicly available database.

Core-genome multi-locus sequence typing (cgMLST) was performed using cgMLSTFinder. V1.2. The cgMLSTFinder utilizes a rapid and accurate K-mer alignment of the reads [23] against EnteroBase to identify the cgMLST schemes based on alleles found in the 3002 *Salmonella* genes [24]. The traditional MLST was performed using PubMLST (https://pubmlst.org/organisms/salmonella-spp, accessed on 30 August 2020), which employs WGS data to identify the sequence types (STs) of bacteria [25].

The in silico antimicrobial resistance (AMR) gene and plasmid detection was conducted using STARAMR 0.7.1 with the following settings: a minimum DNA identity threshold of 95%; and a minimum DNA coverage of 60% for all genome alignments [22]. STARAMR scans bacterial genome contigs against the ResFinder [26], PointFinder [27], and PlasmidFinder [28] databases, provides a summary report of the detected AMR genes and plasmids, and predicts drug resistance based on the identified resistance genes.

Pan-genome analysis and core-genome phylogenetic inference were conducted using Roary, v1.14.5 [29]. The pan-genome analysis and core-genome alignment were performed with a minimum percentage identity of 95% for blastp; a gene was required to be present in at least 99% of the isolates to be considered as core. To ensure standardized genome annotation, all genomes were annotated using prokka, v1.14.5 [30]. The pan-genome visualization was achieved by employing the postprocessing scripts provided by Roary. For the generation of a maximum likelihood phylogeny from the core-genome alignment, IQtree v2.0.3 [31] was utilized, employing a GTR + F + I + G4 substitution model with 1000 bootstrap replications to support the tree nodes. Complementary annotation was performed using emapper, v2.0 [32] with default parameters, relying on eggNOG orthology data [33]. Diamond protein alignment [34] was employed for the sequence searches. For interactive visualization of the phylogenetics, metadata, and gene content, the browser application Phandango was used [35].

Detection of the WGS variants and the prediction of their effects were conducted using Snippy, v4.6.0 (Seemann T., GitHub repository: snippy). The complete genome sequence of SSa strain CFSAN004175 (GCA_001952995.1) served as the reference strain for identifying single-nucleotide variants (SNVs). To enhance the accuracy of phylogenetic reconstructions, recombinant regions within the core-genome alignment were identified and excluded using Gubbins v2.4.1 [36]. Phage regions and Ribosomal RNA were identified using Phastaf, v0.1.0 (Seemann T., GitHub repository: phastaf) and Barrnap, v0.9 (Seemann T., GitHub repository: barrnap), respectively, to facilitate masking in the reference genome. The final set of high-quality core-genome SNPs was extracted from the alignment file using SNP-sites, v2.5.1 [37]. Finally, IQtree was employed for the SNP phylogenetic inference, following previously described methods.

3. Results and Discussion

3.1. Comparative Genomic Analysis of SSa

The MLST analysis identified a total of 49 distinct sequence types (STs) of SSa, with ST-50 being the most prevalent, encompassing 787 genomes, followed by ST-27, comprising 634 genomes (Figure 1). Together, these two STs accounted for 1421 (73%) of the 1954 SSa genomes analyzed in this study (Supplementary Table S1). The other most frequently detected STs included ST-49, ST-95, and ST-680, with 180, 132, and 70 genomes, respectively. The remaining STs had fewer than 30 representative genomes. Our analysis revealed that the strain PP_BR059 belongs to MLST 50 (ST-50). Given the significant role of SSa ST-50 in the genus *Salmonella* epidemiology globally, the high phenotypic virulence exhibited by the PP_BR059 strain underscores the importance of molecular epidemiology for investigating outbreaks, especially when dealing with relatively uncommon organisms and limited knowledge of their diversity and distribution.

Based on the provided metadata on ST-50 sources (Supplementary Table S1), around 100 non-clinical isolates originate from turkey and turkey products. The frequent isolation of ST-50 from turkey (e.g., "Young Turkey," "Turkey Carcass Sponge," "Turkey Meat") in the United States [38] and Europe [39,40] suggests that poultry may be a significant risk factor for human SSa infections worldwide. This aligns with global findings that often implicate poultry and turkey as major sources of *Salmonella* contamination in the food supply chain [41]. Interestingly, environmental sources, such as canal water, river water, and food-processing residues also indicate potential indirect transmission routes (Supplementary Table S1). Additionally, regional distinctions, such as the higher reporting of turkey isolates in the United States and other animal sources in Brazil, may reflect differences in food production systems or surveillance practices.

Interestingly, most ST-50 genomes originated from three BioProjects (Supplementary Table S1). BioProject PRJNA230403 contained 279 SSa clinical strains (stool, urine, rectal swabs, tissues, wounds, and blood) sequenced by PulseNet USA at the Center for Disease Control and Prevention. BioProject PRJNA248792 included 202 SSa strains of clinical origin sequenced by the *Salmonella* Reference Service at the Gastrointestinal Bacteria Reference Unit in the United Kingdom. BioProject PRJNA596817 comprised 95 SSa strains originating from routine public health outbreak investigations in New South Wales (NSW), Australia [42]. In 2019, an outbreak of SSa ST-50 in Scotland involved person-to-person transmission and hospitalizations, primarily affecting children [43]. Although WGS was used to investigate the outbreak [43], the specific genomes identified in the study were not found in pathogen-detection databases, preventing their inclusion in the current analysis. According to the study, 28 isolates of SSa were isolated from patients in Scotland with no travel history between October 2017 and November 2018. The outbreak strain formed a distinct cluster compared to other isolates from Scotland during the same period; no links were found to international isolates within the Enterobase database (containing over 180,000 *Salmonella* genomes).

The SSa ST-50 has also been implicated in various global outbreaks, often associated with food contamination, raw fruits or vegetables, and water [7,44–49] including the 2008 multi-state foodborne outbreak in the United States that affected over 1400 individuals [7]. Finally, SSa strains CFSAN004173, CFSAN004174, and CFSAN004175, isolated from the 2013 multistate cucumber-associated outbreak in the United States, were also classified as ST-50 (Supplementary Table S1). The frequent association of SSa ST50 with outbreaks and isolation from diverse clinical sources strongly suggests that ST50 is likely the predominant type responsible for human SSa infections worldwide.

While most ST-50 genomes in the NCBI database were derived from human clinical sources, over half of the ST-27 genomes (331 out of 634) were sourced from ground turkey in the United States (Supplementary Table S1). These genomes, collected between 2004 and 2018, were part of twelve BioProjects that conducted WGS of SSa as part of multi-state outbreak surveillance routines in the United States (Supplementary Table S1). Furthermore, BioProject PRJNA248792 documented a significant population of ST-27 strains (164 out of 634) originating from human clinical sources in the United Kingdom, recorded from 2013 to 2020. Similarly, BioProject PRJNA230403 included data on 75 ST-27 and 64 ST-95 strains extracted from human stool samples in the United States. All ST-49 genomes contained in BioProject PRJNA248792 were sourced from human clinical cases in the United Kingdom. Lastly, ST-680 was also a highly prevalent sequence type (n = 70) within this SSa collection because it was frequently isolated from ground turkey multi-state outbreak surveillance in the United States (Supplementary Table S1).

In silico resistome analysis showed that SSa exhibits a predominantly pan-susceptible genetic profile with 60% (1149 out of 1954) of the genomes carrying no AMR genes. Pan-susceptibility frequencies were observed across the predominant sequence types as follows: ST-50 with 574 out of 797 (72%); ST-27 with 236 out of 634 (37%); ST-49 with 114 out of 180 (63%); ST-95 with 109 out of 132 (83%); and ST-680 with 37 out of 70 (53%). Collectively, pan-susceptibility in other STs accounted for 79 out of 151 (52%). These results highlight that while most predominant sequence types (STs) exhibit high proportions of pan-susceptible strains, ST-27 stands out, with only 37% of the strains being pan-susceptible. This lower frequency suggests that ST-27, predominantly isolated from turkeys, is more likely to carry resistance genes compared to other STs, indicating a higher potential for AMR within this lineage. A probable reason for this observation is the intense use of antimicrobials in the poultry production chain, which likely exerts a selective pressure that favors the persistence and proliferation of MDR strains in ST-27.

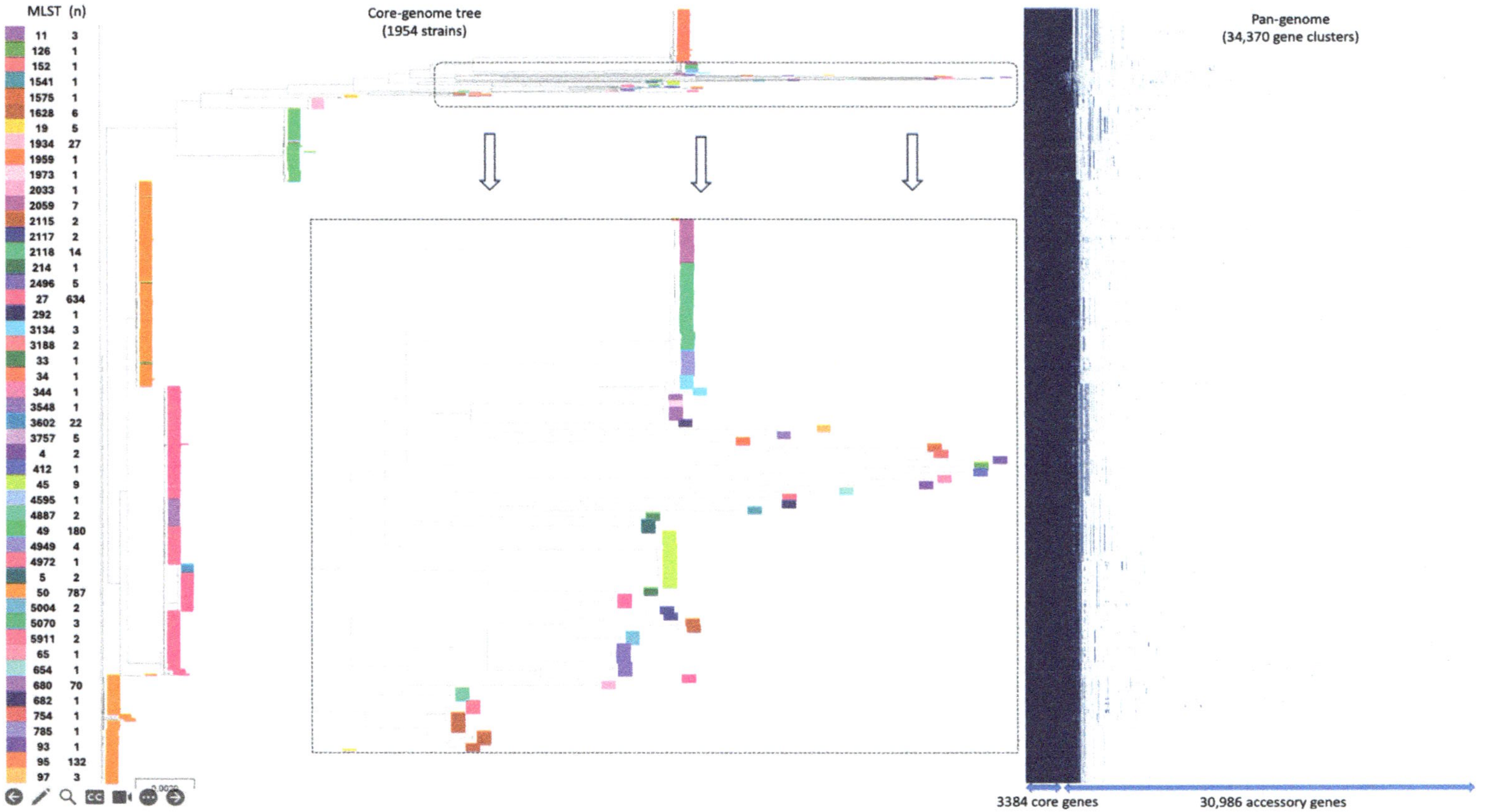

Figure 1. Phylogenetic tree based on the concatenated sequence alignment of core genes identified by Roary, showing the distribution of multi-locus sequence types (MLST) across 1954 *SSa* genomes, including the *PP_BR059* strain sequenced in this study. The corresponding pan-genome matrix on the right illustrates the distribution of core (dark blue) and accessory (light blue) genes across all analyzed genomes.

In contrast, the resistome of the 805 SSa strains, each carrying a minimum of one resistance gene, exhibits a rich diversity encompassing 90 distinct AMR genes (Figure 2 and Supplementary Table S1). Aminoglycosides were the most prevalent class, with resistance genes detected in approximately 30% of strains. This was followed by tetracyclines (19.9%) and β-lactams (19.9%), which also exhibited high prevalences. Conversely, rifamycin, polymyxins, and MLSB (macrolide–lincosamide–streptogramin B) showed significantly lower frequencies, each accounting for less than 2% of strains (Figure 2A). Regarding the resistance genes, *bla*TEM-1 (β-lactamase activity) and *tet* (A and B) (tetracycline resistance) emerge as the most frequently identified genes, present in over 50% of strains. Other highly prevalent genes include *ant* (3″)-Ia (aminoglycoside resistance) and *sul* (1 and 2) (sulfonamide). A long tail of less prevalent genes reflects the diversity and specificity of AMR gene occurrence among the strains with resistance genes that have emerged recently, conferring resistance to the last generation of antimicrobials, such as *bla*CTX-M, TEM- or SHV-type extended-spectrum ß-lactamases (ESBLs), CMY, and mcr, detected in SSa (Figure 2B and Supplementary Table S1). The great concern about the emergence of such resistance genes is their ability to encode resistance against last-resort drugs, such as ESBL, polymyxins, oxazolidinone and phenicol, in the treatment of multidrug-resistant (MDR) enterobacteria [50–52].

Out of the 1954 SSa strains evaluated in this study, 391 (20%) carried resistance genes indicative of an MDR phenotype; seven strains carried the *mcr-1* gene, and seventeen genomes carried the CTX-M gene (Supplementary Table S1). Interestingly, strains carrying these two resistance genes were isolated from different countries, including the United States, Mexico, Singapore, Australia, the United Kingdom, Thailand, Ethiopia, and Canada (Supplementary Table S1). SSa strains exhibiting MDR profiles have also been reported in various countries, including Morocco [53], Ethiopia [54–56], Denmark [40,57], Germany [39], Vietnam [58], Korea [59–61], and Brazil [14,62–64]. Interestingly, only a few reported outbreaks of SSa worldwide have been caused by MDR SSa strains [43,48,65]. Thirty-three percent of the MDR strains (130 out of 391) originated from the outbreak of *Salmonella* infections linked to raw turkey products in 2018 [38]. Additionally, the majority of other MDR genomes (n = 123) were deposited by the Gastrointestinal Bacteria Reference Unit in the United Kingdom and mainly derived from human clinical samples. It is worth noting that there is currently a high incidence of MDR strains within the entire genus *Salmonella* worldwide [66–68], particularly in developing countries with intensive animal production, such as Brazil, where the use of antimicrobials in livestock likely creates substantial selective pressure [69]. The reasons behind the lower frequency of AMR in SSa compared to other NTS *Salmonella* remain unexplained. Accurate determination of AMR is crucial in establishing national and global priorities, guiding public health interventions, and making treatment decisions regarding bacteria capable of causing infectious diseases or contributing to clinical infections, especially those associated with outbreaks.

The strain PP_BR059 also exhibited pan-susceptible phenotypic resistance. The results obtained from the software STARAMR showed agreement between genetic resistance, predicted phenotypic antimicrobial resistance, and phenotypic resistance observed in the disk diffusion test against a panel of 12 antimicrobials (Table 1). All downloaded genomes passed the quality check module of STARAMR (Supplementary Table S1). Interestingly, sixty-seven percent (1310 out of 1954) of the genomes, including the strain PP_BR059, showed no evidence of plasmids (Supplementary Table S1). The most frequently detected plasmids belonged to the incompatibility group I1 (IncI1), present in 369 out of the 1954 genomes (18.88%). Within this incompatibility group, we identified 110 out of the 1954 genomes (5.6%) that displayed no resistance genes, suggesting a pan-susceptible profile. On the other hand, 44 genomes carried IncI1 plasmids with the resistance genes *bla*CMY-2 and *bla*CTX-M, which confer resistance to the ESBLs ceftriaxone and cefoxitin. Although our analysis did not demonstrate that resistance genes are carried on the IncI1 plasmid, these plasmids are commonly detected in enteric bacteria and are also known to contribute to the horizontal transmission of AMR genes among enteric pathogens [13,70]. The combination of plasmids IncFIA (HI1), IncHI1A, and IncHI1B (R27) was the second most common, detected in 48 genomes; the IncX4 plasmid, the third most common, was present in 36 strains. Interestingly, all detected plasmid-mediated colistin resistance *mcr*-1 genes were found in the incompatibility group X4 plasmids (IncX4). Consistent with previous reports [71–73], IncX4, along with IncHI2 and IncI2, accounted for over 90% of reported *mcr*-1-bearing plasmids [74].

Table 1. The panel of antimicrobial drugs used to determine the phenotypic AMR level of the Strain PP_BR059.

Antimicrobial Drug	Susceptible [1] (S)	Intermediary Resistance [1] (I)	Resistant [1] (R)	PP_BR059	*E. coli* [2]	*S. Aureus* [2]	*P. Aeruginosa* [2]
Ampicillin (AM—10 μg)	≥17	14–16	≤13	24 (S)	19 (16–22)	31 (27–35)	R
Chloramphenicol (C—30 μg)	≥18	13–17	≤12	26 (S)	26 (21–27)	21 (19–26)	R
Gentamicin (GM—10 μg)	≥15	13–14	≤12	15 (S)	18 (19–26)	21 (19–27)	25 (16–21)
Amikacin (AN—30 μg)	≥17	15–16	≤14	22 (S)	25 (19–26)	23 (20–26)	26 (18–26)
Ciprofloxacin (CIP—5 μg)	≥21	16–20	≤15	36 (S)	34	22	26
Sulfamethoxazole /Trimethoprim (SXT—23.75 /1.25 μg)	≥16	11–15	≤10	27 (S)	26 (23–29)	27 (24–32)	R
Streptomycin (S—10 μg)	≥15	12–14	≤11	15 (S)	18 (12–20)	20 (14–22)	20
Tetracycline (TE—30 μg)	≥15	12–14	≤11	25 (S)	26 (18–25)	28 (24–30)	13
Amoxicillin /Clavulanic Acid (Amc—20/10 μg)	≥18	14–17	≤13	27 (S)	16 (18–24)	31 (28–36)	R
Nalidixid Acid (NAL—30 μg)	≥19	14–18	≤13	23 (S)	26	16	R
Sulfisoxazole (G—0.25 μg)	≥17	13–16	≤12	19 (S)	18 (15–23)	27 (24–34)	R
Ceftiofur (XNL—30 μg)	≥21	18–20	≤17	23 (S)	24 (26–31)	28 (27–31)	17 (14–18)

[1] Diameter of growth in millimeters. [2] Experiment control strains.

The pan-genome analysis of 1954 SSa genomes identified a total of 34,370 genes. Among these, 3384 genes were identified as core genes present in over 99% of the strains, while 30,986 genes were identified as accessory genes present in at least one strain, and 13,967 unique genes found in individual genomes (Figure 1 and Supplementary Table S2). Our pan-genome analysis aligns with previous studies on the pan-genome of *S.* enterica, based on relatively large datasets of genomes. For instance, Laing et al. identified a conserved core of 3200 genes, and a pan-genome of 25,300 genes from 4893 genomes [75]. Similarly, Park and Andam identified a conserved core of 2636 genes and a pan-genome of 33,257 genes from 297 genomes [76]. Finally, Yin et al. analyzed the genomes of 68 SSa strains and identified 3529 core genes in a pan-genome of 6360 genes [77]. The exact size of the *Salmonella* pan-genome depends on the total number of genomes analyzed and the chosen methodology [78]. However, it is evident that the pan-genome exhibits an "open" structure, similar to other pathogens such as *E. coli* [79]. Consistent with Yin's study, the conserved core genome of SSa identified here has likely reached a stable minimum; any additional strains would not substantially affect the core-gene content (Figure 1). However, the increased frequency of detection of accessory genes suggests the accumulation of unique genes found only in a few or unique isolates. Most of these accessory genes are attributed to mobile genetic elements (MGEs) due to *Salmonella*'s propensity to acquire and lose genetic content for adaptation to different hosts.

The cgMLST analysis revealed extensive diversity within the SSa strains (Supplementary Table S1). The allelic differences in cgMLST resulted in significantly higher discriminatory power when compared with traditional MLST, classifying 1954 genomes into 1437 groups, whereas traditional MLST generated 49 groups (Figure 1 and Supplementary Table S1). In general, most identical cgMLST schemes were found among strains from the same geographic origin, indicating their potential involvement in the same reported outbreak (see geographic location column in Supplementary Table S1). However, we identified some highly similar strains (with identical cgMLST schemes) reported in multiple countries, suggesting the global transmission of SSa, possibly through infected travelers or contaminated foods (highlighted in yellow in Supplementary Table S1). For example, the strains FDA478976-2 and PNUSAS026954 were collected from Mexico in 2008 and the United States in 2017, respectively, from jalapeño peppers and human stool. They were likely involved in the 2008 multi-country and multi-state foodborne outbreak of SSa, which caused over 1400 illnesses in the United States [5,7,44]. Moreover, strains 37779 and FDA8656199-1 were isolated in 2014 from the United Kingdom and the United States, respectively, from human clinical sources and avocados (Supplementary Table S1). Given the multiple potential pathways of transmission, including international travel, imported foods, and animal movements, it is not possible to definitively link these strains to United Kingdom-wide outbreaks based solely on the provided metadata. Lastly, as part of routine public health outbreak investigations in NSW, Australia, between October 2015 and December 2018, 37 isolates belonging to the BioProject PRJNA596817 underwent WGS [42,80]. Interestingly, strain 244421 from the United Kingdom, isolated from a human clinical source, shared an identical cgMLST scheme with these 37 Australian isolates, increasing the likelihood of SSa's potential involvement in the global dissemination of outbreak strains responsible for human clinical manifestations.

We conducted a genome-wide association study focusing on the top five most frequent sequence types (Figure 3). In this analysis, clinical isolates were categorized as strains from human and animal stools, urine, rectal swabs, tissues, wounds, and blood samples. In contrast, environmental isolates were represented by strains from environmental, food, and water sources. The SNPs analysis revealed that the two most prevalent sequence types, ST-50 and ST-27, diverged into two distinct lineages (Figure 3). ST-50 Lineage-1 consisted of 514 isolates, predominantly isolated from the United States (317), the United Kingdom (113), Mexico (48), and all strains from South American countries, including Brazil, Argentina, and Colombia (Supplementary Table S1). This lineage exhibited 37 unique SNPs and showed a balanced distribution between environmental and clinical sources. Conversely, ST-50 Lineage-2 comprised 277 isolates, primarily isolated from Australia (106), the United Kingdom (104), and the United States (57), along with strains from Thailand and Vietnam. Lineage-2 was characterized by the accumulation of 187 SNPs, with most isolates (253) originating from clinical sources. These findings suggest that Lineage-2 diverged from Lineage-1 through the acquisition of novel SNPs. It is plausible that these allelic differences have contributed to the persistence of these strains in clinical environments, as well as to their distinct geographic distribution, which is more prevalent in Europe, Asia, and Oceania compared to the Americas.

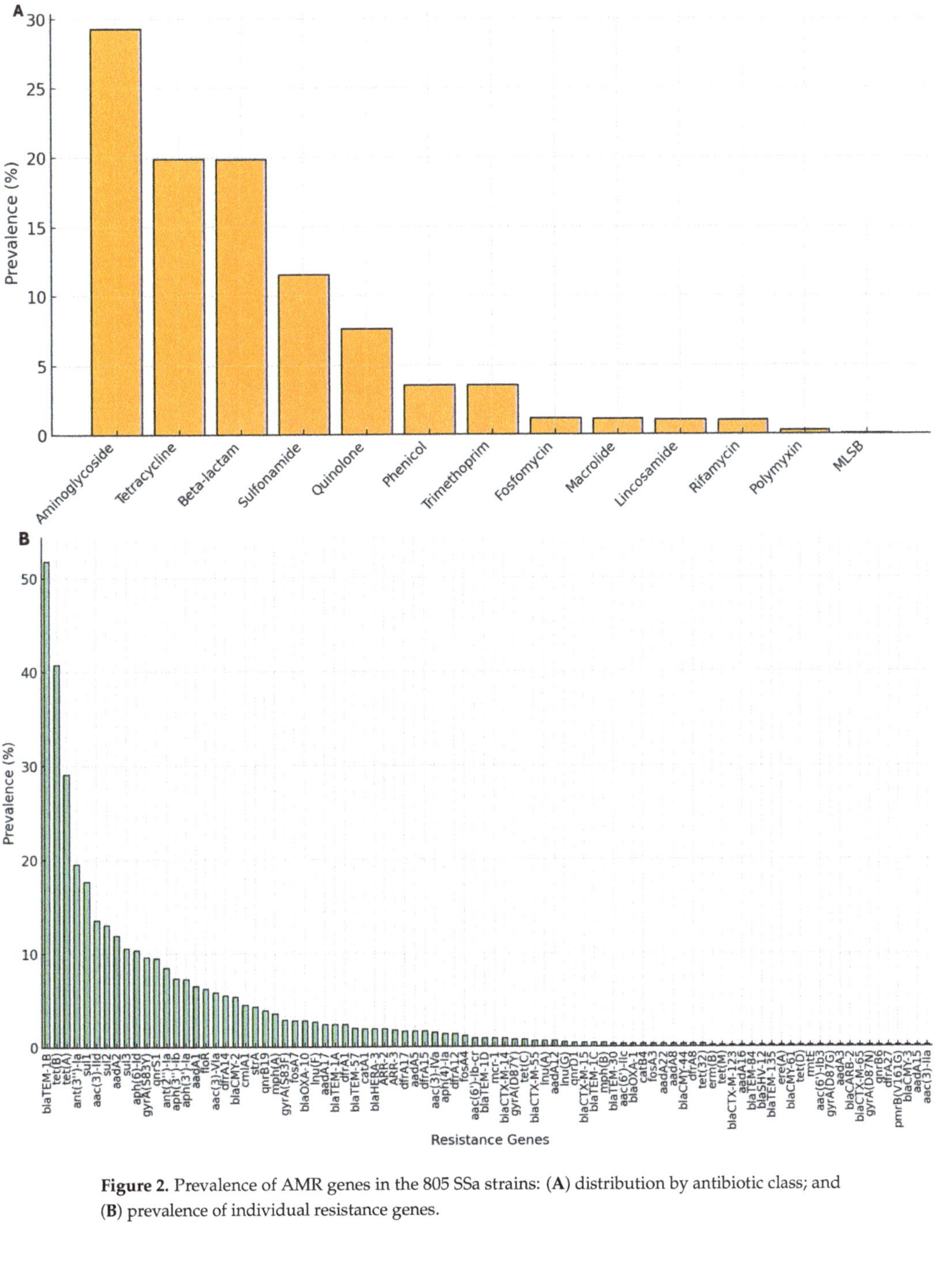

Figure 2. Prevalence of AMR genes in the 805 SSa strains: (**A**) distribution by antibiotic class; and (**B**) prevalence of individual resistance genes.

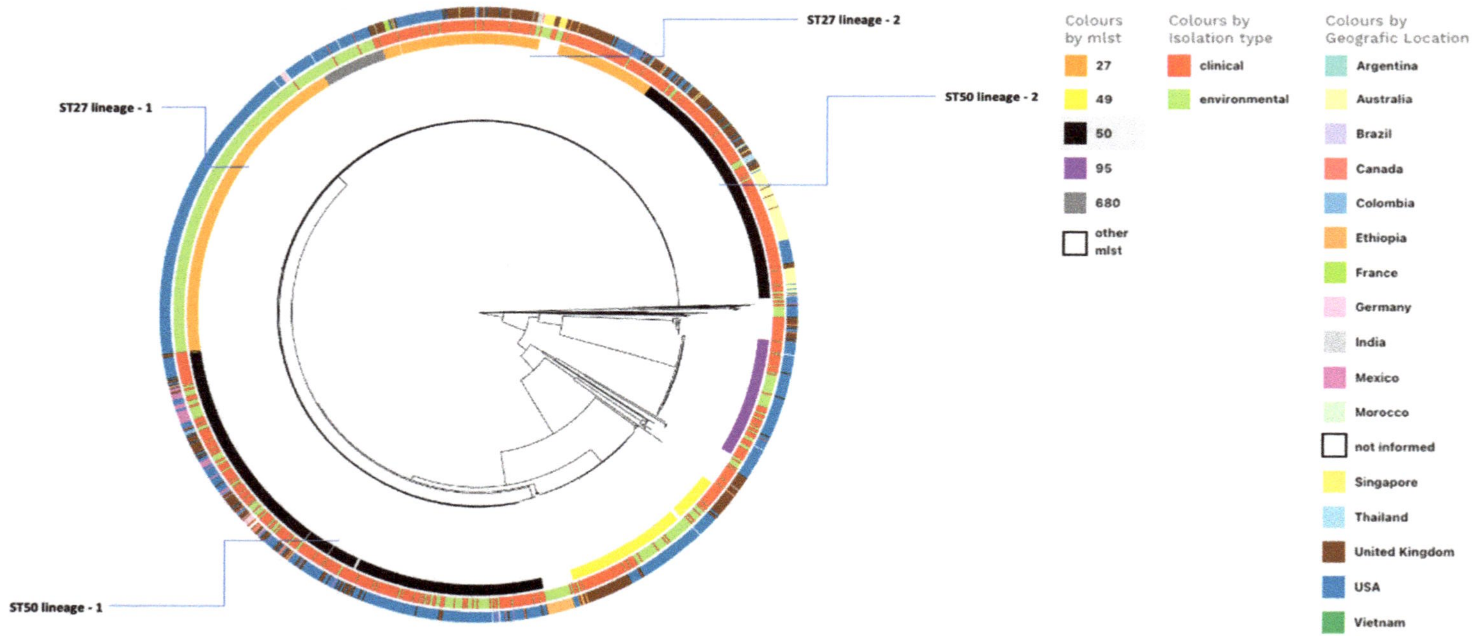

Figure 3. Genome-wide association analysis of single-nucleotide polymorphisms (SNPs) from the 1954 SSa strains, considering the top five most frequent sequence types, the geographic location, and the isolation type. The clinical isolation type is represented by strains originating from sources related to human and animal stools, urine, rectal swabs, tissues, wounds, and blood. The environmental isolation type is represented by strains originating from sources related to the environment, food, and water.

Regarding ST-27, Lineage-1 consisted of 377 isolates, primarily from the United States (349), and 29 other isolates from the United Kingdom, France, and Germany (Supplementary Table S1). This lineage displayed 82 unique SNPs and was mainly associated with environmental sources. In contrast, ST-27 Lineage-2 included 279 isolates, predominantly from the United Kingdom (*n* = 156) and the United States (*n* = 88), along with 35 strains from Australia, Canada, Ethiopia, and India. Lineage-2 was marked by 537 unique SNPs, with most isolates originating from clinical sources. Similar to ST-50, we hypothesize that ST-27 Lineage-1 and Lineage-2 are undergoing divergent evolutions via the accumulation of distinct SNPs, which are likely contributing to the differential niche adaptation in different geographic areas. Additionally, the ST680 cluster, comprising 70 strains, evolved from ST-27 Lineage-1 with a difference of only 6 unique SNPs. While most of these strains originated from environmental sources, six were derived from clinical samples.

These findings offer valuable insights into the phylogenetic relationships and evolutionary dynamics of SSa genomes. The accumulation of unique SNPs in Lineage-2 of both ST-50 and ST-27 suggests ongoing genetic changes that could influence the clinical behavior, geographic distribution, and environmental adaptation of these strains. Further research is needed to clarify the functional implications of these SNPs and their role in the epidemiology and pathogenicity of SSa. Gaining a deeper understanding of the evolutionary dynamics of these strains is crucial for developing effective strategies for the surveillance, prevention, and control of this pathogen and its associated diseases.

3.2. Invasiveness and Virulence Potential of the SSa Strain PP_BR059

To assess the virulence level and invasiveness of the PP_BR059 strain, we compared its invasion and intracellular survival in the human myelogenous leukemia macrophage (THP-1) cell line with the reference strains *Salmonella* Typhimurium LT2 and SL1344. Strain PP_BR059 exhibited a significantly low initial invasion/uptake (3.82%) in cultured human macrophages (Figure 4A); however, this strain showed significantly higher intra-macrophage survival at 2 h (68.72%, Figure 4B) and 20 h (25.68%, Figure 4C) post infection, when compared with the reference strains LT2 (5.69%/23.10%/0.16%) and SL1344 (5.43%/17.77%/2.47%), respectively. *Salmonella enterica* serovars can cause systemic infections by invading non-phagocytic intestinal cells and can survive within phagocytic defense cells such as macrophages and dendritic cells [2,81–86]. SSa is typically associated with outbreaks of NTS gastroenteritis worldwide [4–9,39,45,48,57,87], although some cases of systemic infections have been reported [43,47,88]. Higher intra-macrophage survival, particularly after 20 h post phagocytosis, suggests that strain PP_BR059 is likely more invasive than other strains. Although strain PP_BR059 was isolated from a fecal sample of a case associated with an outbreak, the patient's medical records were unavailable to confirm whether the infection became systemic. We recently reported NTS *Salmonella* Typhimurium in Brazil with high intramacrophage survival rates originating from systemic sites such as blood and cerebrospinal fluid [17]. It is currently unknown if SSa strains circulating in Brazil are also evolving to become more invasive.

Given that strain PP_BR059 exhibited significantly lower phagocytosis rates (Figure 4A), we hypothesized that this strain may also have lower motility compared to other NTS reference strains such as LT2 and SL1344 [89–94]. Surprisingly, strain PP_BR059 showed no impaired motility and did not significantly differ in motility compared to the reference strains (Figure 4D). These findings suggest that the ability of strain PP_BR059 to evade phagocytic engulfment is likely independent of the flagellar motility. To gain a better understanding of the genetic features underlying the above phenotypic characteristics, we performed WGS of strain PP_BR059. The genetic characterization of this strain is presented in the above section of this study.

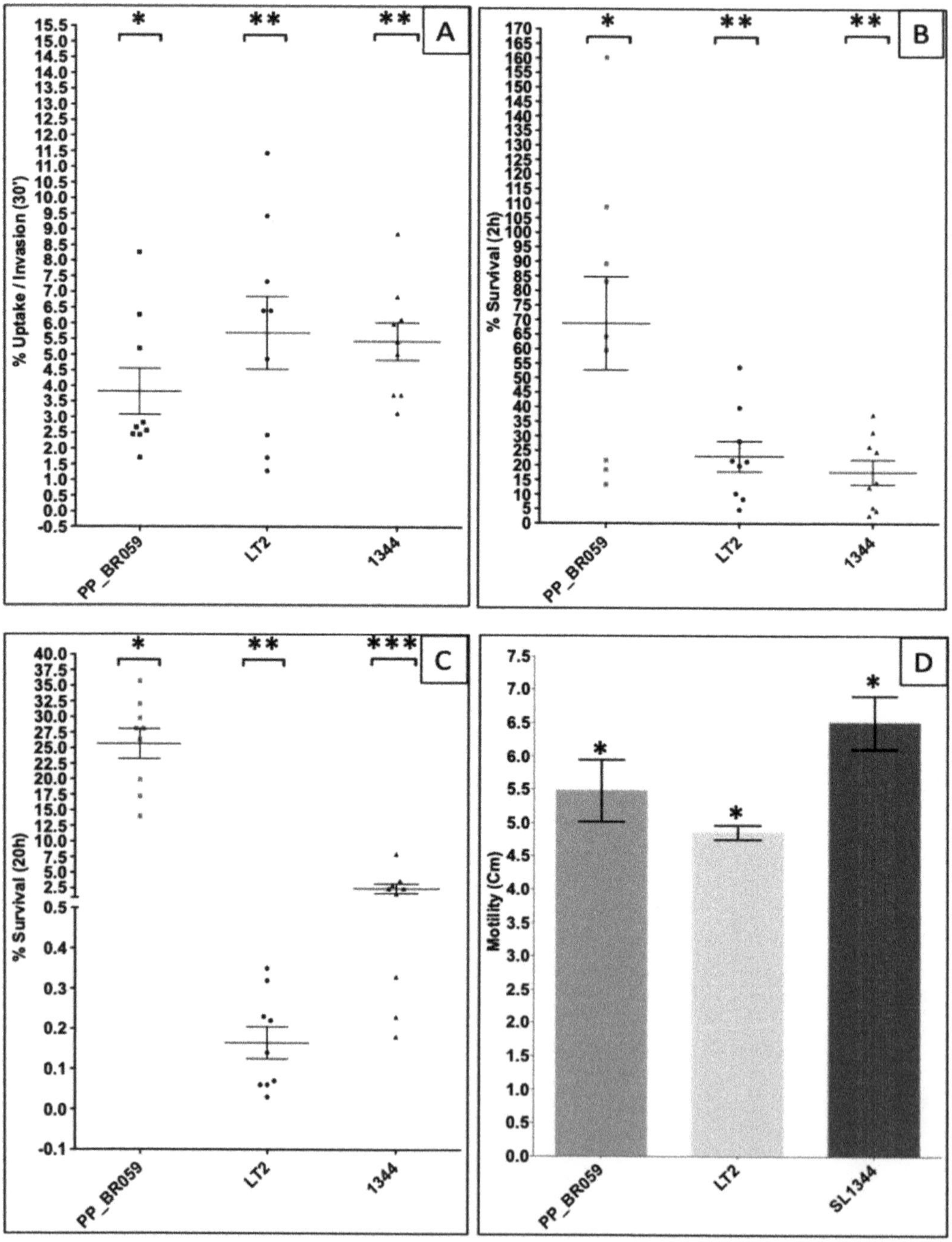

Figure 4. Percent uptake/invasion at 30 min (**A**). Survival at 2 h (**B**); and 20 h (**C**) post infection of the strain PP_BR059 in human macrophages (THP-1 cells). Motility of the strain PP_BR059 compared with the reference ST19 strains LT2 and SL1344 (**D**). Each strain number represents the mean from at least three independent experiments. Statistically significant ($p < 0.05$) differences are indicated by the number of asterisks (*/**/***).

4. Conclusions

In conclusion, this study provides valuable insights into the genomic epidemiology, antimicrobial resistance, and pathogenic potential of *Salmonella enterica* serovar SSa. Our analysis revealed substantial genomic diversity within SSa, with ST-50 and ST-27 accounting for the majority of global isolates. While ST-50 exhibited broad geographic distribution, ST-27 showed evidence of evolutionary divergence, with a subset associated with clinical infections and a higher prevalence of antimicrobial resistance genes. The characterization of strain PP_BR059, isolated from a hospitalized patient in Brazil identified this isolate as ST-50 and showed a significantly higher intra-macrophage survival rate at 2 h and 20 h post-infection. Whole-genome sequencing confirmed that PP_BR059 is pan-susceptible to antimicrobials, consistent with the overall trend of low resistance prevalence among SSa isolates from the global database. These findings underscore the need for continued genomic surveillance to monitor the evolutionary dynamics of emerging non-typhoidal *Salmonella* serovars. Understanding the genetic and phenotypic determinants of virulence and antimicrobial resistance in SSa will be crucial for assessing its public health significance and informing effective control strategies.

Supplementary Materials: The following supporting information can be downloaded at: https://www.mdpi.com/article/10.3390/genes16030254/s1, Table S1: Metadata of SSa isolates used in this study. Table provides detailed information on the isolates analyzed, including their geographic origin, source type (clinical, food, environmental, or animal), sequence type (ST), and associated BioProject accession numbers; Table S2: Pangenome analysis of SSa isolates. Table provides comprehensive pangenome data, including the total number of genes identified across 1,954 SSa genomes. It details the distribution of core, accessory and unique genes. Additionally, it includes the nucleotide sequences of all annotated genes.

Author Contributions: Conceptualization, P.P. and D.H.S.; Methodology, P.P. and D.H.S.; Formal Analysis, P.P.; Investigation, P.P.; Resources, D.H.S., D.d.P.R. and C.A.C.J.; Data Curation, P.P. and D.H.S.; Writing—Original Draft Preparation, P.P.; Writing—Review and Editing, P.P., D.H.S., D.d.P.R. and C.A.C.J.; Visualization, P.P.; Supervision, C.A.C.J.; Project Administration, C.A.C.J.; Funding Acquisition, D.H.S. and C.A.C.J. All authors have read and agreed to the published version of the manuscript.

Funding: This research was funded by Fundação Carlos Chagas Filho de Amparo à Pesquisa do Estado do Rio de Janeiro (FAPERJ) grant numbers E-26/200.891/2021 (Carlos Adam Conte Junior) and E-26/202.227/2018 (Pedro Panzenhagen).

Data Availability Statement: The original contributions presented in this study are included in the article/Supplementary Material. Further inquiries can be directed to the corresponding author.

Conflicts of Interest: The authors declare no conflicts of interest.

References

1. Balasubramanian, R.; Im, J.; Lee, J.-S.; Jeon, H.J.; Mogeni, O.D.; Kim, J.H.; Rakotozandrindrainy, R.; Baker, S.; Marks, F. The Global Burden and Epidemiology of Invasive Non-Typhoidal *Salmonella* Infections. *Hum. Vaccines Immunother.* **2019**, *15*, 1421–1426. [CrossRef]
2. Ao, T.T.; Feasey, N.A.; Gordon, M.A.; Crump, J.A. Global Burden of Invasive Nontyphoidal *Salmonella* Disease, 2010. *Emerg. Infect. Dis.* **2015**, *21*, 941. [CrossRef] [PubMed]
3. CDC. Outbreaks Involving *Salmonella*. In *Reports of Selected Salmonella Outbreak Investigations*; CDC: Atlanta, GA, USA, 2018.
4. Fey, P.D.; Iwen, P.C.; Zentz, E.B.; Briska, A.M.; Henkhaus, J.K.; Bryant, K.A.; Larson, M.A.; Noel, R.K.; Hinrichs, S.H. Assessment of Whole-Genome Mapping in a Well-Defined Outbreak of *Salmonella* Enterica Serotype Saintpaul. *J. Clin. Microbiol.* **2012**, *50*, 3063–3065. [CrossRef] [PubMed]
5. Klontz, K.C.; Klontz, J.C.; Mody, R.K.; Hoekstra, R.M. Analysis of Tomato and Jalapeño and Serrano Pepper Imports into the United States from Mexico before and during a National Outbreak of *Salmonella* Serotype Saintpaul Infections in 2008. *J. Food Prot.* **2010**, *73*, 1967–1974. [CrossRef] [PubMed]

6. Mody, R.K.; Greene, S.A.; Gaul, L.; Sever, A.; Pichette, S.; Zambrana, I.; Dang, T.; Gass, A.; Wood, R.; Herman, K.; et al. National Outbreak of *Salmonella* Serotype Saintpaul Infections: Importance of Texas Restaurant Investigations in Implicating Jalapeño Peppers. *PLoS ONE* **2011**, *6*, e16579. [CrossRef]
7. Taylor, E.; Kastner, J.; Renter, D. Challenges Involved in the *Salmonella* Saintpaul Outbreak and Lessons Learned. *J. Public Health Manag. Pract.* **2010**, *16*, 221–231. [CrossRef] [PubMed]
8. Yao, K.; Muruvanda, T.; Allard, M.W.; Hoffmann, M. Complete Genome Sequences of Three *Salmonella* Enterica Subsp. Enterica Serovar Saintpaul Isolates Associated with a 2013 Multistate Outbreak in the United States. *Genome Announc.* **2017**, *5*, e00456-17. [CrossRef] [PubMed]
9. Boore, A.L.; Jungk, J.; Russo, E.T.; Redd, J.T.; Angulo, F.J.; Williams, I.T.; Cheek, J.E.; Gould, L.H. Added Value of a Household-Level Study during an Outbreak Investigation of *Salmonella* Serotype Saintpaul Infections, New Mexico 2008. *Epidemiol. Infect.* **2013**, *141*, 2068–2073. [CrossRef] [PubMed]
10. Dyda, A.; Nguyen, P.-Y.; Chughtai, A.A.; MacIntyre, C.R. Changing Epidemiology of *Salmonella* Outbreaks Associated with Cucumbers and Other Fruits and Vegetables. *Glob. Biosecurity* **2020**, *2*, 1–13.
11. Mercês, A.F. Top 10 *Salmonella* Serovars Associated with Human Salmonellosis in Brazil (2011–2020). *Gen. Med.* **2022**, *6*, 1306.
12. Rau, R.B.; de Lima-Morales, D.; Wink, P.L.; Ribeiro, A.R.; Barth, A.L. *Salmonella* Enterica Mcr-1 Positive from Food in Brazil: Detection and Characterization. *Foodborne Pathog. Dis.* **2010**, *17*, 202–208. [CrossRef] [PubMed]
13. Rodrigues, I.; Silva, R.; Menezes, J.; Machado, S.; Rodrigues, D.; Pomba, C.; Abreu, D.; Nascimento, E.; Aquino, M.; Pereira, V. High Prevalence of Multidrug-Resistant Nontyphoidal *Salmonella* Recovered from Broiler Chickens and Chicken Carcasses in Brazil. *Braz. J. Poult. Sci.* **2020**, *22*, eRBCA-2019-1206. [CrossRef]
14. Viegas, F.M.; Ramos, C.P.; Xavier, R.G.C.; Lopes, E.O.; Júnior, C.A.O.; Bagno, R.M.; Diniz, A.N.; Lobato, F.C.F.; Silva, R.O.S. Fecal Shedding of *Salmonella* Spp., *Clostridium Perfringens*, and *Clostridioides Difficile* in Dogs Fed Raw Meat-Based Diets in Brazil and Their Owners' Motivation. *PLoS ONE* **2020**, *15*, e0231275. [CrossRef] [PubMed]
15. CLSI. *Performance Standards for Antimicrobial Susceptibility Testing*, 27th ed.; CLSI Supplement M100; Clinical and Laboratory Standards Institute: Wayne, PA, USA, 2017.
16. Shah, D.H.; Zhou, X.; Addwebi, T.; Davis, M.A.; Call, D.R. In Vitro and in Vivo Pathogenicity of *Salmonella* Enteritidis Clinical Strains Isolated from North America. *Arch. Microbiol.* **2011**, *193*, 811–821. [CrossRef] [PubMed]
17. Panzenhagen, P.H.N.; Paul, N.C.; Conte, C.A.; Costa, R.G.; Rodrigues, D.P.; Shah, D.H. Genetically Distinct Lineages of *Salmonella* Typhimurium ST313 and ST19 Are Present in Brazil. *Int. J. Med. Microbiol.* **2018**, *308*, 306–316. [CrossRef] [PubMed]
18. Bolger, A.M.; Lohse, M.; Usadel, B. Trimmomatic: A Flexible Trimmer for Illumina Sequence Data. *Bioinformatics* **2014**, *30*, 2114–2120. [CrossRef]
19. Magoč, T.; Salzberg, S.L. FLASH: Fast Length Adjustment of Short Reads to Improve Genome Assemblies. *Bioinformatics* **2011**, *27*, 2957–2963. [CrossRef]
20. Bankevich, A.; Nurk, S.; Antipov, D.; Gurevich, A.A.; Dvorkin, M.; Kulikov, A.S.; Lesin, V.M.; Nikolenko, S.I.; Pham, S.; Prjibelski, A.D.; et al. SPAdes: A New Genome Assembly Algorithm and Its Applications to Single-Cell Sequencing. *J. Comput. Biol.* **2012**, *19*, 455–477. [CrossRef] [PubMed]
21. Gurevich, A.; Saveliev, V.; Vyahhi, N.; Tesler, G. QUAST: Quality Assessment Tool for Genome Assemblies. *Bioinformatics* **2013**, *29*, 1072–1075. [CrossRef] [PubMed]
22. Bharat, A.; Petkau, A.; Avery, B.P.; Chen, J.C.; Folster, J.P.; Carson, C.A.; Kearney, A.; Nadon, C.; Mabon, P.; Thiessen, J.; et al. Correlation between Phenotypic and In Silico Detection of Antimicrobial Resistance in *Salmonella* Enterica in Canada Using Staramr. *Microorganisms* **2022**, *10*, 292. [CrossRef] [PubMed]
23. Clausen, P.T.L.C.; Aarestrup, F.M.; Lund, O. Rapid and Precise Alignment of Raw Reads against Redundant Databases with KMA. *BMC Bioinform.* **2018**, *19*, 307. [CrossRef] [PubMed]
24. Alikhan, N.-F.; Zhou, Z.; Sergeant, M.J.; Achtman, M. A Genomic Overview of the Population Structure of *Salmonella*. *PLoS Genet.* **2018**, *14*, e1007261. [CrossRef]
25. Larsen, M.V.; Cosentino, S.; Rasmussen, S.; Friis, C.; Hasman, H.; Marvig, R.L.; Jelsbak, L.; Sicheritz-Pontén, T.; Ussery, D.W.; Aarestrup, F.M.; et al. Multilocus Sequence Typing of Total-Genome-Sequenced Bacteria. *J. Clin. Microbiol.* **2012**, *50*, 1355–1361. [CrossRef] [PubMed]
26. Zankari, E.; Hasman, H.; Cosentino, S.; Vestergaard, M.; Rasmussen, S.; Lund, O.; Aarestrup, F.M.; Larsen, M.V. Identification of Acquired Antimicrobial Resistance Genes. *J. Antimicrob. Chemother.* **2012**, *67*, 2640–2644. [CrossRef]
27. Zankari, E.; Allesøe, R.; Joensen, K.G.; Cavaco, L.M.; Lund, O.; Aarestrup, F.M. PointFinder: A Novel Web Tool for WGS-Based Detection of Antimicrobial Resistance Associated with Chromosomal Point Mutations in Bacterial Pathogens. *J. Antimicrob. Chemother.* **2017**, *72*, 2764–2768. [CrossRef]
28. Carattoli, A.; Zankari, E.; García-Fernández, A.; Larsen, M.V.; Lund, O.; Villa, L.; Aarestrup, F.M.; Hasman, H. In Silico Detection and Typing of Plasmids Using PlasmidFinder and Plasmid Multilocus Sequence Typing. *Antimicrob. Agents Chemother.* **2014**, *58*, 3895–3903. [CrossRef] [PubMed]

29. Page, A.J.; Cummins, C.A.; Hunt, M.; Wong, V.K.; Reuter, S.; Holden, M.T.G.; Fookes, M.; Falush, D.; Keane, J.A.; Parkhill, J. Roary: Rapid Large-Scale Prokaryote Pan Genome Analysis. *Bioinformatics* **2015**, *31*, 3691–3693. [CrossRef] [PubMed]
30. Seemann, T. Prokka: Rapid Prokaryotic Genome Annotation. *Bioinformatics* **2014**, *30*, 2068–2069. [CrossRef] [PubMed]
31. Trifinopoulos, J.; Nguyen, L.-T.; von Haeseler, A.; Minh, B.Q. W-IQ-TREE: A Fast Online Phylogenetic Tool for Maximum Likelihood Analysis. *Nucleic Acids Res.* **2016**, *44*, W232–W235. [CrossRef] [PubMed]
32. Huerta-Cepas, J.; Forslund, K.; Coelho, L.P.; Szklarczyk, D.; Jensen, L.J.; Von Mering, C.; Bork, P. Fast Genome-Wide Functional Annotation through Orthology Assignment by eggNOG-Mapper. *Mol. Biol. Evol.* **2017**, *34*, 2115–2122. [CrossRef] [PubMed]
33. Huerta-Cepas, J.; Szklarczyk, D.; Heller, D.; Hernández-Plaza, A.; Forslund, S.K.; Cook, H.; Mende, D.R.; Letunic, I.; Rattei, T.; Jensen, L.J.; et al. eggNOG 5.0: A Hierarchical, Functionally and Phylogenetically Annotated Orthology Resource Based on 5090 Organisms and 2502 Viruses. *Nucleic Acids Res.* **2019**, *47*, D309–D314. [CrossRef]
34. Buchfink, B.; Xie, C.; Huson, D.H. Fast and Sensitive Protein Alignment Using DIAMOND. *Nat. Methods* **2015**, *12*, 59–60. [CrossRef]
35. Hadfield, J.; Croucher, N.J.; Goater, R.J.; Abudahab, K.; Aanensen, D.M.; Harris, S.R. Phandango: An Interactive Viewer for Bacterial Population Genomics. *Bioinformatics* **2018**, *34*, 292–293. [CrossRef]
36. Croucher, N.J.; Page, A.J.; Connor, T.R.; Delaney, A.J.; Keane, J.A.; Bentley, S.D.; Parkhill, J.; Harris, S.R. Rapid Phylogenetic Analysis of Large Samples of Recombinant Bacterial Whole Genome Sequences Using Gubbins. *Nucleic Acids Res.* **2015**, *43*, e15. [CrossRef]
37. Page, A.J.; Taylor, B.; Delaney, A.J.; Soares, J.; Seemann, T.; Keane, J.A.; Harris, S.R. SNP-Sites: Rapid Efficient Extraction of SNPs from Multi-FASTA Alignments. *Microb. Genom.* **2016**, *2*, e000056. [CrossRef] [PubMed]
38. CDC. Multistate Outbreak of Salmonella Reading Infections Linked to Raw Turkey Products. Available online: https://archive.cdc.gov/#/details?url=https://www.cdc.gov/salmonella/reading-07-18/index.html (accessed on 4 January 2025).
39. Beutlich, J.; Rodríguez, I.; Schroeter, A.; Käsbohrer, A.; Helmuth, R.; Guerra, B. A Predominant Multidrug-Resistant *Salmonella* Enterica Serovar Saintpaul Clonal Line in German Turkey and Related Food Products. *Appl. Environ. Microbiol.* **2010**, *76*, 3657–3667. [CrossRef]
40. Cavaco, L.M.; Korsgaard, H.; Sørensen, G.; Aarestrup, F.M. Plasmid-Mediated Quinolone Resistance Due to qnrB5 and qnrS1 Genes in *Salmonella* Enterica Serovars Newport, Hadar and Saintpaul Isolated from Turkey Meat in Denmark. *J. Antimicrob. Chemother.* **2008**, *62*, 632–634. [CrossRef]
41. Cargnel, M.; Filippitzi, M.; Van Cauteren, D.; Mattheus, W.; Botteldoorn, N.; Cambier, L.; Welby, S. Assessing Evidence of a Potential *Salmonella* Transmission across the Poultry Food Chain. *Zoonoses Public Health* **2023**, *70*, 22–45. [CrossRef] [PubMed]
42. Rockett, R.J.; Arnott, A.; Wang, Q.; Howard, P.; Sintchenko, V. Genomic Surveillance Enables Suitability Assessment of *Salmonella* Gene Targets Used for Culture-Independent Diagnostic Testing. *J. Clin. Microbiol.* **2020**, *58*, e00038-20. [CrossRef] [PubMed]
43. Thomson, R.M.; Henderson, H.J.; Smith-Palmer, A. An Outbreak of *Salmonella* Saintpaul in a Scottish Childcare Facility: The Influence of Parental under-Reporting. *BMC Infect. Dis.* **2019**, *19*, 847. [CrossRef] [PubMed]
44. Barton Behravesh, C.; Mody, R.K.; Jungk, J.; Gaul, L.; Redd, J.T.; Chen, S.; Cosgrove, S.; Hedican, E.; Sweat, D.; Chávez-Hauser, L.; et al. 2008 Outbreak of *Salmonella* Saintpaul Infections Associated with Raw Produce. *N. Engl. J. Med.* **2011**, *364*, 918–927. [CrossRef]
45. Beatty, M.E.; LaPorte, T.N.; Phan, Q.; Van Duyne, S.V.; Braden, C. A Multistate Outbreak of *Salmonella* Enterica Serotype Saintpaul Infections Linked to Mango Consumption: A Recurrent Theme. *Clin. Infect. Dis. Off. Publ. Infect. Dis. Soc. Am.* **2004**, *38*, 1337–1338. [CrossRef] [PubMed]
46. Bornemann, R.; Zerr, D.M.; Heath, J.; Koehler, J.; Grandjean, M.; Pallipamu, R.; Duchin, J. An Outbreak of *Salmonella* Serotype Saintpaul in a Children's Hospital. *Infect. Control Hosp. Epidemiol.* **2002**, *23*, 671–676. [CrossRef] [PubMed]
47. Draper, A.D.; Morton, C.N.; Heath, J.N.; Lim, J.A.; Markey, P.G. An Outbreak of *Salmonella* Saintpaul Gastroenteritis after Attending a School Camp in the Northern Territory, Australia. *Commun. Dis. Intell.* **2017**, *41*, E10–E15.
48. Jain, S.; Bidol, S.A.; Austin, J.L.; Berl, E.; Elson, F.; Williams, M.L.; Deasy, M.; Moll, M.E.; Rea, V.; Vojdani, J.D.; et al. Multistate Outbreak of *Salmonella* Typhimurium and Saintpaul Infections Associated with Unpasteurized Orange Juice—United States, 2005. *Clin. Infect. Dis.* **2009**, *48*, 1065–1071. [CrossRef]
49. Munnoch, S.A.; Ward, K.; Sheridan, S.; Fitzsimmons, G.J.; Shadbolt, C.T.; Piispanen, J.P.; Wang, Q.; Ward, T.J.; Worgan, T.L.M.; Oxenford, C.; et al. A Multi-State Outbreak of *Salmonella* Saintpaul in Australia Associated with Cantaloupe Consumption. *Epidemiol. Infect.* **2009**, *137*, 367–374. [CrossRef] [PubMed]
50. Bevan, E.R.; Jones, A.M.; Hawkey, P.M. Global Epidemiology of CTX-M β-Lactamases: Temporal and Geographical Shifts in Genotype. *J. Antimicrob. Chemother.* **2017**, *72*, 2145–2155. [CrossRef] [PubMed]
51. Khan, A.U.; Maryam, L.; Zarrilli, R. Structure, Genetics and Worldwide Spread of New Delhi Metallo-β-Lactamase (NDM): A Threat to Public Health. *BMC Microbiol.* **2017**, *17*, 101. [CrossRef]

52. Wang, Y.; Lv, Y.; Cai, J.; Schwarz, S.; Cui, L.; Hu, Z.; Zhang, R.; Li, J.; Zhao, Q.; He, T.; et al. A Novel Gene, *optrA*, That Confers Transferable Resistance to Oxazolidinones and Phenicols and Its Presence in *Enterococcus Faecalis* and *Enterococcus Faecium* of Human and Animal Origin. *J. Antimicrob. Chemother.* **2015**, *70*, 2182–2190. [CrossRef] [PubMed]
53. Amajoud, N.; Bouchrif, B.; El Maadoudi, M.; Skalli Senhaji, N.; Karraouan, B.; El Harsal, A.; El Abrini, J. Prevalence, Serotype Distribution, and Antimicrobial Resistance of *Salmonella* Isolated from Food Products in Morocco. *J. Infect. Dev. Ctries.* **2017**, *11*, 136–142. [CrossRef] [PubMed]
54. Eguale, T.; Asrat, D.; Alemayehu, H.; Nana, I.; Gebreyes, W.A.; Gunn, J.S.; Engidawork, E. Phenotypic and Genotypic Characterization of Temporally Related Nontyphoidal *Salmonella* Strains Isolated from Humans and Food Animals in Central Ethiopia. *Zoonoses Public Health* **2018**, *65*, 766–776. [CrossRef]
55. Hiko, A.; Irsigler, H.; Bräutigam, L.; Ameni, G.; Fries, R.; Maximilian, B. Antimicrobial Resistance and Genotypic Profiles of *Salmonella* Saintpaul Isolated along Beef Processing and Distribution Continuum. *Heliyon* **2018**, *4*, e01025. [CrossRef]
56. Molla, B.; Miko, A.; Pries, K.; Hildebrandt, G.; Kleer, J.; Schroeter, A.; Helmuth, R. Class 1 Integrons and Resistance Gene Cassettes among Multidrug Resistant *Salmonella* Serovars Isolated from Slaughter Animals and Foods of Animal Origin in Ethiopia. *Acta Trop.* **2007**, *103*, 142–149. [CrossRef] [PubMed]
57. Baggesen, D.L.; Wegener, H.C.; Christensen, J.P. Typing of *Salmonella* Enterica Serovar Saintpaul: An Outbreak Investigation. *APMIS* **1996**, *104*, 411–418. [CrossRef] [PubMed]
58. Nguyen, D.T.A.; Kanki, M.; Nguyen, P.D.; Le, H.T.; Ngo, P.T.; Tran, D.N.M.; Le, N.H.; Dang, C.V.; Kawai, T.; Kawahara, R.; et al. Prevalence, Antibiotic Resistance, and Extended-Spectrum and AmpC β-Lactamase Productivity of *Salmonella* Isolates from Raw Meat and Seafood Samples in Ho Chi Minh City, Vietnam. *Int. J. Food Microbiol.* **2016**, *236*, 115–122. [CrossRef] [PubMed]
59. Lee, K. Diversity of TEM-52 Extended-Spectrum -Lactamase-Producing Non-Typhoidal *Salmonella* Isolates in Korea. *J. Antimicrob. Chemother.* **2003**, *52*, 493–496. [CrossRef]
60. Singh, B.R.; Singh, P.; Agrawal, S.; Teotia, U.; Verma, A.; Sharma, S.; Chandra, M.; Babu, N.; Kant Agarwal, R. Prevalence of Multidrug Resistant *Salmonella* in Coriander, Mint, Carrot, and Radish in Bareilly and Kanpur, Northern India. *Foodborne Pathog. Dis.* **2007**, *4*, 233–240. [CrossRef] [PubMed]
61. Singh, B.R.; Singh, M.; Singh, P.; Babu, N.; Chandra, M.; Agarwal, R.K. Prevalence of Multidrug-Resistant *Salmonella* on Ready-to-Eat Betel Leaves (Paan) and in Water Used for Soaking Betel Leaves in North Indian Cities. *J. Food Prot.* **2006**, *69*, 288–292. [CrossRef] [PubMed]
62. Mendonça, E.P.; de Melo, R.T.; Nalevaiko, P.C.; Monteiro, G.P.; Fonseca, B.B.; Galvão, N.N.; Giombelli, A.; Rossi, D.A. Spread of the Serotypes and Antimicrobial Resistance in Strains of *Salmonella* Spp. Isolated from Broiler. *Braz. J. Microbiol.* **2019**, *50*, 515–522. [CrossRef] [PubMed]
63. Scur, M.C.; da Silva Pinto, F.G.; De Bona, E.A.M.; Weber, L.D.; Alves, L.F.A.; de Moura, A.C. Occurrence and Antimicrobial Resistance of *Salmonella* Serotypes Isolates Recovered from Poultry of Western Paran, Brazil. *Afr. J. Agric. Res.* **2014**, *9*, 823–830. [CrossRef]
64. Wilson, I.G. Antimicrobial Resistance of *Salmonella* in Raw Retail Chickens, Imported Chicken Portions, and Human Clinical Specimens. *J. Food Prot.* **2004**, *67*, 1220–1225. [CrossRef]
65. Hayford, A.E.; Brown, E.W.; Zhao, S.; Mammel, M.K.; Gangiredla, J.; Abbott, J.W.; Friedman, S.L.; Ayers, S.L.; Lewis, J.L.; Lacher, D.W.; et al. Genetic and Resistance Phenotypic Subtyping of *Salmonella* Saintpaul Isolates from Various Food Sources and Humans: Phylogenetic Concordance in Combinatory Analyses. *Infect. Genet. Evol.* **2015**, *36*, 92–107. [CrossRef]
66. Eng, S.-K.; Pusparajah, P.; Ab Mutalib, N.-S.; Ser, H.-L.; Chan, K.-G.; Lee, L.-H. *Salmonella*: A Review on Pathogenesis, Epidemiology and Antibiotic Resistance. *Front. Life Sci.* **2015**, *8*, 284–293. [CrossRef]
67. Hur, J.; Jawale, C.; Lee, J.H. Antimicrobial Resistance of *Salmonella* Isolated from Food Animals: A Review. *Food Res. Int.* **2012**, *45*, 819–830. [CrossRef]
68. Su, L.-H.; Chiu, C.-H.; Chu, C.; Ou, J.T. Antimicrobial Resistance in Nontyphoid *Salmonella* Serotypes: A Global Challenge. *Clin. Infect. Dis.* **2004**, *39*, 546–551. [CrossRef] [PubMed]
69. Rodrigues, G.L.; Panzenhagen, P.; Ferrari, R.G.; dos Santos, A.; Paschoalin, V.M.F.; Conte-Junior, C.A. Frequency of Antimicrobial Resistance Genes in *Salmonella* From Brazil by in Silico Whole-Genome Sequencing Analysis: An Overview of the Last Four Decades. *Front. Microbiol.* **2020**, *11*, 1864. [CrossRef]
70. Smith, H.; Bossers, A.; Harders, F.; Wu, G.; Woodford, N.; Schwarz, S.; Guerra, B.; Rodríguez, I.; van Essen-Zandbergen, A.; Brouwer, M.; et al. Characterization of Epidemic IncI1-Iγ Plasmids Harboring Ambler Class A and C Genes in Escherichia Coli and *Salmonella* Enterica from Animals and Humans. *Antimicrob. Agents Chemother.* **2015**, *59*, 5357–5365. [CrossRef] [PubMed]
71. Cui, M.; Zhang, J.; Gu, Z.; Li, R.; Chan, E.W.; Yan, M.; Wu, C.; Xu, X.; Chen, S. Prevalence and Molecular Characterization of *Mcr-1*-Positive *Salmonella* Strains Recovered from Clinical Specimens in China. *Antimicrob. Agents Chemother.* **2017**, *61*, 10-1128. [CrossRef] [PubMed]

72. Doumith, M.; Godbole, G.; Ashton, P.; Larkin, L.; Dallman, T.; Day, M.; Day, M.; Muller-Pebody, B.; Ellington, M.J.; de Pinna, E.; et al. Detection of the Plasmid-Mediated *Mcr-1* Gene Conferring Colistin Resistance in Human and Food Isolates of *Salmonella Enterica* and *Escherichia Coli* in England and Wales. *J. Antimicrob. Chemother.* **2016**, *71*, 2300–2305. [CrossRef] [PubMed]
73. Poirel, L.; Jayol, A.; Nordmann, P. Polymyxins: Antibacterial Activity, Susceptibility Testing, and Resistance Mechanisms Encoded by Plasmids or Chromosomes. *Clin. Microbiol. Rev.* **2017**, *30*, 557–596. [CrossRef]
74. COMBAT Consortium; Matamoros, S.; van Hattem, J.M.; Arcilla, M.S.; Willemse, N.; Melles, D.C.; Penders, J.; Vinh, T.N.; Thi Hoa, N.; de Jong, M.D.; et al. Global Phylogenetic Analysis of Escherichia Coli and Plasmids Carrying the Mcr-1 Gene Indicates Bacterial Diversity but Plasmid Restriction. *Sci. Rep.* **2017**, *7*, 15364. [CrossRef]
75. Laing, C.R.; Whiteside, M.D.; Gannon, V.P.J. Pan-Genome Analyses of the Species *Salmonella* Enterica, and Identification of Genomic Markers Predictive for Species, Subspecies, and Serovar. *Front. Microbiol.* **2017**, *8*, 1345. [CrossRef]
76. Park, C.J.; Andam, C.P. Distinct but Intertwined Evolutionary Histories of Multiple *Salmonella Enterica* Subspecies. *mSystems* **2020**, *5*, e00515-19. [CrossRef] [PubMed]
77. Yin, Z.; Liu, J.; Du, B.; Ruan, H.-H.; Huo, Y.-X.; Du, Y.; Qiao, J. Whole-Genome-Based Survey for Polyphyletic Serovars of *Salmonella* Enterica Subsp. Enterica Provides New Insights into Public Health Surveillance. *Int. J. Mol. Sci.* **2020**, *21*, 5226. [CrossRef] [PubMed]
78. Jacobsen, A.; Hendriksen, R.S.; Aaresturp, F.M.; Ussery, D.W.; Friis, C. The *Salmonella* Enterica Pan-Genome. *Microb. Ecol.* **2011**, *62*, 487–504. [CrossRef] [PubMed]
79. Touchon, M.; Hoede, C.; Tenaillon, O.; Barbe, V.; Baeriswyl, S.; Bidet, P.; Bingen, E.; Bonacorsi, S.; Bouchier, C.; Bouvet, O.; et al. Organised Genome Dynamics in the Escherichia Coli Species Results in Highly Diverse Adaptive Paths. *PLoS Genet.* **2009**, *5*, e1000344. [CrossRef] [PubMed]
80. Iveson, J.B.; Bradshaw, S.D.; Smith, D.W. The Movement of Humans and the Spread of *Salmonella* into Existing and Pristine Ecosystems. *Microbiol. Aust.* **2017**, *38*, 201–203. [CrossRef]
81. Crump, J.A.; Heyderman, R.S. Invasive *Salmonella* Infections in Africa. *Trans. R. Soc. Trop. Med. Hyg.* **2014**, *108*, 673–675. [CrossRef]
82. Feasey, N.A.; Dougan, G.; Kingsley, R.A.; Heyderman, R.S.; Gordon, M.A. Invasive Non-Typhoidal *Salmonella* Disease: An Emerging and Neglected Tropical Disease in Africa. *Lancet* **2012**, *379*, 2489–2499. [CrossRef]
83. Lan, N.P.H.; Phuong, T.L.T.; Huu, H.N.; Thuy, L.; Mather, A.E.; Park, S.E.; Marks, F.; Thwaites, G.E.; Chau, N.V.V.; Thompson, C.N.; et al. Invasive Non-Typhoidal *Salmonella* Infections in Asia: Clinical Observations, Disease Outcome and Dominant Serovars from an Infectious Disease Hospital in Vietnam. *PLoS Negl. Trop. Dis.* **2016**, *10*, e0004857. [CrossRef]
84. Okoro, C.K.; Barquist, L.; Connor, T.R.; Harris, S.R.; Clare, S.; Stevens, M.P.; Arends, M.J.; Hale, C.; Kane, L.; Pickard, D.J.; et al. Signatures of Adaptation in Human Invasive *Salmonella* Typhimurium ST313 Populations from Sub-Saharan Africa. *PLoS Negl. Trop. Dis.* **2015**, *9*, e0003611. [CrossRef]
85. Ramachandran, G.; Aheto, K.; Shirtliff, M.E.; Tennant, S.M. Poor Biofilm-Forming Ability and Long-Term Survival of Invasive *Salmonella* Typhimurium ST313. *Pathog. Dis.* **2016**, *74*, ftw049. [CrossRef] [PubMed]
86. Vugia, D.J.; Samuel, M.; Farley, M.M.; Marcus, R.; Shiferaw, B.; Shallow, S.; Smith, K.; Angulo, F.J. Invasive *Salmonella* Infections in the United States, FoodNet, 1996–1999: Incidence, Serotype Distribution, and Outcome. *Clin. Infect. Dis.* **2004**, *38*, S149–S156. [CrossRef] [PubMed]
87. Akiyama, T.; Khan, A.A.; Cheng, C.-M.; Stefanova, R. Molecular Characterization of *Salmonella* Enterica Serovar Saintpaul Isolated from Imported Seafood, Pepper, Environmental and Clinical Samples. *Food Microbiol.* **2011**, *28*, 1124–1128. [CrossRef]
88. Daitsu, T.; Maeda, S.; Abe, A.; Kanazawa, C.; Shimizu, Y. Relapse of *Salmonella* Saintpaul Meningitis in a 4-Month-Old Infant. *Pediatr. Int. Off. J. Jpn. Pediatr. Soc.* **2017**, *59*, 233–234. [CrossRef]
89. Jones, B.D.; Lee, C.A.; Falkow, S. Invasion by *Salmonella* Typhimurium Is Affected by the Direction of Flagellar Rotation. *Infect. Immun.* **1992**, *60*, 2475–2480. [CrossRef] [PubMed]
90. Ramachandran, G.; Perkins, D.J.; Schmidlein, P.J.; Tulapurkar, M.E.; Tennant, S.M. Invasive *Salmonella* Typhimurium ST313 with Naturally Attenuated Flagellin Elicits Reduced Inflammation and Replicates within Macrophages. *PLoS Negl. Trop. Dis.* **2015**, *9*, e3394. [CrossRef] [PubMed]
91. Schmitt, C.K.; Ikeda, J.S.; Darnell, S.C.; Watson, P.R.; Bispham, J.; Wallis, T.S.; Weinstein, D.L.; Metcalf, E.S.; O'Brien, A.D. Absence of All Components of the Flagellar Export and Synthesis Machinery Differentially Alters Virulence of *Salmonella* Enterica Serovar Typhimurium in Models of Typhoid Fever, Survival in Macrophages, Tissue Culture Invasiveness, and Calf Enterocolitis. *Infect. Immun.* **2001**, *69*, 5619–5625. [CrossRef]
92. Tomita, T.; Kanegasaki, S. Enhanced Phagocytic Response of Macrophages to Bacteria by Physical Impact Caused by Bacterial Motility or Centrifugation. *Infect. Immun.* **1982**, *38*, 865–870. [CrossRef] [PubMed]

93. Weinstein, D.L.; Carsiotis, M.; Lissner, C.R.; O'Brien, A.D. Flagella Help *Salmonella* Typhimurium Survive within Murine Macrophages. *Infect. Immun.* **1984**, *46*, 819–825. [CrossRef]
94. Yim, L.; Sasías, S.; Martínez, A.; Betancor, L.; Estevez, V.; Scavone, P.; Bielli, A.; Sirok, A.; Chabalgoity, J.A. Repression of Flagella Is a Common Trait in Field Isolates of *Salmonella* Enterica Serovar Dublin and Is Associated with Invasive Human Infections. *Infect. Immun.* **2014**, *82*, 1465–1476. [CrossRef] [PubMed]

genes MDPI

Article

Integrated Whole-Genome Sequencing and In Silico Characterization of *Salmonella* Cerro and Schwarzengrund from Brazil

Nathaly Barros Nunes [1,*], Vinicius Silva Castro [1], Adelino da Cunha-Neto [2,3], Fernanda Tavares Carvalho [3], Ricardo César Tavares Carvalho [3] and Eduardo Eustáquio de Souza Figueiredo [1,2,*]

1 Faculty of Agronomy and Zootechnics, Federal University of Mato Grosso (UFMT), Cuiabá 78060-900, MT, Brazil; viniciuscastro06@gmail.com
2 Faculty of Nutrition, Federal University of Mato Grosso (UFMT), Cuiabá 78060-900, MT, Brazil; adelino.neto@ufmt.br
3 Program in Animal Bioscience, University of Cuiabá (UNIC), Cuiabá 78065-900, MT, Brazil; fefetavares_carvalho@hotmail.com (F.T.C.); ricardo_carvalho88@hotmail.com (R.C.T.C.)
* Correspondence: nathalyb.nunes@outlook.com (N.B.N.); eduardo.figueiredo@ufmt.br (E.E.d.S.F.); Tel.: +55-65-3615-8589 (E.E.d.S.F.)

Abstract

Background: *Salmonella* is a bacterium that causes foodborne infections. This study characterized two strains isolated from cheese and beef in Brazil using whole-genome sequencing (WGS). Objectives: We evaluated their antimicrobial resistance profiles, virulence factors, plasmid content, serotypes and phylogenetic relationships. Methods: DNA was extracted and sequenced on the NovaSeq 6000 platform; the pangenome was assembled using the Roary tool; and the phylogenetic tree was constructed via IQ-TREE. Results and Discussion: For contextualization and comparison, 3493 *Salmonella* genomes of Brazilian origin from NCBI were analyzed. In our isolates, both strains carried the *aac(6′)-Iaa_1* gene, while only Schwarzengrund harbored the *qnrB19_1* gene and the *Col440I_1* plasmid. Cerro presented the islands SPI-1, SPI-2, SPI-3, SPI-4, SPI-5 and SPI-9, while Schwarzengrund also possessed SPI-13 and SPI-14. Upon comparison with other Brazilian genomes, we observed that Cerro and Schwarzengrund represented only 0.40% and 2.03% of the national database, respectively. Furthermore, they revealed that Schwarzengrund presented higher levels of antimicrobial resistance, a finding supported by the higher frequency of plasmids in this serovar. Furthermore, national data corroborated our findings that SPI-13 and SPI-14 were absent in Cerro. A virulence analysis revealed distinct profiles: the *cdtB* and *pltABC* genes were present in the Schwarzengrund isolates, while the *sseK* and *tldE*1 family genes were exclusive to Cerro. The results indicated that the sequenced strains have pathogenic potential but exhibit low levels of antimicrobial resistance compared to national data. The greater diversity of SPIs in Schwarzengrund explains their prevalence and higher virulence potential. Conclusions: Finally, the serovars exhibit distinct virulence profiles, which results in different clinical outcomes.

Keywords: whole-genome sequencing; WGS; *Salmonella* Cerro; *Salmonella* Schwarzengrund; pathogenicity; bacterial virulence; cheese; beef

1. Introduction

Foodborne illnesses caused by pathogens represent a major public health concern worldwide [1]. Among these, salmonellosis stands out as one of the most significant foodborne bacterial zoonoses [2]. The genus *Salmonella* comprises two species: *Salmonella bongori* and *Salmonella enterica* [3]. *S. enterica* currently includes approximately 2600 serotypes [4].

Salmonella has been widely reported in cattle [1] and dairy cows [2]. Its presence in bovine can lead to the contamination of milk or meat through various environmental sources on the farm, including other animals, rodents, wildlife, insects, water, resulting in both direct and indirect infection of humans and other animals [3]. *Salmonella* Cerro has been detected in dairy herds as an asymptomatic carrier, with fecal shedding observed in clinically healthy animals [4]. Although this serotype is rarely associated with human illness [5,6], it has emerged in recent years due to its increasing prevalence among dairy cattle [6].

Salmonella Schwarzengrund is frequently associated with human infections [7] and is also commonly reported in farm animals, food products and agricultural environments, where it has been linked to antimicrobial-resistant infections in humans [8]. In many countries, *S.* Schwarzengrund ranks among the top 15 *Salmonella* serotypes isolated from humans, food sources and animals [9].

Data from the United States Department of Agriculture [10] indicate that the global cattle herd is expected to reach approximately 914 million head in 2025. This report also describes the global distribution of cattle, with Brazil ranking second place in herd size, behind only India among the seven regions analyzed. Between 2014 and 2024, Brazilian beef production increased by 21.88% (in thousand tons). According to the CiCarne Yearbook of the beef production chain, published in March 2025, Brazil is recognized for its high food productivity, particularly in animal-based foods, and it is the world's largest exporter of the most widely consumed meat [11].

The Midwest region is the largest producer of cattle in Brazil. In 2024, Mato Grosso state led all states in beef exports, shipping 161,98 thousand tons abroad, with China and the United States as the main destinations [12]. Regarding milk production, recent data indicate that the Central-West region ranks third place among Brazilian regions, accounting for approximately 11.3% of the raw milk purchased by dairies. Despite these high levels of bovine productivity and exports, there are still limited studies on pathogenic bacteria in this sector, especially *Salmonella*.

In Brazil, several studies have utilized whole-genome sequencing (WGS) to investigate pathogenic bacteria, including various *Salmonella* serotypes [13–17]. Given the limitations of conventional serology, WGS has become a critical tool for characterizing foodborne pathogens and predicting genotypic antimicrobial resistance [18,19]. Insights into pathogen genomes are essential for disease prevention, management and treatment [20]. WGS enables the identification of resistance genes, virulence factors, mobile genetic elements and relevant mutations [21,22].

In a previous study, Cunha Neto et al. [23] analyzed 225 cheese samples from Brazilian dairies, and Muller et al. [24] examined 107 beef samples from various slaughterhouses in Mato Grosso, Brazil. In both cases, the isolates were phenotypically identified as *Salmonella* Schwarzengrund. However, it is well known that traditional isolation and serological methods have limited accuracy and may allow cross-reactions, which can hinder the complete characterization of bacterial strains. Therefore, the aim of this study was to sequence the first two genomes of *Salmonella* strains isolated from cheese and beef in Mato Grosso state, Brazil, to confirm their serovar designations and to genotypically characterize their antimicrobial resistance profile, virulence determinants and phylogenetic relationships in the context of other Brazilian *Salmonella* genomes.

2. Materials and Methods

2.1. Strain Recovery and DNA Extraction

The strains used in this study were previously identified as *Salmonella* Schwarzengrund in the studies by [23,24]. For DNA extraction, a single colony was resuspended in brain heart infusion (BHI) broth (KASVI®, laboratories Conda S.A, Madrid, Spain) and incubated overnight at 35 °C. Genomic DNA was then extracted and quantified using Qiagen DNeasy Blood and Tissue Kit (Qiagen, Hilden, Germany) and QUBIT 2.0 fluorometer (Invitrogen®, Carlsbad, CA, USA), respectively, following the manufacturer's instructions.

2.2. Whole-Genome Sequencing and Data Processing

The sequencing library was prepared using the NEBNext Ultra II DNA library Prep Kit (New England Biolabs, Ipswich, MA, USA). Sequencing was performed on the Illumina NovaSeq 6000 platform (Illumina Inc., San Diego, CA, USA) by GenOne Biotechnologies (Rio de Janeiro, Brazil), achieving a minimum coverage depth of 100×.

The quality of raw sequencing reads was assessed using FastQC and further processed with fastp. De novo genome assembly was performed using Shovill, which employs SPAdes v.1.1.0 (https://github.com/tseemann/shovill) (accessed on 16 March 2025). The genome sequences generated in this study were deposited in the National Center for Biotechnology Information (NCBI) database under BioProject accession number PRJNA1154933.

In addition, all sequences available in the NCBI Pathogen Detection database (https://www.ncbi.nlm.nih.gov/pathogens/) (accessed on 15 March 2025) were downloaded after applying filters for *Salmonella* and Brazil. Thus, a total of 3493 sequences were retrieved on (17 March 2025) and subjected to serotyping using SeqSero2 tool v1.3.1 [25] based on the assembled contigs processed on a Linux Mint system. Of these, 71 assemblies were identified as *Salmonella enterica* serovar Schwarzengrund and 14 assemblies as *Salmonella enterica* serovar Cerro. Supplementary Materials Table S1 containing the accession numbers and associated metadata for each analyzed strain is available.

2.3. Genome Quality, Annotation, Virulence and Resistance Gene Determination

Genome quality was first assessed using QUAST (v. 5.3.0) with default parameters, generating PDF reports. In addition, NCBI assemblies were evaluated using the seqkit tool (v2.10.0), specifically the L50 metric. The L50 is defined as the minimum number of largest contigs whose combined length comprises at least 50% of the total genome assembly. In this study, the highest L50 value observed was 68, which nevertheless indicated that the assemblies from the Pathogen Detection database were of sufficient quality for downstream analyses. Supplementary Materials Table S1 is provided with N50, L50 and additional sequencing quality metrics.

Genome annotation was then performed with Prokka (1.14.6) [26], using GFF3 output format and setting a minimum contig size of 200 bp. For specialized gene annotation, including virulence genes, ABRicate v. 1.0.1 was employed with the VirulenceFinder Database [27]. Antimicrobial resistance genes were identified by running ABRIcate again, this time using the Resfinder database (updated on 22 March 2024) [28]. All ABRicate analyses were performed using the default parameters (80% coverage and identity). Pathogenicity islands were detected using SPIFinder 2.0, applying thresholds of 95% identity and 60% coverage [29–31]. For plasmid analysis, we used the PlasmidFinder (Version 2.1.6; Database version: 18 January 2023) [32]. All the analyses described above were carried out on a total of 84 strains retrieved from the NCBI Pathogen Detection database and the 2 newly sequenced strains in this study, totaling 86 strains.

2.4. Phylogenetic Tree Comparison

For a genome comparison, the GFF3 files generated by Prokka (as described above) were used to construct a pangenome from our set of 86 serotypes. The Roary tool (v3.13.0) [33] was run with a minimum sequence identity of 95% and a gene presence threshold of 99% of isolates to define the core genome. We then generated two figures: one showing the accessory genome (all non-core genes) and another depicting the pangenome based on core-genome genes.

After obtaining the accessory-genome alignment, a maximum-likelihood phylogenetic tree was constructed using IQ-TREE v2.4.0 [34], with ultrafast bootstrap analysis of 1.000 replicates. For the pangenome analysis, Single nucleotide polymorphisms (SNPs) were extracted from the core genome alignment using SNP, and a phylogenetic tree was constructed with FastTree 2.1.10 [35].

For the metadata of the tree figures, manual curation was performed by the authors, and the data were distributed into the following groups and depicted in a phylogenetic tree: carcass/meat/organ; clinical/fluids; environmental sources; fecal/swab sample; food/feed or others; type of isolation (clinical or environmental); serotype (Cerro or Schwarzengrund); presence or absence of genes in the pathogenicity islands; and resistance profile to some classes of antibiotics.

2.5. Data Processing and Visualization

The phylogenetic tree was visualized using the Microreact web platform, enabling interactive exploration of the tree alongside metadata [36]. Data processing and analysis, including the generation of the heatmap and grouped bar chart, were performed in Python (v. 3.10.11) using Pandas (1.5.2), Numpy (1.24.2), Matplotlib (3.6.3) and Seaborn (0.12.1).

Reports obtained from Resfinder, PlasmidFinder and VFDB were merged and converted into binary matrices (0 = absence; 1 = presence), considering only genes detected with an identity greater than 80%. For heatmap generation, these binary values were transformed into percentages (1 = 100%; 0 = 0%), and frequencies were calculated and visualized by color intensities. The heatmap illustrating the average percentage distribution of pathogenicity islands was generated with Seaborn to facilitate group comparisons. For antimicrobial resistance gene analysis, the Resfinder results were processed and visualized as grouped bar charts using Matplotlib, highlighting the incidence of resistance genes across serotypes.

3. Results

3.1. Serovar Determination and Sequence Classification

*Salmonella*_R5748 was identified as serovar Cerro, while *Salmonella*_R10633 was classified as Schwarzengrund. To assess the prevalence of these serovars in Brazil, we queried the Pathogen Detection database (https://www.ncbi.nlm.nih.gov/pathogens/) (accessed on 21 April 2025) for additional *Salmonella* sequences of Brazilian origin. Serotype assignments were performed using SeqSero2 [25]. In total, 3493 sequences were retrieved from NCBI; of these, 71 (2,03%) were identified as Schwarzengrund and 14 (0.40%) as Cerro.

Based on the metadata, the distribution of Schwarzengrund isolates was as follows: 49.3% (35/71) were obtained from carcass/meat/organs; 36.6% (26/71) from environmental sources; 5.6% (4/71) from food or animal feed (including *Salmonella*_R10633); 4.2% (3/71) from fecal samples or anal junction swabs; 2.8% (2/71) from other or undetermined sources; and 1.4% (1/71) from clinical specimens. For Cerro, 28.6% (4/14) of the isolates were obtained from carcass/meat/organs; 21.4% (3/14) from environmental sources; 21.4% (3/14) from fecal/swab samples; 14.3% (2/14) from food or animal feed; and 14.3% (2/14) from other/undetermined sources.

Notably, both [23,24] identified these isolates as *Salmonella* Schwarzengrund using conventional serology. This discrepancy between in vitro serological results and in silico genomic analysis represents the first major inconsistency observed in this study.

3.2. Detection of Resistance Genes and Plasmids in Salmonella Cerro and Schwarzengrund

RESfinder and PlasmidFinder were applied to the WGS data to identify resistance genes and plasmids in *S.* Cerro (*Salmonella*_R5748) and *S.* Schwarzengrund (*Salmonella*_R10633). Both strains harbored the *aac(6′)-Iaa_1*, which confers resistance to the aminoglycosides amikacin (AMI) and tobramycin (TOB). In the *S.* Schwarzengrund isolate, the *qnr*B19_1 gene was also identified, encoding resistance to the fluoroquinolone ciprofloxacin (CIP). Thus, resistance in these sequenced strains was associated with the aminoglycoside and quinolone classes, although few genes were identified. No plasmid was detected in the *S.* Cerro (*Salmonella*_R5748), whereas *S.* Schwarzengrund (*Salmonella*_R10633) harbored the *Col440I_1* plasmid (Table 1).

Table 1. Resistance genes and plasmids found in the sequenced serotypes through WGS.

Strain	***S.* Cerro (*Salmonella*_R5748)**	***S.* Schwarzengrund (*Salmonella*_R10633)**	**Classes of Antibiotics/Function**
Resistance genes	*Aac(6′)-Iaa_1*	*Aac(6′)-Iaa_1* *qnrB19_1*	Aminoglicosides Quinolones
Plasmids	None	*Col440I_1*	They play a role in antimicrobial resistance and virulence

To compare the resistance profiles of our isolates with other Cerro and Schwarzengrund strains available in NCBI, we re-analyzed publicly available genomes using Resfinder and summarized the antibiotic-specific resistance data in Figure 1. In this bar graph, the *aac(6′)-Iaa_1* gene, conferring resistance to the aminoglycosides amikacin (AMI) and tobramycin (TOB), was present in 100% of both serotypes. For antibiotics outside the aminoglycosides class, *S.* Cerro exhibited resistance frequencies of approximately 10–40% across the agents evaluated. In contrast, *S.* Schwarzengrund displayed higher resistance rates (20–40%) for most antibiotic genes and a notably high frequency (~85%) of ciprofloxacin (CIP) resistance, primarily due to the presence of the *qnrB19_1* gene.

Regarding plasmids, we conducted a comparative genomic annotation of Brazilian genomes from the Cerro and Schwarzengrund serovars and observed greater genomic plasticity in Schwarzengrund. This finding aligns with our isolates, where the *Col440I_1* plasmid was detected exclusively in this serovar. Although the overall frequency of *Col440I_1* was similar between Cerro and Schwarzengrund, other plasmids, such as *Col* (pHAD28), were more frequently identified in both serovars, despite being absent in our isolates (Figure 2). The *Col440I_1* plasmid belongs to the *ColE* family and functions as a cryptic plasmid involved in initiating plasmid DNA replication within the host cell [37]. In contrast, *Col* (pHAD28) has been described as originating from *Klebsiella pneumoniae* and is widely implicated in the dissemination of aminoglycoside resistance, often in association with genes from the *aac(6′)* family [38].

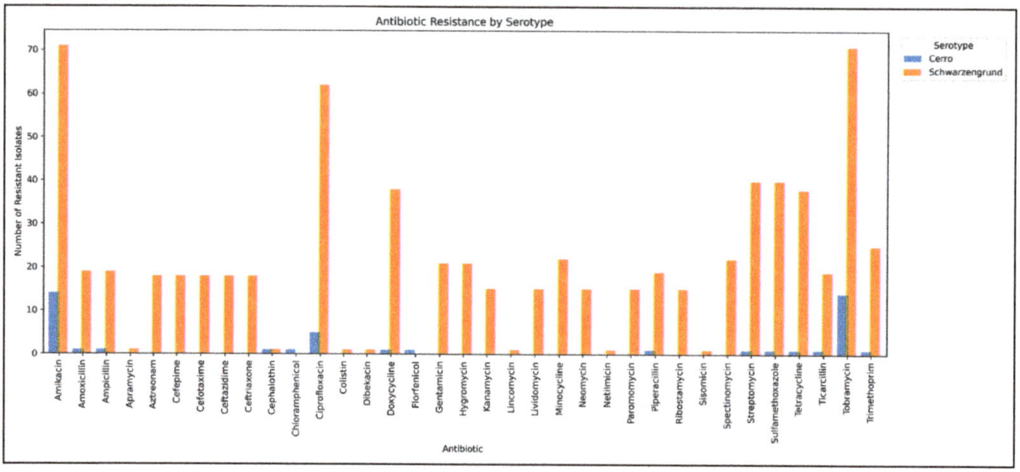

Figure 1. Frequency (%) of antimicrobial resistance genes in *S.* Cerro and *S.* Schwarzengrund isolates. Legend: Bar graph showing the resistance profile of Cerro (blue) and Schwarzengrund (orange). The x-axis indicates the antibiotics, and the y-axis indicates the frequency of antibiotic resistance.

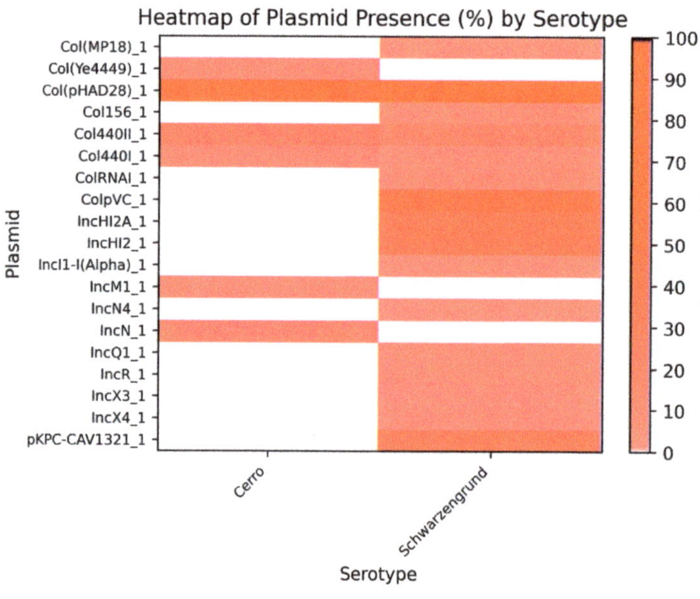

Figure 2. Heatmap of plasmid prevalence in *Salmonella* Cerro and Schwarzengrund. Legend: Heatmap indicates the presence (%) of plasmids in Cerro and Schwarzengrund. The color gradient ranges from white (0% presence) to black (100% presence).

3.3. Phylogeny of Salmonella *Cerro and Schwarzengrund and SPI Determinants*

A total of 3493 *Salmonella* genomes were downloaded from the NCBI database, from which all the available isolates belonging to the Cerro and Schwarzengrund serovars were identified. To infer their evolutionary relationships, we constructed a maximum-likelihood phylogenetic tree based on the core genome, incorporating 85 genomes (71 Schwarzengrund and 14 Cerro) (Figure 3). In this tree, isolates from the Cerro and Schwarzengrund serovars form distinct clades that correlate with serotype, source and isolation type. Branch annotations further indicate the presence or absence of key virulence genes located within *Salmonella* pathogenicity islands (SPIs), as highlighted in the figure.

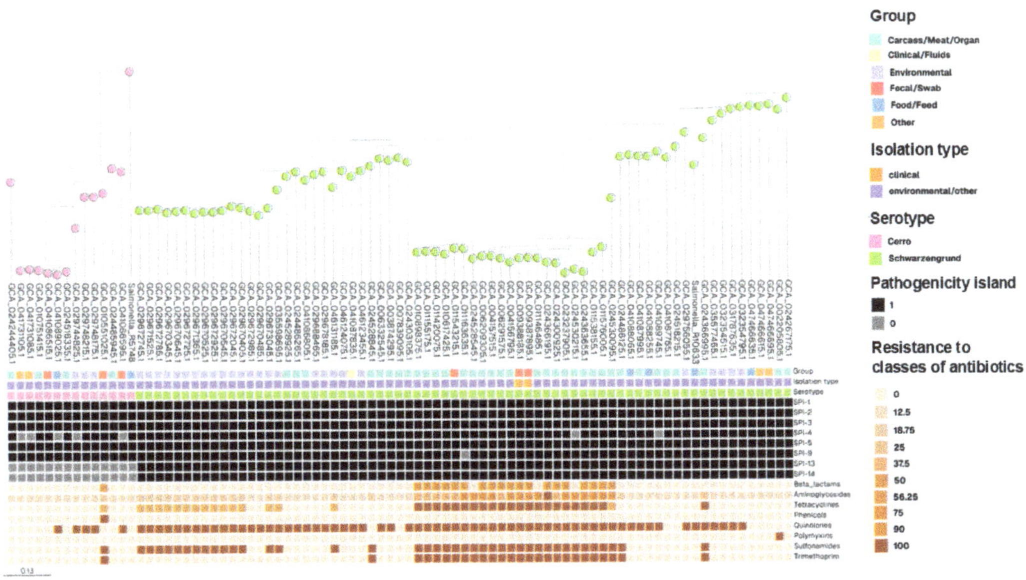

Figure 3. Phylogenetic tree of sequenced *Salmonella* Cerro and *Salmonella* Schwarzengrund serotypes present in the NCBI database, their virulence and frequency of antibiotic classes. Legend: The strains sequenced in this study are named *Salmonella*_R5748 and *Salmonella*_R10633. Beta_lactams: Amoxicillin, Ampicillin, Aztreonam, Cefepime, Cefotaxime, Ceftazidime, Ceftriaxone, Cephalothin, Piperacillin, Ticarcillin. Aminoglycosides: Amikacin, Apramycin, Gentamicin, Kanamycin, Neomycin, Netilmicin, Paromomycin, Tobramycin, Streptomycin, Dibekacin, Ribostamycin, Sisomicin, Spectinomycin, Hygromycin, Lividomycin. Tetracyclines: Tetracycline, Doxycycline, Minocycline. Phenicols: Chloramphenicol, Florfenicol. Quinolones: Ciprofloxacin. Polymyxins: Colistin. Sulfonamides: Sulfamethoxazole. Trimethoprim: Trimethoprim. Pathogenicity island: (1) Present, (0) Absent.

The clades observed in this study clearly demonstrate the separation between the Cerro and Schwarzengrund serovars, particularly in relation to pathogenicity islands. All Cerro isolates lacked islands 13 and 14, whereas these islands were consistently present in Schwarzengrund strains (Figure 3). In addition, a higher overall resistance to antibiotic classes was observed in Schwarzengrund compared to Cerro strain, which may partially explain the greater number of sequenced genomes available for this serovar.

Regarding the strains sequenced in this study, *Salmonella*_R5748 (Cerro) showed a notable phylogenetic distance from the other Cerro isolates available in the NCBI database. Although the reference strains are of the same geographical origin (Brazil) and have a similar antimicrobial resistance profile (low resistance to antibiotic classes), hypotheses may explain this phylogenetic distance, such as the origin of contamination. The Cerro strain in this study was isolated from food, while the others were isolated from other contaminated matrices; only one strain (GCA_024516335.1) had the same contaminated origin (food), but it was phylogenetically distant. In contrast, *Salmonella*_R10633 (Schwarzengrund) clustered closely with an environmentally derived isolate (GCA_029756055.1), sharing a similar profile in terms of pathogenicity island content and antibiotic resistance patterns.

In Figure 4, the pangenome of *Salmonella* Cerro (14 strains) and *S.* Schwarzengrund (71 strains) is contrasted. In *S.* Cerro, 3730 gene clusters are shared by more than 99% of genomes (core genome), with no clusters identified in the soft core. Also, 685 clusters belong to the shell, and 1695 are found in less than 15% of genomes (cloud). For *S.* Schwarzengrund, 3879 clusters constitute the core, 251 the soft core, 716 the shell and 2801 the cloud. The core-genome SNP phylogeny reveals two major subclades within the Cerro lineage and at least three well-supported lineages in Schwarzengrund. The adjacent presence/absence heatmap illustrates that the patterns of accessory gene gain and loss closely correspond to the phylogenetic clades, suggesting that differential acquisition of accessory genes contributes to lineage divergence and may reflect specific adaptative processes.

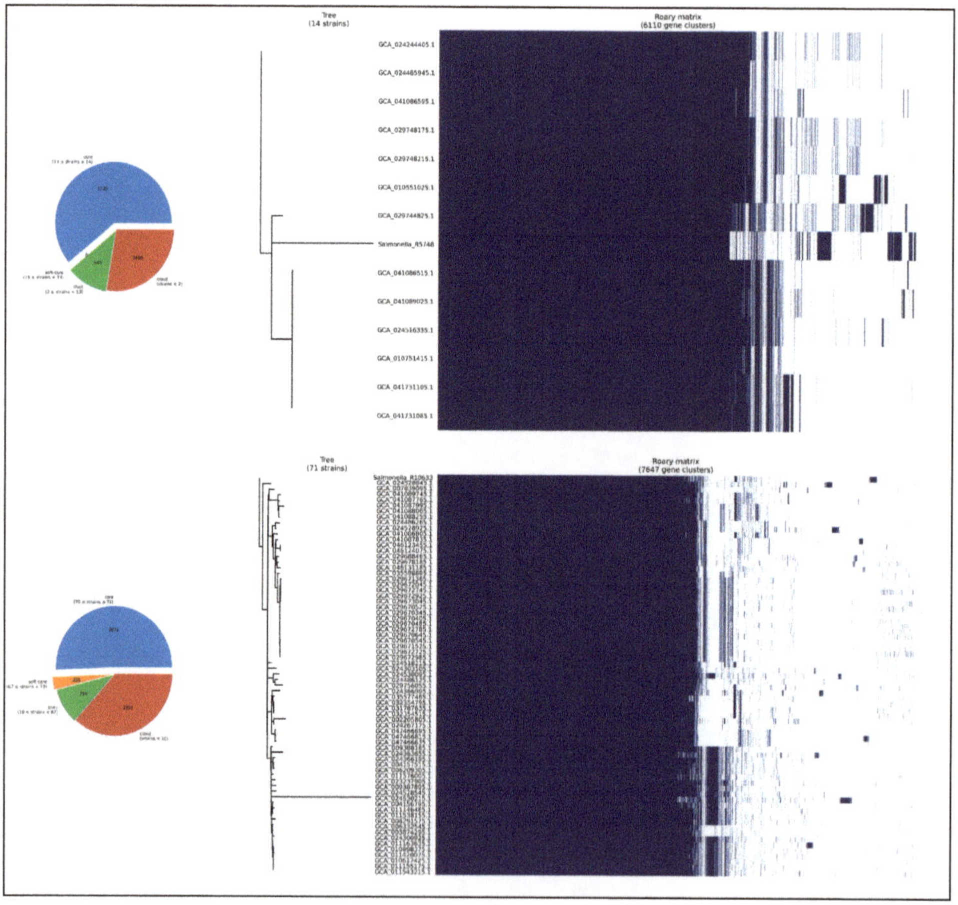

Figure 4. Pangenome composition and core-genome SNP phylogeny of *Salmonella* Cerro and Schwarzengrund isolates. Legend: The figures on the left show a pie chart of pangenome composition, dividing gene clusters into core (present in >99% of strains; in blue), soft core (95–99%; orange), shell (15–95%; green) and cloud (<15%; red). On the right, a core-genome SNP phylogeny (dendrogram) is displayed alongside a binary presence/absence matrix (blue = gene present; white = gene absent).

3.4. Salmonella Pathogenicity Islands and Virulence Genes

Expanding the analysis of pathogenicity islands, Figure 5 shows that there is a considerable difference between the two serovars, especially in the presence of SPI-4, SPI-13 and SPI-14. These islands are mainly associated with key virulence functions: SPI-4 contributes to host cell adherence and cytotoxicity; SPI-13 is involved in macrophage internalization and intracellular survival; and SPI-14 plays a role in systemic virulence and bile salt tolerance, facilitating survival during passage through the gastrointestinal tract and subsequent invasion of internal organs.

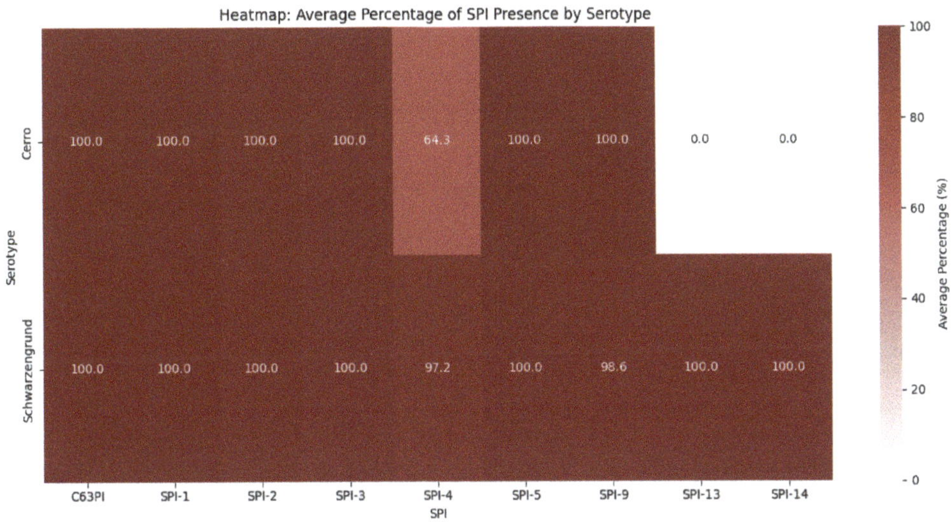

Figure 5. Heatmap of the frequency of SPIs in relation to serovars. Legend: Heatmap showing the frequency of pathogenicity islands (SPIs) under the serotypes in shades of red.

Finally, we assessed the virulence gene profiles of Cerro and Schwarzengrund isolates to identify potential differences between the groups. Genomic annotation using the VFDB database revealed a total of 139 virulence genes across all genomes. Among these, 116 genes were conserved across 100% of the isolates, while only 23 genes showed variation between the two serovars (Figure 6).

The results shown in Figure 6 indicate the presence of genes from the *pltABC* family in all Schwarzengrund genomes, while none of the Cerro strains carried these genes. The *pltABC* gene cluster encodes a tripartite exotoxin, originally identified in *Salmonella typhi*, which contributes to host cell damage and disruption of the immune response. Similarly, the *cdtB* gene was detected exclusively in Schwarzengrund isolates. On the other hand, *S.* Cerro genomes consistently harbored the *sseK1* and *sseK2* genes (responsible for favoring intracellular survival and systemic dissemination), as well as the *tlde1* gene (responsible for encoding a toxin with an antimicrobial effect on competing bacterial cells).

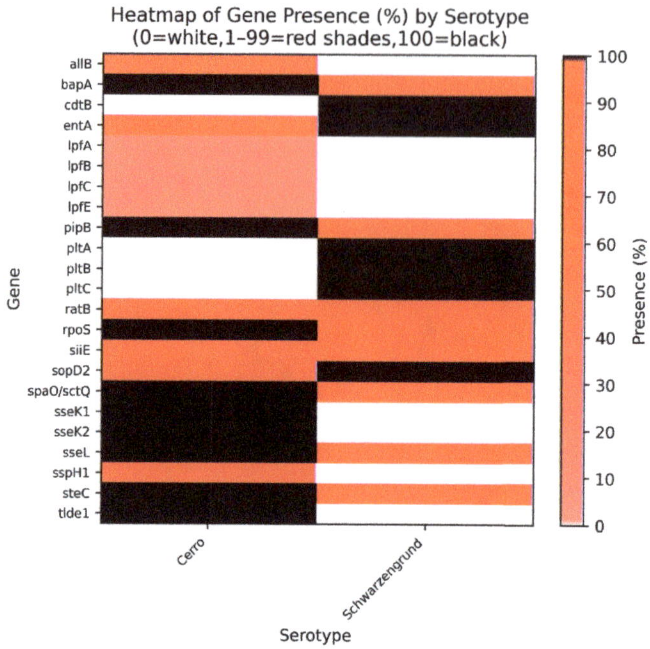

Figure 6. Heatmap of the percentage presence of virulence genes in *Salmonella* Cerro and Schwarzengrund isolates. Legend: Heatmap of the percentage presence of selected virulence genes in isolates of *Salmonella* serotypes Cerro and Schwarzengrund. The vertical axis lists the genes analyzed; the horizontal axis lists the serotypes. The color gradient ranges from white (0% presence) to black (100% presence). *Salmonella*_R5748 (Cerro) demonstrated the presence of the following genes: *allB, bapA, entA, lpfA, lpfB, lpfC, lpfE, pipB, rpoS, siiE, spaO/sctQ, sseK1, sseK2, sseL, steC tlde1*, while exhibiting absence of the following genes: *cdtB, pltA, pltB, pltC, ratB, sopD2, sspH1*. *Salmonella*_R10633 (Schwarzengrund) demonstrated the presence of the following genes: *bapA, cdtB, entA, pipB, pltA, pltB, pltC, ratB, rpoS, siiE, sopD2, spaO/sctQ, sseK1, sspH1, tlde1*, while exhibiting absence of the following genes: *allB, lpfA, lpfB, lpfC, lpfE, sseK2, sseL, steC*.

4. Discussion

4.1. Serovar Determination and Sequence Classification

Serotyping via agglutination remains the initial step in the characterization of *Salmonella* [39], although in silico platforms based on WGS are increasingly employed for faster and more reliable serotype prediction [40]. In the present study, we observed discrepancy between traditional serology and WGS for *Salmonella*_R5748, which was initially classified as Schwarzengrund via serology but later confirmed as Cerro by SeqSero. According to [41], an agreement rate was observed between traditional serotyping and WGS using SeqSero; however, 7.7% of the isolates showed discordant results, and 5.9% remained untyped by the molecular tool. Thus, these findings underscore the transformative potential of WGS-based tools in public health microbiology [40,42]. The current global trend is to use whole-genome sequencing in outbreak detection and routine surveillance of pathogenic bacteria [43]. Based on the studies highlighting the importance of WGS, it is valuable to emphasize that this tool could easily make a significant contribution to surveillance systems in Brazil.

Salmonella Cerro and Schwarzengrund have been reported worldwide across a variety of food matrices [6,9,44,45], and dozens of isolates have had their genomes sequenced and made publicly available in the NCBI database. In contrast, only a limited number of studies in Brazil have performed WGS to characterize *Salmonella* isolates [13,15]. To our knowledge, this is the first study is Mato Grosso to apply WGS for the characterization of *Salmonella* serotypes isolated from both cheese and beef, thereby addressing a critical gap in regional surveillance and food safety research.

S. Schwarzengrund is a relatively uncommon cause of human salmonellosis worldwide [46], but its incidence has increased in recent decades in countries such as Thailand, Slovakia, New Zealand and Venezuela [46]. Its prevalence has also risen in Japan, where isolates have been recovered from chicken meat [46,47]. In China, poultry meat remains the primary source of contamination, along with cases in humans and animals [48]. Reflecting these trends, nearly half (49.3%) of all *S.* Schwarzengrund genomes in the NCBI Pathogen Detection database originate from carcasses, meat and organs. In the present study, we identified that there is a global convergence regarding the source of isolation that has been reported in the literature with our findings, since the sequenced Schwarzengrund strain was isolated from beef.

Poultry, especially chicken, has been identified as the main vehicle for *S.* Schwarzengrund infections [49,50]. In Brazil, *S.* Schwarzengrund has similarly been isolated from poultry and chicken meat [51,52], and more recently from beef [24], highlighting its relevance across multiple food matrices.

S. Cerro is rarely implicated in human salmonellosis [53], although it has been associated with clinical disease in cattle in the United States [54]. In Brazil, *S.* Cerro has been recovered from poultry [55], swine [56] and wild boar [57]. Despite this diversity of hosts, its prevalence in Brazilian cattle remains poorly characterized; however, one study did report the detection of *S.* Cerro in veal calves [58]. The relatively low incidence of human infection caused by *S.* Cerro, especially compared to *S.* Schwarzengrund, likely contributes to its under-representation in the NCBI Pathogen Detection database.

Salmonella primarily colonizes the intestinal tract of humans and animals, including cattle [59]. Salmonellosis in cattle is a global concern, with serotypes such as *S. Dublin* and *S. Typhimurium* commonly implicated [60]. Notably, outbreaks linked to *Salmonella* transmission through raw milk in Denmark and raw meat in Ireland have been documented [61,62]. These incidents highlight the importance of ensuring food quality control measures to ensure consumer safety.

Salmonella enterica frequently colonizes both dairy products and livestock, serving as vehicles and reservoirs for human exposure [63,64]. Human salmonellosis is most commonly attributed to the consumption of contaminated poultry meat and eggs, as well as dairy and beef products [2]. In our metadata analysis, only 5.6% (4/71) of the *S.* Schwarzengrund isolates originated from food sources or animal feed, highlighting the under-representation of these reservoirs in publicly available genomic data.

Salmonella colonization in cattle can result in the contamination of both milk and meat at the farm level [65]. Cattle often carry the pathogen asymptomatically, introducing it into slaughterhouses and posing significant food safety risks through cross-contamination during processing. In this study, we focused on strains isolated from cheese and beef. Notably, no other *Salmonella* isolates from cheese were present in the current metadata, underscoring the novelty and relevance of our findings.

4.2. Detection of Resistance Genes and Plasmids in Salmonella Cerro and Schwarzengrund

Antimicrobial resistance can be transmitted to humans through the food chain, either via the ingestion of resistant bacteria or via horizontal gene transfer to other microorganisms [66]. Resistance among foodborne pathogens represents a serious global public health threat [67,68]. In our study, we detected the *aac(6′)-Iaa_1* gene, associated with resistance to aminoglycosides, specifically amikacin (AMI) and tobramycin (TOB), in both sequenced serotypes. This same behavior was observed in reference strains available in the NCBI database, as both Cerro and Schwarzengrund harbored genes conferring resistance to this antibiotic class.

Although approximately 85 aminoglycoside-modifying enzymes have been identified, only a subset are commonly selected and responsible for the majority of aminoglycoside resistance. Among them, *aac(6′)-I* is particularly notable [69]. The same authors suggest that intrinsic resistance to aminoglycosides may be explained by the mechanism of antibiotic uptake, which depends on bacterial respiration. This process generates an electrochemical potential across the cytoplasmic membrane. However, a reduced or absent membrane potential, often observed in anaerobic bacteria, may account for resistance in these organisms [70].

The *qnrB19* gene was previously identified by [71] in the WGS of *Salmonella enterica* isolated from duck carcasses. Similarly, ref. [14] detected the *qnrB19* gene in serotypes circulating within the Brazilian food production chain. Their study was also the first in Brazil to report the presence of this gene in *S.* Schwarzengrund isolates from poultry and environmental samples. In addition, the *qnrB19* gene has been reported in other *Salmonella* serotypes in Brazil, with [72] documenting its presence in isolates from both food and human sources. In our phylogenetic analysis, we detected that most Schwarzengrund isolates have genes that confer resistance to the quinolone class (high frequency). In 2024, the World Health Organization [73] published a list of bacterial pathogens of public health importance, in which *Salmonella* Tiphi is in the high group category (difficult to treat and showing increasing trends of antimicrobial resistance), with an association of resistance to fluoroquinolones.

We observed that the *S.* Schwarzengrund serotype exhibited resistance to ciprofloxacin (CIP), a quinolone antibiotic, mediated by the *qnrB19_1* gene. Genes in the qnr family encode pentapeptide-repeat proteins that bind DNA gyrase, thereby blocking quinolone activity [13]. *Salmonella* spp. can serve as both recipients or donors of resistance determinants, facilitating horizontal gene transfer and posing significant food safety risks to human health [74]. Although CIP remains the drug of choice for treating complicated gastrointestinal infections, resistance among *Salmonella* isolates has been widely reported [75,76]. In addition, our findings reveal distinct resistome profiles for *S.* Cerro and *S.* Schwarzengrund: *S.* Cerro exhibited a lower frequency of antibiotic resistance genes across most antibiotic classes, while *S.* Schwarzengrund harbored a broader and more diverse resistome. Plasmids can play a significant role in antimicrobial resistance or virulence, with *Col440I_1* playing a role in initiating plasmid DNA replication within the host cell [37]. This leads us to hypothesize that they may be acting more directly in the relationship with MDR strains.

Mato Grosso ranked among Brazil's top beef-exporting states in 2024, supplying predominantly China and the United States [12]. This highlights the critical importance of maintaining herd health to ensure the safety of animal-derived foods. However, the routine use of veterinary antibiotics (VAs) and synthetic growth promoters (SGPs) can compromise dietary sustainability and lead to the accumulation of residues in meat, milk and eggs [77]. Moreover, China (23%), the United States (13%), Brazil (13%), India (3%) and Germany (3%) together account for the highest levels of antimicrobial consumption in food animal production systems [78].

Salmonella gastroenteritis is typically treated with quinolones and fluoroquinolones, especially in elderly patients [79]. Ciprofloxacin is considered the drug of choice for salmonellosis and has been associated with the presence of the *qnrB19* gene in non-human isolates [80]. The *qnr* genes are typically plasmid-mediated, conferring resistance to quinolones, including ciprofloxacin, commonly used in veterinary medicine [81]. In our study, *S.* Schwarzengrund displayed elevated resistance across multiple antibiotics

(Figure 1), with approximately 85% of the isolates resistant to CIP. The *qnrB19* gene was identified by [14] as the most prevalent resistance determinant in their study. Also, ref. [82] reported high levels of fluoroquinolone resistance among several serotypes, including *S.* Schwarzengrund. According to [83], such resistance may also result from mutations in the *gyrA* gene.

No plasmids were detected in the *S.* Cerro isolates in this study, whereas *S.* Schwarzengrund harbored a *Col440I_1* replicon. Although *Col440I_1*-type plasmids have been reported in both serotypes in other Brazilian isolates, our data demonstrate their presence exclusively in *S.* Schwarzengrund (Figure 2), highlighting their greater genomic plasticity. *Col440I_1* plasmids have been reported in *S.* Schwarzengrund by [22,84] in Chile. According to [85], plasmids can play a critical role in the antimicrobial resistance and virulence of *S.* Schwarzengrund. Consistent with these findings, we observed a higher prevalence of plasmids in *S.* Schwarzengrund compared to *S.* Cerro.

4.3. Phylogeny of *Salmonella* Cerro and Schwarzengrund and SPI Determinants

The phylogenetic analysis revealed a clear division between *Salmonella* Cerro and *Salmonella* Schwarzengrund, consistent with their classification as distinct serotypes. In Brazil, little is known about the occurrence of *S.* Cerro in cattle, and this serotype has not been reported in the literature as a direct cause of human salmonellosis [53]. Notably, within the *S.* Cerro clade, the strain sequenced in this study was phylogenetically more distant from the other Cerro isolates, a divergence further supported by the pangenome analysis. This distinct clustering may be partially explained by its limited antibiotic resistance (Figure 1), although a consistent resistance pattern across antibiotic classes was still observed among the Cerro strains.

According to [86], several studies have sought to explain why *S.* Cerro is frequently associated with livestock but rarely causes human disease. In the study by [53], the authors evaluated the diversity of *Salmonella* subtypes in dairy cattle and found that *S.* Cerro lacked virulence genes located in pathogenicity islands 10, 12 and 13. Consistent with these findings, Figure 5 shows that all *S.* Cerro genomes available in the NCBI database lack SPI-13 virulence genes, a pattern not observed in *S.* Schwarzengrund.

The clades and branches of the *S.* Schwarzengrund serotype form distinct isolation groups, with farm animals frequently identified as the sources of contamination. The isolate sequenced in this study did not cluster within any well-defined group but was phylogenetically closest to the GCA_029756055.1 strain, which grouped with strains recovered from environmental water. *S.* Schwarzengrund infections have increased globally in recent years [85]. Moreover, ref. [87] had documented infections caused by this serotype in both humans and birds across various Western countries. Notably, all strains clustering within our isolate were obtained in Brazil from poultry and environmental sources; none originated from beef.

Finally, upon comparison of the resistance frequencies across antibiotic classes, we observed that *S.* Schwarzengrund obtained from the NCBI database exhibited broad resistance profiles, with many strains classified as multidrug-resistant (Figure 3). In contrast, the strain sequenced in this study showed resistance only to the quinolone class. *S.* Cerro isolates, by comparison, displayed uniformly low resistance frequencies across all antibiotic classes.

4.4. Virulence Genes in Pathogenicity Islands

Pathogenicity islands are genomic regions acquired through horizontal gene transfer that contribute to microbial virulence [88]. Ref. [89] provided an early overview of their evolution in *Salmonella enterica* serotypes, initially describing twelve pathogenicity islands. Subsequent research has expanded this number to seventeen known islands in

Salmonella [88,90]. Ref. [91] further elucidated how these islands enhance the pathogenic potential of different *Salmonella* serovars. In particular, ref. [92] identified several virulence genes in *Salmonella* serotypes that are implicated in transmission and infection via type III secretion system (T3SS), which plays a critical role in host invasion. The pathogenicity of *S.* Schwarzengrund is driven by a wide array of virulence factors, notably the T3SS, encoded by *Salmonella* islands I and II [85,93]. SPI-II activity is essential for the formation and maintenance of the *Salmonella*-containing vacuole (SCV), an intracellular niche that supports bacterial survival and replication [88].

Salmonella pathogenicity islands (SPIs) harbor a diverse array of virulence genes, most of which are chromosomally located [14]. These islands are horizontally acquired genomic segments that play a central role in *Salmonella* pathogenesis [94]. Of the approximately 23 SPIs described to date [95], SPIs 1 and 2 are considered the most critical virulence determinants in *Salmonella* [96].

Pathogenicity island I (SPI-1) is the most extensively characterized among the *Salmonella* pathogenicity islands [97]. According to [94], SPI-I encodes effector proteins that are secreted via type III secretion system, facilitating the invasion of intestinal epithelial cells. In contrast, SPI-2, SPI-3 and SPI-4 are primarily involved in promoting bacterial growth and survival within the host during the systemic phase of the infection [97].

Salmonella pathogenicity island 4 (SPI-4) encodes six main genes: *sii*A–F. Among them, *Sii*C, *Sii*D and *Sii*F form a type I secretion system responsible for secreting *Sii*E [88]. In our study, *Sii*E was detected in all sequences analyzed. This gene mediates adhesion to bovine intestinal epithelial cells [98]. Also, genes of the *plt*ABC operon encode an exotoxin that damages host cells and subverts immune defenses [99]. These virulence genes were present in our *S.* Schwarzengrund isolate but were absent from all *S.* Cerro strains examined.

In *S.* Cerro, the effector genes *sse*K1 and *sse*K2 were detected in 100% of the isolates and were absent from all *S.* Schwarzengrund strains. These genes are located on *Salmonella* pathogenicity island II [100].

Until recently, the contributions of the *sse*K2 gene to *Salmonella* virulence has not been well understood. Refs. [99,100] examined the role of *sse*K2 in *Salmonella enterica* and demonstrated that deletion of this gene significantly attenuates pathogenicity both in vivo and in vitro. *Sse*K2 is a novel translocated effector protein originally described in *S. typhimurium* and is highly conserved across *Salmonella* strains [100].

We observed that Cerro lacked virulence genes located on pathogenicity islands 13 and 14. Similarly, ref. [53] reported the absence of SPI-13 in *S.* Cerro. The exact number of genes that constitute SPI-13 remains unclear [101,102]. This island has also been observed in other *S. enterica* subspecies [103,104], as well as in *S.* Gallinarum [105]. According to [106], an in silico analysis revealed that most genes within SPI-13 encode proteins involved in bacterial metabolism, leading to the hypothesis that the absence of this island may result in a nutritional disadvantage, potentially impairing *Salmonella* virulence in specific hosts.

In contrast, *S.* Schwarzengrund harbored genes located on pathogenicity islands 13 and 14, a finding also observed by [107]. Genes associated with SPI-14 play a central role in the activation of SPI-1 genes, primarily through a regulator encoded within SPI-14 (STM14_1008) known as low-oxygen-induced factor A (*LoiA*), which functions as a key virulence determinant [108].

4.5. Limitations

Although the data obtained for the *Salmonella* Cerro and *Salmonella* Schwarzengrund serovars are relevant, it is important to acknowledge some limitations. Given the number of strains sequenced (N = 2) and the lack of phenotypic validation, the observed percentages of Cerro and Schwarzengrund strains may not accurately reflect their true prevalence due to potential selection bias in sequencing efforts. Serotypes such as S. Typhimurium, S. Typhi and S. Enteritidis are more frequently studied and sequenced, which does not necessarily imply that they are more prevalent in the environment than Cerro and Schwarzengrund. We encourage further studies focusing on these less characterized serotypes, as the findings related to resistance genes, virulence factors, plasmid presence and pathogenicity islands presented here represent potential capabilities rather than definitive resistance or virulence profiles.

Finally, it is important to note that the strains sequenced in this study were isolated from food sources, which may affect comparisons with clinical isolates in public databases. Clinical strains are generally better characterized and tend to exhibit higher resistance profiles than environmental or food-derived isolates.

5. Conclusions

In conclusion, *Salmonella* Cerro and *Salmonella* Schwarzengrund represent a small fraction of the strains isolated and sequenced in the NCBI database. Our analysis revealed that S. Cerro exhibited a lower antimicrobial resistance profile but carried virulence gene clusters with potential threat to human health. Notably, the S. Cerro virulome lacked genes located on SPI-13 and SPI-14, whereas S. Schwarzengrund harbored genes associated with both islands. A virulome analysis indicated that most S. Schwarzengrund strains were multidrug-resistant (MDR); however, the strain sequenced in this study showed resistance only to aminoglycoside and quinolone. Although only one plasmid was detected in our S. Schwarzengrund isolate, a broader diversity of plasmids was observed in both serotypes across the dataset, underscoring the potential for future dissemination of resistance and virulence genes through horizontal gene transfer.

Supplementary Materials: The following supporting information can be downloaded at: https://www.mdpi.com/article/10.3390/genes16080880/s1, Table S1: Quality of sequences.

Author Contributions: Conceptualization, N.B.N., V.S.C., A.d.C.-N. and E.E.d.S.F.; Methodology, N.B.N. and V.S.C.; Validation, N.B.N., V.S.C., A.d.C.-N. and F.T.C.; Formal Analysis, N.B.N. and V.S.C.; Investigation, N.B.N., V.S.C. and F.T.C.; Writing—Original Draft, N.B.N.; Writing—Review and Editing, N.B.N., V.S.C., A.d.C.-N. and R.C.T.C.; Visualization, N.B.N. and V.S.C.; Supervision, E.E.d.S.F.; Project Administration, E.E.d.S.F.; Funding Acquisition, R.C.T.C. and E.E.d.S.F. All authors have read and agreed to the published version of the manuscript.

Funding: This work was supported by the Fundação Coordenação de Aperfeiçoamento de Pessoal de Nível Superior (CAPES), Brazil, under process codes 88887.828908/2023-00 and 88887.692875/2022-00, by the Conselho Nacional de Desenvolvimento Científico e Tecnológico (CNPq)/Brazil, under process code 310181/2021-6, and E.E.d.S.F. by Fundação de Apoio a Pesquisa de Mato Grosso (FAPEMAT) under process code PDPGCENTROESTE3084571P.

Institutional Review Board Statement: Not applicable.

Informed Consent Statement: Not Applicable.

Data Availability Statement: The original contributions presented in this study are included in the article and Supplementary Materials. Further inquiries can be directed to the corresponding author.

Conflicts of Interest: The authors declare no conflicts of interest.

References

1. Ehuwa, O.; Jaiswal, A.K.; Jaiswal, S. *Salmonella*, Food Safety and Food Handling Practices. *Foods* **2021**, *10*, 907. [CrossRef] [PubMed]
2. Bonifait, L.; Thépault, A.; Baugé, L.; Rouxel, S.; Le Gall, F.; Chemaly, M. Occurrence of *Salmonella* in the Cattle Production in France. *Microorganisms* **2021**, *9*, 872. [CrossRef] [PubMed]

3. Wu, S.; Hulme, J.P. Recent Advances in the Detection of Antibiotic and Multi-Drug Resistant *Salmonella*: An Update. *Int. J. Mol. Sci.* **2021**, *22*, 3499. [CrossRef] [PubMed]
4. Dos Santos Bersot, L.; Quintana Cavicchioli, V.; Viana, C.; Konrad Burin, R.C.; Camargo, A.C.; de Almeida Nogueira Pinto, J.P.; Nero, L.A.; Destro, M.T. Prevalence, Antimicrobial Resistance, and Diversity of *Salmonella* along the Pig Production Chain in Southern Brazil. *Pathogens* **2019**, *8*, 204. [CrossRef]
5. Hughes, L.E.; Gibson, E.A.; Roberts, H.E.; Davies, E.T.; Davies, G.; Sojka, W.J. Bovine Salmonellosis in England and Wales. *Br. Vet. J.* **1971**, *127*, 225–238. [CrossRef]
6. Salaheen, S.; Sonnier, J.; Kim, S.W.; Haley, B.J.; Van Kessel, J.A.S. Interaction of *Salmonella enterica* with Bovine Epithelial Cells Demonstrates Serovar-Specific Association and Invasion Patterns. *Foodborne Pathog. Dis.* **2020**, *17*, 608–610. [CrossRef]
7. Van Kessel, J.A.S.; Karns, J.S.; Wolfgang, D.R.; Hovingh, E.; Schukken, Y.H. Dynamics of *Salmonella* Serotype Shifts in an Endemically Infected Dairy Herd. *Foodborne Pathog. Dis.* **2012**, *9*, 319–324. [CrossRef]
8. Rodriguez-Rivera, L.D.; Cummings, K.J.; Loneragan, G.H.; Rankin, S.C.; Hanson, D.L.; Leone, W.M.; Edrington, T.S. *Salmonella* Prevalence and Antimicrobial Susceptibility Among Dairy Farm Environmental Samples Collected in Texas. *Foodborne Pathog. Dis.* **2016**, *13*, 205–211. [CrossRef]
9. Tewari, D.; Sandt, C.H.; Miller, D.M.; Jayarao, B.M.; M'ikanatha, N.M. Prevalence of *Salmonella* Cerro in Laboratory-Based Submissions of Cattle and Comparison with Human Infections in Pennsylvania, 2005–2010. *Foodborne Pathog. Dis.* **2012**, *9*, 928–933. [CrossRef]
10. Vugia, D.J.; Samuel, M.; Farley, M.M.; Marcus, R.; Shiferaw, B.; Shallow, S.; Smith, K.; Angulo, F.J. Emerging Infections Program FoodNet Working Group. Invasive *Salmonella* Infections in the United States, FoodNet, 1996–1999: Incidence, Serotype Distribution, and Outcome. *Clin. Infect. Dis.* **2004**, *38*, S149–S156. [CrossRef]
11. Keelara, S.; Scott, H.M.; Morrow, W.M.; Hartley, C.S.; Griffin, D.L.; Gebreyes, W.A.; Thakur, S. Comparative Phenotypic and Genotypic Characterization of Temporally Related Nontyphoidal *Salmonella* Isolated from Human Clinical Cases, Pigs, and the Environment in North Carolina. *Foodborne Pathog. Dis.* **2014**, *11*, 156–164. [CrossRef]
12. Chen, M.H.; Wang, S.W.; Hwang, W.Z.; Tsai, S.J.; Hsih, Y.C.; Chiou, C.S.; Tsen, H.Y. Contamination of *Salmonella* Schwarzengrund Cells in Chicken Meat from Traditional Markets in Taiwan and Comparison of Their Antibiograms with Those of Human Isolates. *Poult. Sci.* **2010**, *89*, 359–365. [CrossRef] [PubMed]
13. Publication: USDA ARS. Available online: https://www.ars.usda.gov/research/publications/publication/?seqNo115=127846 (accessed on 5 June 2025).
14. Biscola, P.H.N.; Malafaia, G.C. Cicarne Yearbook of the Beef Production Chain: 2024–2025.—Portal Embrapa. Available online: https://www.embrapa.br/busca-de-publicacoes/-/publicacao/1174114/anuario-cicarne-da-cadeia-produtiva-da-carne-bovina-2024{-}{-}-2025 (accessed on 4 June 2025).
15. IBGE. Estatística Da Produção Pecuária. 2025, 83. Available online: https://biblioteca.ibge.gov.br/index.php/biblioteca-catalogo?view=detalhes&id=73087 (accessed on 4 June 2025).
16. Almeida, F.; Seribelli, A.A.; Medeiros, M.I.C.; Rodrigues, D.D.P.; de MelloVarani, A.; Luo, Y.; Allard, M.W.; Falcão, J.P. Phylogenetic and antimicrobial resistance gene analysis of *Salmonella Typhimurium* strains isolated in Brazil by whole genome sequencing. *PLoS ONE* **2018**, *13*, e0201882. [CrossRef] [PubMed]
17. Monte, D.F.; Lincopan, N.; Berman, H.; Cerdeira, L.; Keelara, S.; Thakur, S.; Fedorka-Cray, P.J.; Landgraf, M. Genomic Features of High-Priority *Salmonella enterica* Serovars Circulating in the Food Production Chain, Brazil, 2000–2016. *Sci. Rep.* **2019**, *9*, 11058. [CrossRef] [PubMed]
18. Rodrigues, G.L.; Panzenhagen, P.; Ferrari, R.G.; dos Santos, A.; Paschoalin, V.M.F.; Conte-Junior, C.A. Frequency of Antimicrobial Resistance Genes in *Salmonella* From Brazil by in Silico Whole-Genome Sequencing Analysis: An Overview of the Last Four Decades. *Front. Microbiol.* **2020**, *11*, 1864. [CrossRef]
19. Tiba-Casas, M.R.; Sacchi, C.T.; Gonçalves, C.R.; Almeida, E.A.; Soares, F.B.; de Jesus Bertani, A.M.; Fernandes, S.A.; de Paula Eduardo, M.B.; Camargo, C.H. Molecular Analysis of Clonally Related *Salmonella* Typhi Recovered from Epidemiologically Unrelated Cases of Typhoid Fever, Brazil. *Int. J. Infect. Dis.* **2019**, *81*, 191–195. [CrossRef]
20. Vilela, F.P.; dos Prazeres Rodrigues, D.; Allard, M.W.; Falcão, J.P. Genomic Analyses of Drug-Resistant *Salmonella enterica* Serovar Heidelberg Strains Isolated from Meat and Related Sources between 2013 and 2017 in the South Region of Brazil. *Curr. Genet.* **2023**, *69*, 141–152. [CrossRef]
21. Lakicevic, B.; Jankovic, V.; Pietzka, A.; Ruppitsch, W. Wholegenome Sequencing as the Gold Standard Approach for Control of *Listeria Monocytogenes* in the Food Chain. *J. Food Prot.* **2023**, *86*, 100003. [CrossRef]
22. Li, I.-C.; Wu, H.-H.; Chen, Z.-W.; Chou, C.-H. Prevalence of IncFIB Plasmids Found among *Salmonella enterica* Serovar Schwarzengrund Isolates from Animal Sources in Taiwan Using Whole-Genome Sequencing. *Pathogens* **2021**, *10*, 1024. [CrossRef]
23. Baert, L.; McClure, P.; Winkler, A.; Karn, J.; Bouwknegt, M.; Klijn, A. Guidance Document on the Use of Whole Genome Sequencing (WGS) for Source Tracking from a Food Industry Perspective. *Food Control* **2021**, *130*, 108148. [CrossRef]

24. Tagg, K.A.; Francois Watkins, L.; Moore, M.D.; Bennett, C.; Joung, Y.J.; Chen, J.C.; Folster, J.P. Novel Trimethoprim Resistance Gene DfrA34 Identified in *Salmonella* Heidelberg in the USA. *J. Antimicrob. Chemother.* **2019**, *74*, 38–41. [CrossRef]
25. da Cunha-Neto, A.; Carvalho, L.A.; Castro, V.S.; Barcelos, F.G.; Carvalho, R.C.T.; Rodrigues, D.d.P.; Conte-Junior, C.A.; de Figueiredo, E.E.S. *Salmonella* Anatum, *S.* Infantis and *S.* Schwarzengrund in Brazilian Cheeses: Occurrence and Antibiotic Resistance Profiles. *Int. J. Dairy Technol.* **2020**, *73*, 296–300. [CrossRef]
26. Muller, B.; Cunha-Neto, A.; Castro, V.S.; Carvalho, R.C.T.; Carvalho Teixeira, L.A.; Rodrigues, D.D.P.; De Souza Figueiredo, E.E. *Salmonella* Schwarzengrund, Akuafo, and O:16 Isolated from Vacuum-Packaged Beef Produced in the State of Mato Grosso, Brazil. *J. Infect. Dev. Ctries.* **2021**, *15*, 1876–1882. [CrossRef]
27. Zhang, S.; den Bakker, H.C.; Li, S.; Chen, J.; Dinsmore, B.A.; Lane, C.; Lauer, A.C.; Fields, P.I.; Deng, X. SeqSero2: Rapid and Improved *Salmonella* Serotype Determination Using Whole-Genome Sequencing Data. *Appl. Environ. Microbiol.* **2019**, *85*, e01746-19. [CrossRef]
28. Seemann, T. Prokka: Rapid Prokaryotic Genome Annotation. *Bioinformatics* **2014**, *30*, 2068–2069. [CrossRef]
29. Chen, L.; Zheng, D.; Liu, B.; Yang, J.; Jin, Q. VFDB 2016: Hierarchical and Refined Dataset for Big Data Analysis—10 Years On. *Nucleic Acids Res.* **2016**, *44*, D694–D697. [CrossRef] [PubMed]
30. Zankari, E.; Hasman, H.; Cosentino, S.; Vestergaard, M.; Rasmussen, S.; Lund, O.; Aarestrup, F.M.; Larsen, M.V. Identification of Acquired Antimicrobial Resistance Genes. *J. Antimicrob. Chemother.* **2012**, *67*, 2640–2644. [CrossRef]
31. Camacho, C.; Coulouris, G.; Avagyan, V.; Ma, N.; Papadopoulos, J.; Bealer, K.; Madden, T.L. BLAST+: Architecture and Applications. *BMC Bioinform.* **2009**, *10*, 421. [CrossRef] [PubMed]
32. Clausen, P.T.L.C.; Aarestrup, F.M.; Lund, O. Rapid and Precise Alignment of Raw Reads against Redundant Databases with KMA. *BMC Bioinform.* **2018**, *19*, 307. [CrossRef] [PubMed]
33. Yoon, S.H.; Park, Y.-K.; Kim, J.F. PAIDB v2.0: Exploration and Analysis of Pathogenicity and Resistance Islands. *Nucleic Acids Res* **2015**, *43*, D624–D630. [CrossRef]
34. Carattoli, A.; Zankari, E.; García-Fernández, A.; Voldby Larsen, M.; Lund, O.; Villa, L.; Møller Aarestrup, F.; Hasman, H. In Silico Detection and Typing of Plasmids Using PlasmidFinder and Plasmid Multilocus Sequence Typing. *Antimicrob. Agents Chemother.* **2014**, *58*, 3895–3903. [CrossRef]
35. Page, A.J.; Cummins, C.A.; Hunt, M.; Wong, V.K.; Reuter, S.; Holden, M.T.G.; Fookes, M.; Falush, D.; Keane, J.A.; Parkhill, J. Roary: Rapid Large-Scale Prokaryote Pan Genome Analysis. *Bioinformatics* **2015**, *31*, 3691–3693. [CrossRef] [PubMed]
36. Nguyen, L.-T.; Schmidt, H.A.; von Haeseler, A.; Minh, B.Q. IQ-TREE: A Fast and Effective Stochastic Algorithm for Estimating Maximum-Likelihood Phylogenies. *Mol. Biol. Evol.* **2015**, *32*, 268–274. [CrossRef] [PubMed]
37. Price, M.N.; Dehal, P.S.; Arkin, A.P. FastTree 2—Approximately Maximum-Likelihood Trees for Large Alignments. *PLoS ONE* **2010**, *5*, e9490. [CrossRef] [PubMed]
38. Argimón, S.; Abudahab, K.; Goater, R.J.E.; Fedosejev, A.; Bhai, J.; Glasner, C.; Feil, E.J.; Holden, M.T.G.; Yeats, C.A.; Grundmann, H.; et al. Microreact: Visualizing and Sharing Data for Genomic Epidemiology and Phylogeography. *Microb. Genom.* **2016**, *2*, e000093. [CrossRef]
39. González-Santamarina, B.; García-Soto, S.; Dang-Xuan, S.; Abdel-Glil, M.Y.; Meemken, D.; Fries, R.; Tomaso, H. Genomic Characterization of Multidrug-Resistant *Salmonella* Serovars Derby and Rissen From the Pig Value Chain in Vietnam. *Front. Vet. Sci.* **2021**, *8*, 705044. [CrossRef]
40. Shelenkov, A.; Mikhaylova, Y.; Voskanyan, S.; Egorova, A.; Akimkin, V. Whole-Genome Sequencing Revealed the Fusion Plasmids Capable of Transmission and Acquisition of Both Antimicrobial Resistance and Hypervirulence Determinants in Multidrug-Resistant *Klebsiella Pneumoniae* Isolates. *Microorganisms* **2023**, *11*, 1314. [CrossRef]
41. Diep, B.; Barretto, C.; Portmann, A.-C.; Fournier, C.; Karczmarek, A.; Voets, G.; Li, S.; Deng, X.; Klijn, A. *Salmonella* Serotyping; Comparison of the Traditional Method to a Microarray-Based Method and an in Silico Platform Using Whole Genome Sequencing Data. *Front. Microbiol.* **2019**, *10*, 2554. [CrossRef]
42. Zhang, S.; Yin, Y.; Jones, M.B.; Zhang, Z.; Deatherage Kaiser, B.L.; Dinsmore, B.A.; Fitzgerald, C.; Fields, P.I.; Deng, X. *Salmonella* Serotype Determination Utilizing High-Throughput Genome Sequencing Data. *J. Clin. Microbiol.* **2015**, *53*, 1685–1692. [CrossRef]
43. Leekitcharoenphon, P.; Nielsen, E.M.; Kaas, R.S.; Lund, O.; Aarestrup, F.M. Evaluation of Whole Genome Sequencing for Outbreak Detection of Salmonella Enterica. *PLoS ONE* **2014**, *9*, e87991. [CrossRef]
44. Köser, C.U.; Ellington, M.J.; Cartwright, E.J.P.; Gillespie, S.H.; Brown, N.M.; Farrington, M.; Holden, M.T.G.; Dougan, G.; Bentley, S.D.; Parkhill, J.; et al. Routine Use of Microbial Whole Genome Sequencing in Diagnostic and Public Health Microbiology. *PLoS Pathog.* **2012**, *8*, e1002824. [CrossRef] [PubMed]
45. Chapagain, P.P.; Kessel, J.S.V.; Karns, J.S.; Wolfgang, D.R.; Hovingh, E.; Nelen, K.A.; Schukken, Y.H.; Grohn, Y.T. A Mathematical Model of the Dynamics of *Salmonella* Cerro Infection in a US Dairy Herd. *Epidemiol. Infect.* **2008**, *136*, 263–272. [CrossRef] [PubMed]

46. Duc, V.M.; Shin, J.; Nagamatsu, Y.; Fuhiwara, A.; Toyofuku, H.; Obi, T.; Chuma, T. Increased *Salmonella* Schwarzengrund Prevalence and Antimicrobial Susceptibility of Salmonella Enterica Isolated from Broiler Chickens in Kagoshima Prefecture in Japan between 2013 and 2016. *J. Vet. Med. Sci.* **2020**, *82*, 585–589. [CrossRef]
47. Aarestrup, F.M.; Hendriksen, R.S.; Lockett, J.; Gay, K.; Teates, K.; McDermott, P.F.; White, D.G.; Hasman, H.; Sørensen, G.; Bangtrakulnonth, A.; et al. International Spread of Multidrug-Resistant *Salmonella* Schwarzengrund in Food Products. *Emerg. Infect. Dis.* **2007**, *13*, 726–731. [CrossRef]
48. Sasaki, Y.; Ikeda, T.; Yonemitsu, K.; Kuroda, M.; Ogawa, M.; Sakata, R.; Uema, M.; Momose, Y.; Ohya, K.; Watanabe, M.; et al. Antimicrobial Resistance Profiles of *Campylobacter Jejuni* and *Salmonella* Spp. Isolated from Enteritis Patients in Japan. *J. Vet. Med. Sci.* **2023**, *85*, 463–470. [CrossRef]
49. Matsui, K.; Nakazawa, C.; Thiri Maung Maung Khin, S.; Iwabuchi, E.; Asai, T.; Ishihara, K. Molecular Characteristics and Antimicrobial Resistance of *Salmonella enterica* Serovar Schwarzengrund from Chicken Meat in Japan. *Antibiotics* **2021**, *10*, 1336. [CrossRef]
50. Du, X.; Jiang, X.; Ye, Y.; Guo, B.; Wang, W.; Ding, J.; Xie, G. Next Generation Sequencing for the Investigation of an Outbreak of *Salmonella* Schwarzengrund in Nanjing, China. *Int. J. Biol. Macromol.* **2018**, *107*, 393–396. [CrossRef]
51. Bangtrakulnonth, A.; Pornreongwong, S.; Pulsrikarn, C.; Sawanpanyalert, P.; Hendriksen, R.S.; Wong, D.M.A.L.F.; Aarestrup, F.M. *Salmonella* Serovars from Humans and Other Sources in Thailand, 1993–2002. *Emerg. Infect. Dis.* **2004**, *10*, 131–136. [CrossRef]
52. Tsai, H.-J.; Hsiang, P.-H. The Prevalence and Antimicrobial Susceptibilities of *Salmonella* and *Campylobacter* in Ducks in Taiwan. *J. Vet. Med. Sci.* **2005**, *67*, 7–12. [CrossRef]
53. Tejada, T.S.; Silva, C.S.J.; Lopes, N.A.; Silva, D.T.; Agostinetto, A.; Silva, E.F.; Menezes, D.B.; Timm, C.D. DNA Profiles of *Salmonella* Spp. Isolated from Chicken Products and From Broiler and Human Feces. *Braz. J. Poult. Sci.* **2016**, *18*, 693–700. [CrossRef]
54. Voss-Rech, D.; Vaz, C.S.L.; Alves, L.; Coldebella, A.; Leão, J.A.; Rodrigues, D.P.; Back, A. A Temporal Study of *Salmonella enterica* Serotypes from Broiler Farms in Brazil. *Poult. Sci.* **2015**, *94*, 433–441. [CrossRef]
55. Rodriguez-Rivera, L.D.; Wright, E.M.; Siler, J.D.; Elton, M.; Cummings, K.J.; Warnick, L.D.; Wiedmann, M. Subtype Analysis of *Salmonella* Isolated from Subclinically Infected Dairy Cattle and Dairy Farm Environments Reveals the Presence of Both Human- and Bovine-Associated Subtypes. *Vet. Microbiol.* **2014**, *170*, 307–316. [CrossRef]
56. Kovac, J.; Cummings, K.J.; Rodriguez-Rivera, L.D.; Carroll, L.M.; Thachil, A.; Wiedmann, M. Temporal Genomic Phylogeny Reconstruction Indicates a Geospatial Transmission Path of *Salmonella* Cerro in the United States and a Clade-Specific Loss of Hydrogen Sulfide Production. *Front. Microbiol.* **2017**, *8*, 737. [CrossRef]
57. Pulido-Landínez, M.; Sánchez-Ingunza, R.; Guard, J.; do Nascimento, V.P. Assignment of Serotype to Salmonella enterica Isolates Obtained from Poultry and Their Environment in Southern Brazil. *Lett. Appl. Microbiol.* **2013**, *57*, 288–294. [CrossRef] [PubMed]
58. de Azevedo, E.C.; Martins, B.T.F.; Tiba Casas, M.R.; Possebon, F.S.; Araújo Junior, J.P.; Nero, L.A.; Yamatogi, R.S. Multidrug Resistance and Virulence Profiles of *Salmonella* Isolated from Swine Lymph Nodes. *Microb. Drug Resist.* **2021**, *27*, 562–570. [CrossRef] [PubMed]
59. Carraro, P.E.; de Barbosa, F.O.; Benevides, V.P.; Casas, M.R.T.; Berchieri Junior, A.; Bürger, K.P. Prevalence and Antimicrobial Resistance of *Salmonella* Spp. Isolated from Free-Ranging Wild Boars in the State of São Paulo, Brazil. *Cienc. Rural* **2022**, *52*, e20210263. [CrossRef]
60. Gabana, A.D.A.; Núncio, A.S.P.; Lopes, B.C.; de Oliveira, J.A.; da Silva Monteiro, L.; de Menezes Coppola, M.; Furian, T.Q.; Borges, K.A.; Rodrigues, L.B.; Mayer, F.Q. Different Multidrug-Resistant *Salmonella* Spp. Serovars Isolated from Slaughter Calves in Southern Brazil. *Curr. Microbiol.* **2022**, *80*, 11. [CrossRef]
61. Gutema, F.D.; Agga, G.E.; Abdi, R.D.; De Zutter, L.; Duchateau, L.; Gabriël, S. Prevalence and Serotype Diversity of *Salmonella* in Apparently Healthy Cattle: Systematic Review and Meta-Analysis of Published Studies, 2000–2017. *Front. Vet. Sci.* **2019**, *6*, 102. [CrossRef]
62. Barrow, P.A.; Jones, M.A.; Thomson, N. Salmonella. In *Pathogenesis of Bacterial Infections in Animals*; John Wiley & Sons, Ltd.: Hoboken, NJ, USA, 2010; pp. 231–265. ISBN 978-0-470-95820-9.
63. Ågren, E. Salmonella in Swedish Cattle. 2017. Available online: https://publications.slu.se/?file=publ/show&id=104094 (accessed on 16 April 2025).
64. Kemal, J. A Review on the Public Health Importance of Bovine Salmonellosis. *J Vet. Sci Technol* **2014**, *5*, 175. [CrossRef]
65. Gould, L.H.; Mungai, E.; Barton Behravesh, C. Outbreaks Attributed to Cheese: Differences Between Outbreaks Caused by Unpasteurized and Pasteurized Dairy Products, United States, 1998–2011. *Foodborne Pathog. Dis.* **2014**, *11*, 545–551. [CrossRef]
66. Koohmaraie, M.; Scanga, J.A.; De la zerda, M.J.; Koohmaraie, B.; Tapay, L.; Beskhlebnaya, V.; Mai, T.; Greeson, K.; Samadpour, M. Tracking the Sources of *Salmonella* in Ground Beef Produced from Nonfed Cattle. *J. Food Prot.* **2012**, *75*, 1464–1468. [CrossRef]
67. Fossler, C.P.; Wells, S.J.; Kaneene, J.B.; Ruegg, P.L.; Warnick, L.D.; Bender, J.B.; Godden, S.M.; Halbert, L.W.; Campbell, A.M.; Zwald, A.M.G. Prevalence of *Salmonella* spp. on Conventional and Organic Dairy Farms. *J. Am. Vet. Med. Assoc.* **2004**, *225*, 567–573. [CrossRef] [PubMed]

68. Nhung, N.T.; Van, N.T.B.; Cuong, N.V.; Duong, T.T.Q.; Nhat, T.T.; Hang, T.T.T.; Nhi, N.T.H.; Kiet, B.T.; Hien, V.B.; Ngoc, P.T.; et al. Antimicrobial Residues and Resistance against Critically Important Antimicrobials in Non-Typhoidal *Salmonella* from Meat Sold at Wet Markets and Supermarkets in Vietnam. *Int. J. Food Microbiol.* **2018**, *266*, 301–309. [CrossRef] [PubMed]
69. Carroll, L.M.; Wiedmann, M.; den Bakker, H.; Siler, J.; Warchocki, S.; Kent, D.; Lyalina, S.; Davis, M.; Sischo, W.; Besser, T.; et al. Whole-Genome Sequencing of Drug-Resistant *Salmonella enterica* Isolates from Dairy Cattle and Humans in New York and Washington States Reveals Source and Geographic Associations. *Appl. Environ. Microbiol.* **2017**, *83*, e00140-17. [CrossRef] [PubMed]
70. World Health Organization. *Antimicrobial Resistance: Global Report on Surveillance*; World Health Organization: Geneva, Switzerland, 2014; ISBN 978-92-4-156474-8.
71. Nguyen, T.T.; Le, H.V.; Vu Thi Hai, H.; Nguyen Tuan, T.; Nguyen, H.M.; Pham Xuan, D.; Tran Thi Thanh, H.; Le Thi, H.H. Whole-Genome Analysis of Antimicrobial-Resistant *Salmonella enterica* Isolated from Duck Carcasses in Hanoi, Vietnam. *Curr. Issues Mol. Biol.* **2023**, *45*, 2213–2229. [CrossRef]
72. Moura, Q.; Fernandes, M.R.; Silva, K.C.; Monte, D.F.; Esposito, F.; Dropa, M.; Noronha, C.; Moreno, A.M.; Landgraf, M.; Negrão, F.J.; et al. Virulent Nontyphoidal *Salmonella* Producing *CTX-M* and *CMY-2* β-Lactamases from Livestock, Food and Human Infection, Brazil. *Virulence* **2018**, *9*, 281–286. [CrossRef]
73. World Health Organization. *WHO Bacterial Priority Pathogens List 2024 Bacterial Pathogens of Public Health Importance, to Guide Research, Development, and Strategies to Prevent and Control Antimicrobial Resistance*, 1st ed.; World Health Organization: Geneva, Switzerland, 2024; ISBN 978-92-4-009346-1.
74. McMillan, E.A.; Gupta, S.K.; Williams, L.E.; Jové, T.; Hiott, L.M.; Woodley, T.A.; Barrett, J.B.; Jackson, C.R.; Wasilenko, J.L.; Simmons, M.; et al. Antimicrobial Resistance Genes, Cassettes, and Plasmids Present in *Salmonella enterica* Associated With United States Food Animals. *Front. Microbiol.* **2019**, *10*, 832. [CrossRef]
75. Qian, H.; Cheng, S.; Liu, G.; Tan, Z.; Dong, C.; Bao, J.; Hong, J.; Jin, D.; Bao, C.; Gu, B. Discovery of Seven Novel Mutations of *GyrB*, *ParC* and *ParE* in *Salmonella Typhi* and *Paratyphi* Strains from Jiangsu Province of China. *Sci Rep* **2020**, *10*, 7359. [CrossRef]
76. Randall, L.P.; Coldham, N.G.; Woodward, M.J. Detection of Mutations in *Salmonella enterica gyrA*, *GyrB*, *ParC* and *ParE* Genes by Denaturing High Performance Liquid Chromatography (DHPLC) Using Standard HPLC Instrumentation. *J. Antimicrob. Chemother.* **2005**, *56*, 619–623. [CrossRef]
77. Gonzalez Ronquillo, M.; Angeles Hernandez, J.C. Antibiotic and Synthetic Growth Promoters in Animal Diets: Review of Impact and Analytical Methods. *Food Control* **2017**, *72*, 255–267. [CrossRef]
78. Van Boeckel, T.P.; Brower, C.; Gilbert, M.; Grenfell, B.T.; Levin, S.A.; Robinson, T.P.; Teillant, A.; Laxminarayan, R. Global Trends in Antimicrobial Use in Food Animals. *Proc. Natl. Acad. Sci. USA* **2015**, *112*, 5649–5654. [CrossRef]
79. Hopkins, K.L.; Davies, R.H.; Threlfall, E.J. Mecanismos de Resistência Às Quinolonas Em *Escherichia Coli* e *Salmonella*: Desenvolvimentos Recentes. *Int. J. Antimicrob. Agents* **2005**, *25*, 358–373. [CrossRef]
80. Casas, M.R.T.; Camargo, C.H.; Soares, F.B.; da Silveira, W.D.; Fernandes, S.A. Presence of Plasmid-Mediated Quinolone Resistance Determinants and Mutations in Gyrase and Topoisomerase in *Salmonella enterica* Isolates with Resistance and Reduced Susceptibility to Ciprofloxacin. *Diagn. Microbiol. Infect. Dis.* **2016**, *85*, 85–89. [CrossRef]
81. Salah, F.D.; Soubeiga, S.T.; Ouattara, A.K.; Sadji, A.Y.; Metuor-Dabire, A.; Obiri-Yeboah, D.; Banla-Kere, A.; Karou, S.; Simpore, J. Distribution of Quinolone Resistance Gene (*Qnr*) in ESBL-Producing *Escherichia Coli* and *Klebsiella* spp. in Lomé, Togo. *Antimicrob Resist. Infect. Control* **2019**, *8*, 104. [CrossRef]
82. Heisig, P. High-Level Fluoroquinolone Resistance in a *Salmonella Typhimurium* Isolate Due to Alterations in Both *GyrA* and *GyrB* Genes. *J. Antimicrob. Chemother.* **1993**, *32*, 367–377. [CrossRef] [PubMed]
83. Casin, I.; Breuil, J.; Darchis, J.P.; Guelpa, C.; Collatz, E. Fluoroquinolone Resistance Linked to GyrA, GyrB, and ParC Mutations in *Salmonella enterica Typhimurium* Isolates in Humans. *Emerg. Infect. Dis.* **2003**, *9*, 1455–1457. [CrossRef] [PubMed]
84. Moreno-Switt, A.I.; Pezoa, D.; Sepúlveda, V.; González, I.; Rivera, D.; Retamal, P.; Navarrete, P.; Reyes-Jara, A.; Toro, M. Transduction as a Potential Dissemination Mechanism of a Clonal *Qnr*B19-Carrying Plasmid Isolated From *Salmonella* of Multiple Serotypes and Isolation Sources. *Front. Microbiol.* **2019**, *10*, 2503. [CrossRef]
85. Felix, M.A.; Han, J.; Khajanchi, B.K.; Sanad, Y.M.; Zhao, S.; Foley, S.L. *Salmonella enterica* Serovar Schwarzengrund: Distribution, Virulence, and Antimicrobial Resistance. *Microorganisms* **2025**, *13*, 92. [CrossRef]
86. Cohn, A.R.; Orsi, R.H.; Carroll, L.M.; Liao, J.; Wiedmann, M.; Cheng, R.A. *Salmonella enterica* Serovar Cerro Displays a Phylogenetic Structure and Genomic Features Consistent with Virulence Attenuation and Adaptation to Cattle. *Front. Microbiol.* **2022**, *13*, 1005215. [CrossRef]
87. Li, I.-C.; Wu, R.; Hu, C.-W.; Wu, K.-M.; Chen, Z.-W.; Chou, C.-H. Comparison of Conventional Molecular and Whole-Genome Sequencing Methods for Differentiating *Salmonella enterica* Serovar Schwarzengrund Isolates Obtained from Food and Animal Sources. *Microorganisms* **2021**, *9*, 2046. [CrossRef]
88. Kombade, S.; Kaur, N.; Kombade, S.; Kaur, N. *Pathogenicity Island in Salmonella spp.—A Global Challenge*; IntechOpen: London, UK, 2021; ISBN 978-1-83969-018-1.

89. Hensel, M. Evolution of Pathogenicity Islands of *Salmonella enterica*. *Int. J. Med. Microbiol.* **2004**, *294*, 95–102. [CrossRef]
90. Riquelme, S.; Varas, M.; Valenzuela, C.; Velozo, P.; Chahin, N.; Aguilera, P.; Sabag, A.; Labra, B.; Álvarez, S.A.; Chávez, F.P.; et al. Relevant Genes Linked to Virulence Are Required for *Salmonella Typhimurium* to Survive Intracellularly in the Social Amoeba Dictyostelium Discoideum. *Front. Microbiol.* **2016**, *7*, 1305. [CrossRef]
91. Nieto, P.A.; Pardo-Roa, C.; Salazar-Echegarai, F.J.; Tobar, H.E.; Coronado-Arrázola, I.; Riedel, C.A.; Kalergis, A.M.; Bueno, S.M. New Insights about Excisable Pathogenicity Islands in *Salmonella* and Their Contribution to Virulence. *Microbes Infect.* **2016**, *18*, 302–309. [CrossRef] [PubMed]
92. Pornsukarom, S.; Patchanee, P.; Erdman, M.; Cray, P.F.; Wittum, T.; Lee, J.; Gebreyes, W.A. Comparative Phenotypic and Genotypic Analyses of *Salmonella* Rissen That Originated from Food Animals in Thailand and United States. *Zoonoses Public Health* **2015**, *62*, 151–158. [CrossRef]
93. Han, J.; Aljahdali, N.; Zhao, S.; Tang, H.; Harbottle, H.; Hoffmann, M.; Frye, J.G.; Foley, S.L. Infection Biology of *Salmonella enterica*. *EcoSal Plus* **2024**, *12*, eesp-0001. [CrossRef] [PubMed]
94. Fàbrega, A.; Vila, J. *Salmonella enterica* Serovar *Typhimurium* Skills To Succeed in the Host: Virulence and Regulation. *Clin. Microbiol. Rev.* **2013**, *26*, 308–341. [CrossRef] [PubMed]
95. Sévellec, Y.; Vignaud, M.-L.; Granier, S.A.; Lailler, R.; Feurer, C.; Le Hello, S.; Mistou, M.-Y.; Cadel-Six, S. Polyphyletic Nature of *Salmonella enterica* Serotype Derby and Lineage-Specific Host-Association Revealed by Genome-Wide Analysis. *Front. Microbiol.* **2018**, *9*, 891. [CrossRef]
96. Sousa, A.D.A. Caracterização Fenotípica e Genética de Resistência Antimicrobiana em Salmonella enterica. Master's Thesis, Fernando Pessoa University, Ponte de Lima, Portugal, 2023.
97. Marcus, S.L.; Brumell, J.H.; Pfeifer, C.G.; Finlay, B.B. *Salmonella* Pathogenicity Islands: Big Virulence in Small Packages. *Microbes Infect.* **2000**, *2*, 145–156. [CrossRef]
98. Ibarra, J.A.; Steele-Mortimer, O. *Salmonella*—the Ultimate Insider. *Salmonella* ate Intracellular Survival. *Cell Microbiol.* **2009**, *11*, 1579–1586. [CrossRef]
99. Sheehan, J.R.; Sadlier, C.; O'Brien, B. Bacterial Endotoxins and Exotoxins in Intensive Care Medicine. *BJA Educ.* **2022**, *22*, 224–230. [CrossRef]
100. Kujat Choy, S.L.; Boyle, E.C.; Gal-Mor, O.; Goode, D.L.; Valdez, Y.; Vallance, B.A.; Finlay, B.B. SseK1 and SseK2 Are Novel Translocated Proteins of *Salmonella enterica* Serovar Typhimurium. *Infect. Immun.* **2004**, *72*, 5115–5125. [CrossRef]
101. Zhang, X.; He, L.; Zhang, C.; Yu, C.; Yang, Y.; Jia, Y.; Cheng, X.; Li, Y.; Liao, C.; Li, J.; et al. The Impact of *Sse*K2 Deletion on *Salmonella enterica* Serovar Typhimurium Virulence in vivo and in vitro. *BMC Microbiol.* **2019**, *19*, 182. [CrossRef]
102. Espinoza, R.A.; Silva-Valenzuela, C.A.; Amaya, F.A.; Urrutia, Í.M.; Contreras, I.; Santiviago, C.A.; Espinoza, R.A.; Silva-Valenzuela, C.A.; Amaya, F.A.; Urrutia, Í.M.; et al. Differential Roles for Pathogenicity Islands SPI-13 and SPI-8 in the Interaction of *Salmonella* Enteritidis and Salmonella Typhi with Murine and Human Macrophages. *Biol. Res.* **2017**, *50*, 5. [CrossRef]
103. Desai, P.T.; Porwollik, S.; Long, F.; Cheng, P.; Wollam, A.; Bhonagiri-Palsikar, V.; Hallsworth-Pepin, K.; Clifton, S.W.; Weinstock, G.M.; McClelland, M. Evolutionary Genomics of *Salmonella enterica* Subspecies. *MBio* **2013**, *4*, e00579-12. [CrossRef] [PubMed]
104. Haneda, T.; Ishii, Y.; Danbara, H.; Okada, N. Genome-Wide Identification of Novel Genomic Islands That Contribute to *Salmonella* Virulence in Mouse Systemic Infection. *FEMS Microbiol. Lett.* **2009**, *297*, 241–249. [CrossRef] [PubMed]
105. Shah, D.H.; Lee, M.-J.; Park, J.-H.; Lee, J.-H.; Eo, S.-K.; Kwon, J.-T.; Chae, J.-S. Identification of *Salmonella* Gallinarum Virulence Genes in a Chicken Infection Model Using PCR-Based Signature-Tagged Mutagenesis. *Microbiology* **2005**, *151 Pt 12*, 3957–3968. [CrossRef] [PubMed]
106. Elder, J.R.; Chiok, K.L.; Paul, N.C.; Haldorson, G.; Guard, J.; Shah, D.H. The *Salmonella* Pathogenicity Island 13 Contributes to Pathogenesis in Streptomycin Pre-Treated Mice but Not in Day-Old Chickens. *Gut Pathog.* **2016**, *8*, 16. [CrossRef]
107. Benevides, V.P.; Saraiva, M.M.S.; Ferreira, V.A.; Funnicelli, M.I.G.; Rodrigues Alves, L.B.; Almeida, A.M.; Christensen, H.; Olsen, J.E.; Berchieri Junior, A. Unveiling the Genomic Landscape of Understudied *Salmonella enterica* Serovars from Poultry and Human: Implications for Food Safety. *Curr. Microbiol.* **2025**, *82*, 279. [CrossRef]
108. Jiang, L.; Feng, L.; Yang, B.; Zhang, W.; Wang, P.; Jiang, X.; Wang, L. Signal Transduction Pathway Mediated by the Novel Regulator LoiA for Low Oxygen Tension Induced *Salmonella* Typhimurium Invasion. *PLoS Pathog.* **2017**, *13*, e1006429. [CrossRef]

genes

MDPI

Article

Comparative Genomics and Functional Profiling Reveal Lineage-Specific Metabolic Adaptations in Globally Emerging Fluoroquinolone-Resistant *Salmonella* Kentucky ST198

Juned Ahmed [1], Rachel C. Soltys [2], Smriti Shringi [1], Jean Guard [3], Bradd J. Haley [4] and Devendra H. Shah [1,*]

[1] School of Veterinary Medicine, Texas Tech University, Amarillo, TX 79106, USA; junahmed@ttu.edu (J.A.)
[2] Animal Disease Diagnostic Laboratory, Department of Comparative Pathobiology, College of Veterinary Medicine, Purdue University, West Lafayette, IN 47907, USA
[3] United States National Poultry Research Center, Agriculture Research Service, United States Department of Agriculture, Athens, GA 30605, USA; jean.guard@yahoo.com
[4] Environmental Microbial and Food Safety Laboratory, Beltsville Agricultural Research Center, Agricultural Research Service, United States Department of Agriculture, Beltsville, MD 20705, USA; bradd.haley@usda.gov
* Correspondence: devendra.shah@ttu.edu

Abstract

Background: *Salmonella* Kentucky comprises two major lineages, ST152 and fluoroquinolone-resistant (Flu^R) ST198, which have diverged genotypically and phenotypically along distinct evolutionary and epidemiological trajectories. ST198 is linked to global human disease, while ST152 is primarily animal-associated in the U.S. We hypothesized that lineage-specific metabolic adaptations contribute to their differing host associations and pathogenicity. Methods: We performed comparative metabolic profiling of ST198 (n = 3) and ST152 (n = 4) strains across 948 substrates and environmental conditions. Growth assays tested the ability of these lineages and other non-typhoidal *Salmonella* (NTS) serovars (n = 5) to utilize *myo*-inositol and lactulose as sole carbon sources. Comparative genomic analyses of 294 ST198, 173 ST152, and 1300 other NTS serovars identified nutrient utilization genes. Results: ST198 exhibited significantly higher respiratory activity and broader metabolic versatility across carbon, nitrogen/sulfur sources, and stress conditions. The canonical *iol* gene cluster for *myo*-inositol catabolism was conserved in ST198 but absent in ST152, which nonetheless showed weak growth on *myo*-inositol, suggesting an alternative metabolic pathway for *myo*-inositol may exist. We also report for the first time that, despite lineage-specific differences in metabolic efficiency, multiple NTS serovars, including *S.* Kentucky, can metabolize lactulose, a synthetic disaccharide traditionally associated with beneficial gut microbes. These results suggest the potential existence of a novel lactulose metabolic pathway in NTS. Conclusions: These findings highlight ST198's metabolic adaptability and reveal novel metabolic capacities in NTS. A mechanistic understanding of nutrient utilization pathways, particularly of *myo*-inositol and lactulose, will provide novel insights into the mechanisms underlying nutrient metabolism that likely modulate the ecological success and pathogenic potential of NTS in human and animal hosts.

Keywords: *Salmonella* Kentucky; ST152; ST198; metabolism

1. Introduction

Non-typhoidal *Salmonella* (NTS), a genetically and antigenically diverse enteric pathogen with >1500 serotypes, is a leading cause of foodborne enteric illness and a significant contributor to antimicrobial resistance (AMR) worldwide, responsible for ~150 million infections and 60,000 deaths each year [1]. Among the various NTS serovars, *Salmonella* Kentucky (*S.* Kentucky) is frequently isolated from food animals in the United States, particularly chicken, turkey, swine, and cattle [2–4]. Two predominant genetic lineages of *S.* Kentucky, ST198 and ST152, exhibit distinct epidemiological and genetic characteristics [3–9]. ST152 is frequently isolated from U.S. food animals but is rarely associated with human illness. ST198 comprises two sub-clades: fluoroquinolone-susceptible (Flu^S) ST198.1, primarily isolated from U.S. food animals [2,4,10], and fluoroquinolone-resistant (Flu^R) ST198.2, predominantly isolated from food animals in North Africa, the Middle East, Europe, and South Asia, and often identified as a cause of human illnesses, including U.S. travelers returning from international travel [2,4,5,8,9,11].

The Flu^R ST198.2 sub-clade (hereafter referred to as Flu^R ST198) has been increasingly detected in human clinical cases globally, posing a significant public health threat [7, 12–14]. In North America, Flu^R ST198 is commonly associated with travel-associated illnesses, particularly among travelers returning from North Africa, the Middle East, and South Asia, and it is often linked to the consumption of poultry or poultry-related products [2,4,9]. Notably, Flu^R ST198 harbors two mutations in *gyrA* and three mutations in *parC* in the quinolone resistance-determining region (QRDR), which is concerning given that ciprofloxacin remains the first-line antibiotic for the treatment of NTS infections in humans worldwide [6,15]. Recently, Flu^R ST198 of *S.* Kentucky was implicated in a nosocomial infection in a hospitalized American black bear at a veterinary teaching hospital in New York [16], highlighting the potential risk this lineage poses to animal health, food safety, and public health in the USA.

We previously reported that Flu^R ST198 has undergone microevolution driven by the accumulation of lineage-specific mutations [2,9]. Similar microevolutionary changes in other *Salmonella* serotypes and multidrug-resistant (MDR) lineages of NTS serovars, such as *S.* Typhimurium ST213/ST313 and *S.* Enteritidis ST11, exhibit enhanced metabolic capacities, improved host adaptation, and pathogenicity [17–19]. However, whether lineage-specific mutations among ST152 and Flu^R ST198 have resulted in distinct metabolic capacities remains unknown. Understanding these differences could provide critical insights into the mechanisms driving emergence, pathogenicity, and niche-specific adaptation of Flu^R ST198.

In this study, we hypothesized that the globally emerging Flu^R ST198 and domestically prevalent ST152 lineages have genetically evolved to exhibit distinct metabolic capabilities. To test this hypothesis, we conducted metabolic fingerprinting of seven epidemiologically distinct strains of ST152 ($n = 4$) and Flu^R ST198 ($n = 3$) by measuring respiratory activity (RA) across 948 substrates and environmental conditions. Our results demonstrate that Flu^R ST198 consistently exhibits significantly higher ($p < 0.01$) RA than ST152 when utilizing several carbon and nitrogen/sulfur compounds as sole sources of energy. Notably, Flu^R ST198 showed markedly increased RA and growth in media supplemented with two gut-associated nutrients, lactulose and *myo*-inositol, commonly used as dietary supplements. We report that the *myo*-inositol-mediated metabolic fitness of Flu^R ST198 and several other NTS serovars is likely due to the conserved carriage of the *myo*-inositol catabolism gene cluster (*iolCatGC*), which is completely absent in the ST152 lineage and a few other NTS serovars. Furthermore, we report that different NTS, including Flu^R ST198, can efficiently metabolize lactulose as a sole energy source for growth in a serotype-independent manner, due to an unknown genetic mechanism. Together, these findings suggest that Flu^R ST198 has acquired distinct metabolic traits that may modulate its fitness, pathogenic potential, and capacity for global dissemination.

2. Materials and Methods

2.1. Strain Selection and Growth

A subset of seven well-characterized epidemiologically and genetically distinct strains of S. Kentucky ST152 ($n = 4$) and ST198 ($n = 3$) were used in this study [9]. All ST198 strains in this study were resistant to fluoroquinolones (Flu^R), exhibited multidrug resistance, and were isolated from individuals with travel histories to Asia, Africa, or the Middle East (Table 1). Conversely, all ST152 strains were fluoroquinolone-sensitive (Flu^S), with three strains isolated from poultry sources and one isolated from a human case without travel history outside the USA (Table 1). Five additional NTS strains representing clinically relevant serovars (S. Typhimurium, S. Hadar, S. Enteritidis, S. Montevideo, and S. Schwarzengrund) were also tested in individual growth kinetics experiments in this study [20]. All strains were revived from −80 °C storage by streaking on tryptic soy agar (TSA) followed by incubation at 37 °C for 16 h. Colonies grown on TSA were used for Phenotype Microarray and/or growth assays.

Table 1. Wild-type S. Kentucky and other NTS strains used in this study.

Serovar	ST Type	Strain ID	Source	Resistance	Isolation Year	Travel History	Reference
S. Kentucky	ST198	PU61	Human	Flu^R	2004	Egypt, the Middle East, India, Tanzania, Ethiopia, Ivory Coast, or Morocco	Soltys et al., 2021 [9]
S. Kentucky	ST198	PU128	Human	Flu^R	2013		
S. Kentucky	ST198	PU131	Human	Flu^R	2013		
S. Kentucky	ST152	P55	Poultry	Flu^S	2012	Not available	
S. Kentucky	ST152	P72	Poultry	Flu^S	2012	Not available	
S. Kentucky	ST152	PU116	Human	Flu^S	2012	No travel outside North America	
S. Kentucky	ST152	PU363	Poultry	Flu^S	2005	Not available	
S. Typhimurium	Unknown	21611	Unknown	Not tested	Unknown	Not available	Burin and Shah, 2021 [20]
S. Hadar	Unknown	27032	Human	Not tested	2014	Not available	
S. Enteritidis	Unknown	MD15	Human	Not tested	2010	Not available	
S. Montevideo	Unknown	DHS-R6	Unknown	Not tested	Unknown	Not available	
S. Schwarzengrund	Unknown	26682	Human	Not tested	2014	Not available	

2.2. Comparative Metabolic Profiling of ST152 and Flu^R ST198

We performed comparative metabolic profiling of ST152 ($n = 4$) and Flu^R ST198 ($n = 3$) strains across 948 metabolic substrates and environmental conditions using Phenotype Microarray (PM) (Biolog, Hayward, CA, USA). The PM analysis included eight metabolic panels for carbon (PM-1-2), nitrogen (PM-3, PM-6-8), sulfur (PM-4), phosphorus (PM-4), and nutrient supplements (PM-5) as energy sources. Additionally, two sensitivity panels (PM-9 and PM-10) evaluated the sensitivity of strains to osmolytes, toxicities, and various pH conditions. Testing was performed following previously established protocols [21–23]. Each strain was inoculated onto PM-1 to PM-10 plates and incubated at 42 °C for 48 h. Respiratory activity (RA) was measured by the reduction in tetrazolium violet dye, with color change indicating cellular respiration and electron transport chain function. The data from PM tests were organized in Microsoft Excel (Table S1) and analyzed via hierarchical clustering (Euclidean distance, complete linkage) for each nutrient/condition using RStudio (R version 4.4.1). The RA differences between ST152 and Flu^R ST198 were compared using Student's t-test, and a mean RA difference ≥ 50 units between ST152 and Flu^R ST198 with $p < 0.01$ was considered statistically significant [21–23]. The heatmaps for hierarchical clustering were generated using the pheatmap package within RStudio (version 4.4.1) [24].

2.3. Benchtop Growth Assay

To confirm the differential metabolic fitness traits between ST152 and Flu^R ST198 strains detected by the PM array, each strain was tested in a benchtop growth assay for its ability to utilize two representative carbon compounds, including *myo*-inositol and lactulose, at mammalian (37 °C) and avian (42 °C) physiological temperatures. Briefly, bacterial colonies were revived by streaking on TSA and incubated overnight at 37 °C. Next, colonies were resuspended in M9 broth (BD, Sparks, MD, USA) to a 0.5 McFarland standard, followed by serial 10-fold dilutions to achieve approximately 200 colony-forming units (CFU)/mL. Approximately 360 CFU of each strain was inoculated into 1.8 mL of M9 media (200 CFU/mL) in 96-well blocks supplemented with varying concentrations of lactulose (1, 10, and 20 mM) and *myo*-inositol (1, 10, and 20 mM) and incubated at 42 °C for 96 h. At 24 h intervals, 100 µL from each culture was withdrawn, followed by plating of 10-fold dilutions on TSA and incubation for 16 h at 37 °C to determine the CFUs at each time point. In each of the experiments, M9 broth supplemented with glucose (1, 10, and 20 mM) was used as a positive control, M9 broth without any nutrient supplementation was used as a negative control, and glucose (20 mM) without bacterial inoculation was used as a sterility control. To determine the growth of ST152 and Flu^R ST198 at mammalian physiological temperature (37 °C), a single concentration of lactulose (20 mM) or *myo*-inositol (20 mM) was used, and the CFU determinations were conducted up to 72 h. Growth in each metabolite and each concentration at both temperatures was tested in duplicates in three independent experiments. Growth assay results were expressed as $\log_{10}$ CFU/mL. The growth data were organized and analyzed using Microsoft Excel (Version 2408) and RStudio (R version 4.4.1). A $\log_{10}$ difference was calculated as the difference in the $\log_{10}$ CFU between the end-time-point (72 or 96 h) and the starting inoculum (0 h). Doubling times (Td) were calculated as previously described [22] using the formula $t/(3.3 \times \log(b/B))$, where t is the end time-point for Td calculation (96 h for 42 °C growth data and 72 h for 37 °C growth data), b is CFU at the endpoint (72 h or 96 h), and B is the initial inoculum (CFU at 0 h). Log_{10} and Td differences were calculated and compared using Student's t-test. A p-value of <0.05 was considered statistically significant. To determine the ability of other NTS serovars to utilize *myo*-inositol and lactulose for growth, we tested five NTS strains representing the following serovars (*S.* Typhimurium, *S.* Infantis, *S.* Hadar, *S.* Enteritidis, *S.* Montevideo, and *S.* Schwarzengrund). These NTS strains were grown in M9 media supplemented with lactulose (20 mM) and *myo*-inositol (20 mM) by incubating at 37 °C for 72 h. The growth of each strain was determined by calculating the $\log_{10}$ increase at 72 h post-incubation.

2.4. Comparative Genomics Analysis

To identify and inventory lineage-specific genes likely associated with the differential metabolic potential of ST198 and ST152 lineages, pan-genome analysis of the 7 epidemiologically distinct strains of *S.* Kentucky ST152 (n = 4) and Flu^R ST198 (n = 3) used in this study was performed using PanExplorer followed by gene functional enrichment analysis [25]. Follow-up comparative genomics analysis focused on determining the genetic basis of *myo*-inositol and lactulose-mediated differential metabolic fitness of ST152 and Flu^R ST198. To determine the genetic basis of differential utilization of *myo*-inositol between ST152 and Flu^R ST198 strains, we compared the genomes of the ST152 and Flu^R ST198 strains used in this study to identify the *iol* gene cluster (*iolCatGC*) known to contribute to *myo*-inositol metabolism and growth of *S.* Typhimurium in the presence of *myo*-inositol as the sole energy source [26,27], using MultiGeneBlast version 1.1.13 [28]. To confirm whether differential carriage of the *iol* gene cluster was consistent across ST152 and ST198 lineages, we expanded genetic screening to the publicly available whole-genome sequences of 294 ST198 strains and 173 ST152 strains along with a few other STs, including ST314

($n = 14$), ST2132 ($n = 3$), ST19 ($n = 2$), and ST32, ST64, ST166, ST318, ST639, and ST1679 ($n = 1$ each) (Table S2), using the compare region viewer tool of the Bacterial and Viral Bioinformatics Resource Center (BV-BRC) [29]. Finally, to determine carriage of the *iol* gene cluster among other NTS serovars, a total of 1300 complete high-quality genomes representing 106 different NTS serovars (Table S2) were evaluated using the compare region viewer tool of BV-BRC.

Lactulose metabolism is widespread among gut microbes and is typically attributed to glycoside hydrolases (GHs) that hydrolyze the β-1,4-glycosidic bond of lactulose into fermentable monosaccharides (e.g., fructose, galactose) [30–33]. However, lactulose metabolism by NTS has not been described, and its genetic basis remains unknown. Thus, to identify any gene encoding a putative GH protein likely involved in lactulose metabolism in *Salmonella,* we first searched through the genome of *S.* Kentucky strain PU131, a reference strain of Flu^R ST198 [15]. We then searched the candidate proteins against the CAZy database to identify putative glycoside transporters and β-galactosidases that can hydrolyze the β-1,4-glycosidic bond in lactulose and metabolize it to monosaccharides [31,32,34]. To determine the carriage of putative transporter and β-galactosidase among other *Salmonella* genomes, we screened the whole-genome sequences of 294 ST198 strains and 173 ST152 strains along with a few other STs, including ST314 ($n = 14$) and other STs ($n = 11$) of *S.* Kentucky and 1300 complete high-quality genomes, representing 106 NTS serovars (other than *S.* Kentucky) (Table S3).

2.5. RT-qPCR

RT-qPCR was conducted to confirm the upregulation of *myo*-inositol inducible genes (*iolT1* and *reiD*) in NTS when grown in the presence of *myo*-inositol as the sole source of energy. Total RNA was extracted from a representative strain of *S.* Kentucky ST198 (PU61), which carries the *iol* gene cluster and also exhibited higher growth in the presence of *myo*-inositol as the sole energy source. Briefly, the PU61 strain was grown overnight in TSA at 37 °C. Next, ~10^8 CFU of this strain was resuspended in M9 media containing 20 mM of *myo*-inositol or glucose (control) and incubated at 37 °C for 6 h. After 6 h, ~10^9 CFUs were harvested by centrifugation at 10,000 RCF for 10 min at 4 °C for RNA extraction. Total RNA was extracted using the Direct-zol RNA Miniprep Plus Kit (Zymo Research, Irvine, CA, USA) following the manufacturer's instructions with minor modifications, which included in-tube DNase treatment using Turbo DNase (Invitrogen, Waltham, MA, USA) followed by RNA clean up using the RNA Clean & Concentrator-25 (Zymo Research). An equal concentration (~820 ng) of total RNA was used to prepare cDNA using the High-Capacity cDNA Reverse Transcription Kit (Applied Biosystems, Waltham, MA, USA) according to the manufacturer's protocol. Finally, 2 μL of cDNA and 800 nM of pre-designed primers (Table S4) for inositol catabolic genes (*iolT1* and *reiD*) were used for qPCR using the PowerUp SYBR Master Mix (Applied Biosystems) with the following conditions: 1 cycle at 50 °C for 2 min, 1 cycle at 95°C for 2 min, 40 cycles at 95°C for 15 s, and 40 cycles at 60°C for 1 min. Cq values were obtained and compared with glucose as the reference condition. Each sample was tested in triplicate. Gene expression was analyzed using the ΔCq method, which represents the difference in Cq between the test condition (*myo*-inositol) and the control condition (glucose). Log_2 fold change was computed as $\log_2 (2^{-\Delta Cq})$, and the calculation was performed in Python (version 3.11.13) using the NumPy package. The null hypothesis tested was that the mean $\Delta Cq = 0$ (i.e., no change in expression relative to glucose). A p-value < 0.05 was considered statistically significant.

3. Results and Discussion

3.1. Differences in Metabolic Fitness Between ST152 and Flu^R ST198 Lineages

The metabolic fitness differences between ST152 and Flu^R ST198 lineages were evaluated by measuring the respiratory activity (RA) of multiple epidemiologically distinct *S.* Kentucky strains across 948 metabolic conditions (Table S1). Hierarchical clustering revealed distinct metabolic profiles, with Flu^R ST198 lineage strains exhibiting broader and more efficient metabolic capacities via utilization of various energy sources compared to the ST152 lineage strains (Figures 1 and S1). Statistical analysis showed significantly higher RA ($p < 0.01$) for Flu^R ST198 strains in 17 out of 948 (1.79%) conditions, including six carbon sources, six nitrogen/sulfur sources, and five stress conditions (Table S1). The largest RA differences for Flu^R ST198 strains were observed with *myo*-inositol (RA difference: 247.9, p-value < 0.01) and lactulose (RA difference: 152.9, p-value < 0.01) as sole energy sources (Figure 1). In addition, Flu^R ST198 strains also displayed significantly higher RA (RA difference > 50, p-value < 0.01) when utilizing glycine, mono-methyl succinate, dextrin, and laminarin as the sole energy sources. Conversely, ST152 strains exhibited significantly higher RA with melibionic acid as the sole energy source (RA difference: 213.6, p-value < 0.01) (Figure 1). For nitrogen/sulfur metabolism, Flu^R ST198 strains outperformed ST152 strains in utilizing S-Methyl-L-Cysteine, His-Trp, Phe-Pro, Val-Ser, Gly-Gly-Gly, and His-His as sole energy sources, with no nitrogen/sulfur substrates favoring ST152 (Figure 2A). Flu^R ST198 strains also demonstrated great resilience under environmental stress, with higher RA under five of the six stress conditions tested, notably 4% sodium formate, pH 4.5 + L-Alanine, pH 4.5 + L-Asparagine, and pH 4.5 with or without L-Proline (Figure 2B). Conversely, ST152 strains exhibited significantly higher RA when grown with 6% NaCl with L-Carnitine (Figure 2B).

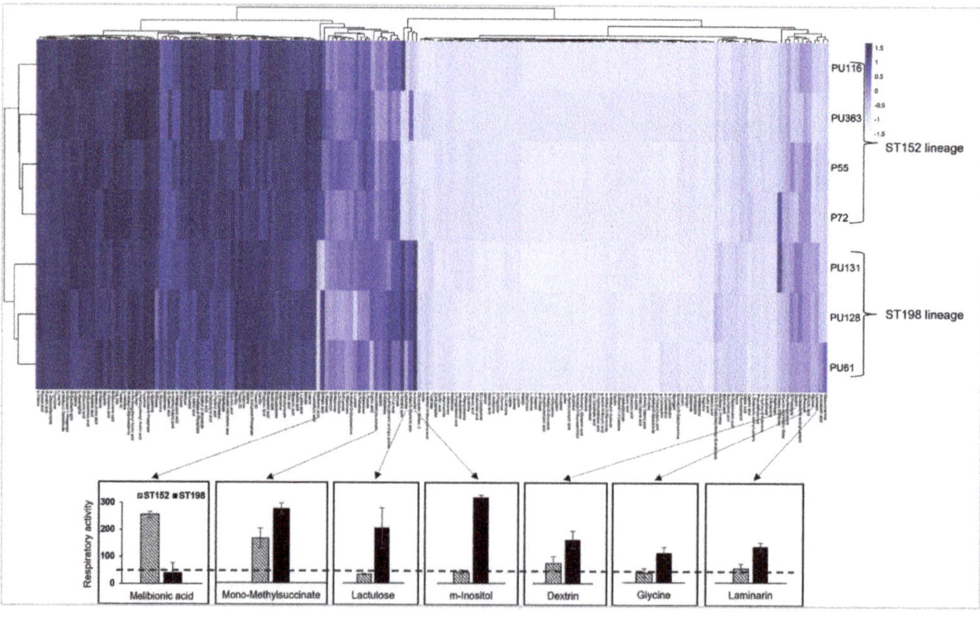

Figure 1. Hierarchical clustering showing distinct metabolic profiles of ST152 and Flu^R ST198 strains based on the differences in respiratory activity (RA) in the presence of 190 different carbon compounds as sole energy sources. The *X*-axis shows the clusters of carbon sources. The *Y*-axis shows the clusters formed by strains within each lineage tested in this study. The dark blue indicates higher RA, and the

light blue to white color indicates lower RA, with the bar graphs below showing seven carbon sources that are significantly differentially utilized by ST152 and Flu^R ST198 strains with an RA difference of >50 at a p-value of ≤ 0.01. The Y-axis in each bar graph shows RA activity in the presence of each carbon source, and the dotted line represents the RA value cutoff of 50.

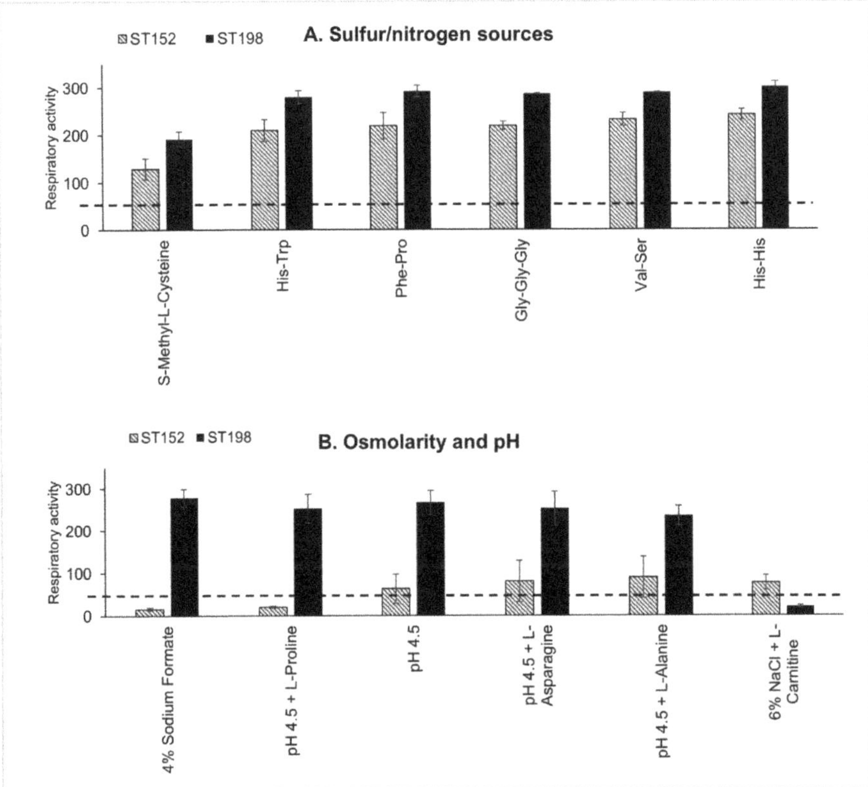

Figure 2. Difference in the respiratory activity (RA) between ST152 and Flu^R ST198 strains. ST152 and Flu^R ST198 strains showed significant differences ($p < 0.01$) in RA for six different nitrogen/sulfur compounds as sole energy source (**A**) and in response to six stressors (**B**). The dotted lines indicate the RA value cutoff of 50.

These data suggest that globally emerging Flu^R ST198 lineage exhibits enhanced nutritional fitness, utilizing diverse carbon sources that likely contribute to its ecological success and pathogenic potential. Similar lineage-specific metabolic divergence has been reported in other NTS serovars, such as *S.* Typhimurium ST213/ST313 and *S.* Enteritidis ST11, where lineage-specific genome degradation has been linked to distinct metabolic capacities, host adaptation, and pathogenicity [17–19,35,36]. For instance, the ST313 lineage of *S.* Typhimurium exhibits distinct metabolic profiles, particularly the enhanced ability to utilize several carbon compounds as sole energy sources, which likely facilitate its adaptation to host-associated niches and contribute to the ability to cause extraintestinal invasive disease [17]. Therefore, it is likely that the metabolic differences observed between ST152 and Flu^R ST198 reflect lineage-specific adaptations that modulate persistence in food-production systems and host environments. Overall, these results indicate that the Flu^R ST198 lineage possesses an expanded and more efficient capacity to utilize a range of carbon and nitrogen/sulfur substrates and enhanced tolerance and survival under a range of stressful environmental conditions, which may confer a metabolic fitness advantage in diverse or host-associated environments.

3.2. Lineage- and Temperature-Dependent Differences in Myo-Inositol Utilization by ST152 and Flu^R ST198 Strains

To validate the differential utilization of *myo*-inositol, growth kinetics assays were performed in media supplemented with varying concentrations (1 mM, 10 mM, and 20 mM) of *myo*-inositol at avian physiological temperature (42 °C). Flu^R ST198 strains exhibited significantly increased growth (1.96 ± 1.12 to 1.88 ± 0.2 $\log_{10}$ CFU/mL increase) across all concentrations of *myo*-inositol tested, with most consistent growth by all strains achieved at 20 mM concentration (Figure 3A,C). In contrast, the ST152 strains showed a consistent decline in viable counts across all concentrations at 42 °C, indicating an inability to effectively utilize *myo*-inositol as a sole energy source under avian physiological conditions.

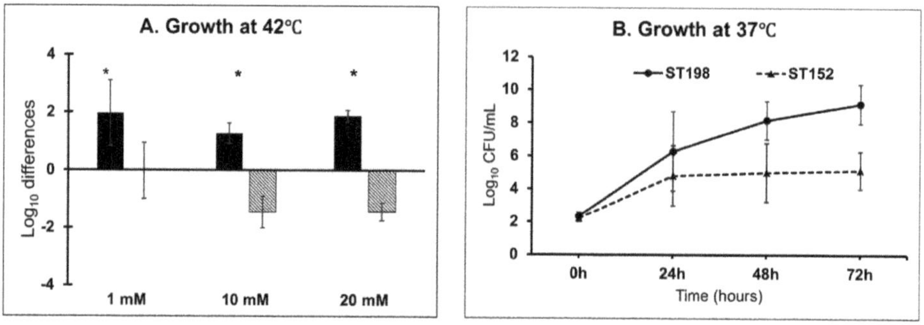

C. Doubling time (Td)				
	42°C		37°C	
Nutrients	Td of ST152 (h)	Td of ST198 (h)	Td of ST152 (h)	Td of ST198 (h)
myo-inositol (20 mM)	−40.43 ± 20.14	6.24 ± 1.16*	8.81 ± 4.87	3.25 ± 0.43*

Figure 3. The $\log_{10}$ difference in growth of Flu^R ST198 (black bars) strains was significantly higher than ST152 (hatched bar) using *myo*-inositol as the sole energy source at three different concentrations at 42 °C (**A**) and using a 20 mM concentration at 37 °C (**B**). The doubling time of Flu^R ST198 was also significantly lower ($p < 0.05$) than ST152 at both 42 °C and 37 °C (**C**). * Indicates significant difference between Flu^R ST198 and ST152 under same nutrient and temperature ($p < 0.05$).

To further evaluate temperature-dependent effects, growth assays were repeated at the mammalian physiological temperature (37 °C) using 20 mM *myo*-inositol, the concentration that supported maximal ST198 growth and has been used in previous in vitro and infection studies [37,38]. At 37 °C, Flu^R ST198 strains reached significantly higher cell densities (9.15 ± 1.19 $\log_{10}$ CFU/mL 72 h post-incubation with a Td of 3.25 ± 0.43 h) compared to ST152 strains (5.14 ± 1.14 $\log_{10}$ CFU/mL 72 h post-incubation with a Td of 8.81 ± 4.87 h) (Figure 3B,C). These results suggest that ST152 strains possess a markedly reduced capacity to metabolize *myo*-inositol, although they are capable of residual growth at 37 °C. The absence of growth at 42 °C but limited growth at 37 °C in ST152 strains may reflect temperature-sensitive regulation or partial functionality of pathways involved in *myo*-inositol utilization. Together, these findings indicate that Flu^R ST198 strains can efficiently use *myo*-inositol at both avian and mammalian physiological temperatures, whereas ST152 strains exhibit impaired and temperature-dependent utilization.

3.3. Lineage- and Temperature-Dependent Differences in Lactulose Utilization by ST152 and Flu^R ST198 Strains

Growth kinetics assays in media supplemented with lactulose (1 mM, 10 mM, and 20 mM) at the avian physiological temperature (42 °C) revealed striking lineage-specific differences in lactulose utilization as the sole source of energy for growth. Flu^R ST198 strains exhibited a clear dose-dependent increase in growth, with viable counts rising from 1.92 $\pm$ 0.66 $\log_{10}$ CFU/mL at 1 mM to 3.45 $\pm$ 0.49 $\log_{10}$ CFU/mL at 10 mM and 4.54 $\pm$ 0.03 $\log_{10}$ CFU/mL at 20 mM, indicating efficient metabolism of lactulose as a sole energy source (Figure 4A). In contrast, ST152 strains showed a consistent decline in growth, from 2.53 $\pm$ 0.40 $\log_{10}$ CFU/mL at 1 mM to 1.12 $\pm$ 1.47 $\log_{10}$ CFU/mL at 10 mM, and a net loss of viability (-1.86 ± 0.20 $\log_{10}$ CFU/mL) at 20 mM, suggesting either an inability to catabolize lactulose or potential toxic effects at higher concentrations at the avian physiological temperature.

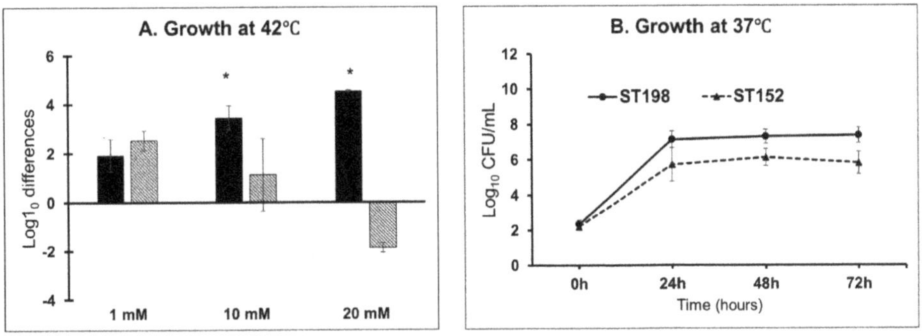

C. Doubling time (Td)

	42°C		37°C	
Nutrients	Td of ST152 (h)	Td of ST198 (h)	Td of ST152 (h)	Td of ST198 (h)
Lactulose (20 mM)	−47.56 ± 54.02	6.28 ± 0.06*	6.25 ± 1.39	4.35 ± 0.2*

Figure 4. The $\log_{10}$ CFU/mL differences in growth of Flu^R ST198 (black bars) strains were significantly higher than ST152 (hatched bar) under lactulose as the sole energy source at 10 mM and 20 mM concentrations at 42 °C (**A**) and at 20 mM concentration at 37 °C incubation temperature (**B**). The doubling time of Flu^R ST198 was also significantly lower ($p < 0.05$) than ST152 at both 42 °C and 37 °C (**C**). * Indicates significant difference between Flu^R ST198 and ST152 under same nutrient and temperature ($p < 0.05$).

Follow-up assays at mammalian physiological temperature (37 °C) using 20 mM lactulose demonstrated that strains within both lineages were capable of growth. However, ST152 strains exhibited reduced proliferation compared to ST198 ($p < 0.05$), reaching 5.80 $\pm$ 0.63 $\log_{10}$ CFU/mL at 72 h with a doubling time of 6.25 $\pm$ 1.39 h, while Flu^R ST198 strains reached 7.36 $\pm$ 0.43 $\log_{10}$ CFU/mL with a faster doubling time of 4.35 $\pm$ 0.2 h (Figure 4B,C).

Taken together, these results highlight the superior metabolic flexibility of ST198, which can efficiently utilize lactulose across both avian and mammalian physiological temperatures. In contrast, ST152 shows temperature-dependent impairment in lactulose utilization with significantly reduced growth at avian physiological temperature; however, this effect is minimized at mammalian physiological temperature. These metabolic constraints suggest lactulose may differentially modulate ecological adaptation in different hosts, impacting fitness, environmental persistence, or zoonotic potential.

3.4. Pan-Genome and Comparative Gene Functional Analysis of ST152 and Flu^R ST198 Strains

The comparative genomic analysis of four strains of the ST152 lineage and three strains of the Flu^R ST198 lineage used in this study revealed a pan-genome consisting of 5464 genes (Figure 5A). A total of 280 (5.1%) genes were identified as ST198-lineage specific, whereas 264 (4.8%) genes were identified as ST152-lineage specific (Figure 5A). The functional distribution of ST152 and Flu^R ST198 lineage-specific genes revealed that each lineage carried a unique repertoire of genes, with the majority of genes encoding bacterial metabolism, hypothetical proteins, and proteins of unknown functions (Figure 5B,C). The differential carriage of the broad repertoire of genes involved in metabolic functions corroborates the differences in the metabolic phenotypes of these two lineages observed in this study. The results of the pan-genome analysis suggest that lineage-specific mutation and genome degradation may directly impact the metabolic adaptation of the sequence types in different niches, including the host and the environment. Our follow-up investigations focused on determining the linkage between genotype and phenotype for the differential ability of ST152 and Flu^R ST198 strains to utilize *myo*-inositol and lactulose as sole carbon sources.

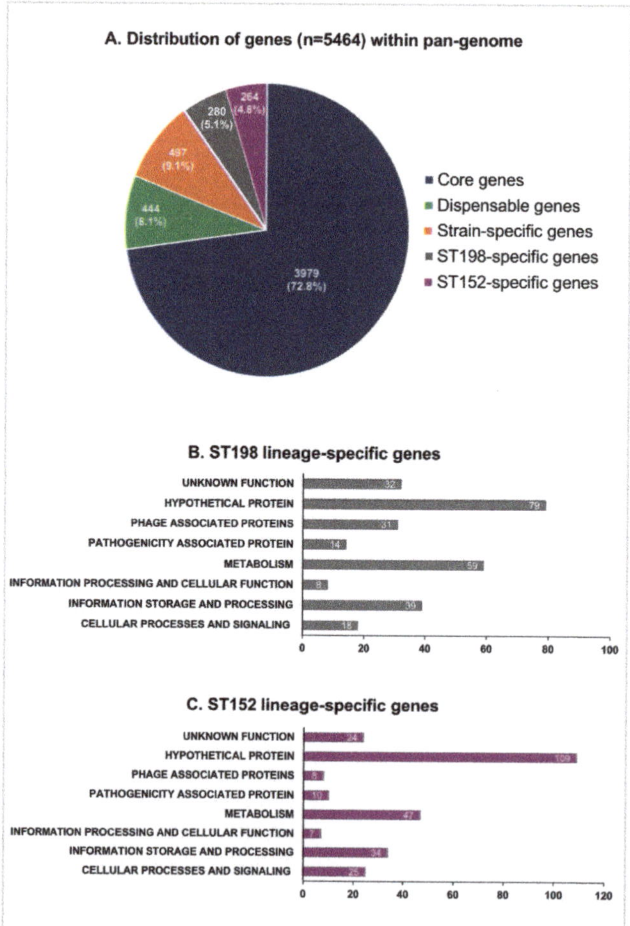

Figure 5. Distribution of genes (n = 5464) (**A**) within the pan-genome of epidemiologically distinct Flu^R ST198 (n = 3) (**B**) and ST152 (n = 4) strains (**C**).

3.4.1. Phenotype-to-Genotype Association for *Myo*-Inositol Metabolism in *S.* Kentucky ST198 and ST152 Lineages and Across NTS

A recent large-scale genomic survey encompassing 193,753 genomes from 24,812 commensal and pathogenic bacterial species, including *Salmonella* spp., found that *myo*-inositol utilization is widespread. This is largely attributed to the presence of a conserved *iol* gene cluster, which mediates *myo*-inositol metabolism [27,39]. Previous studies have confirmed that *S.* Typhimurium can utilize *myo*-inositol as a sole carbon source due to this canonical *iol* cluster [26,27]. In our study, strains belonging to the *S.* Kentucky ST152 lineage exhibited significantly impaired growth in *myo*-inositol minimal media compared to Flu^R ST198 strains, suggesting differences in the presence or functionality of the *iol* gene cluster. To investigate this genotype–phenotype association, we screened the genomes of both ST152 and Flu^R ST198 strains for the canonical *iol* gene cluster. All Flu^R ST198 genomes analyzed carried the complete canonical *iol* gene cluster (Figure 6A), consistent with their robust growth in *myo*-inositol (Figure 3). In contrast, none of the ST152 strains carried the cluster, correlating with their impaired growth (Figure 3).

We extended this analysis to 294 ST198 and 173 ST152 genomes to assess lineage-wide conservation. The canonical *iol* gene cluster was conserved in all ST198 genomes but absent from all ST152 genomes (Figure 7A). Other *S.* Kentucky sequence types (ST314, ST2132, ST64, ST166, and ST318) also lacked this cluster, whereas it was present in ST198, ST19, ST32, ST639, and ST1679 strains (Figure 7A). To further support this functional link, we examined the expression of two key *myo-inositol-responsive* genes [40,41], *iolT1* (transport) and *reiD* (regulatory), in the presence of *myo*-inositol. Both genes were upregulated compared to glucose conditions (Figure 8), aligning with their location in the *iol* cluster of the Flu^R ST198 genome (PU61) and the observed growth phenotype (Figure 3). These findings demonstrate that the canonical *iol* gene cluster is differentially distributed among *S.* Kentucky lineages and that its presence strongly correlates with the ability to utilize *myo*-inositol as a sole energy source.

To evaluate the broader relevance of this observation across NTS serovars, we analyzed 1792 genomes from 107 NTS serovars (Table S2). The canonical *iol* gene cluster was conserved in 45 (42%) serovars, absent in 54 (50%)—including *S.* Enteritidis, a clinically significant serovar, and variably present within 8 (8%) other serovars (Figure 7B). To validate these genomic predictions, representative strains were tested for growth in *myo*-inositol media. Serovars carrying the *iol* cluster, including *S.* Kentucky ST198 (7.8 $\pm$ 0.16 logs), *S.* Typhimurium (7.7 $\pm$ 0.73 logs), and *S.* Hadar (6.04 $\pm$ 0.31 logs), showed significantly higher growth than those lacking the cluster: *S.* Kentucky ST152 (1.7 $\pm$ 0.23 logs), *S.* Enteritidis (1.2 $\pm$ 0.07 logs), *S.* Montevideo (1.30 $\pm$ 0.11 logs), and *S.* Schwarzengrund (1.08 $\pm$ 0.11 logs) (Figure 9A). These findings affirm that *myo*-inositol metabolism in NTS is primarily dependent on the presence of the canonical *iol* gene cluster. Interestingly, strains lacking the cluster still demonstrated residual growth, albeit inefficiently, suggesting the potential presence of non-canonical or alternative metabolic routes for *myo*-inositol utilization. However, the molecular basis of these alternative pathways remains to be elucidated.

Myo-inositol, a polyol found in various foods, plays vital roles in cellular processes, including signaling, motility, membrane trafficking, and phagocytosis [27,42]. *Myo*-inositol is commonly used as a dietary supplement and for the treatment and management of several clinical conditions [43,44], and its intake ranges from 0.3 to 2.6 g/day in Western diets and up to 4.5 g/day globally [45,46], with therapeutic doses reaching 4–18 g/day [47,48]. Although typically associated with beneficial gut microbiota, recent evidence shows that several pathogenic bacterial genomes, including NTS, harbor the *iol* cluster [39]. Interestingly, *S.* Dublin has been shown to secrete a *myo*-inositol polyphosphate 4-phosphatase that disrupts host inositol phosphate signaling, contributing to diarrhea within 30 min of infection [49,50]. Our findings, showing widespread *iol* cluster carriage and associated metabolic capacity, suggest that *myo*-inositol utilization may confer a niche-specific metabolic fitness advantage to ST198 and other NTS serovars that harbor the *iol* gene cluster. Future in vivo studies are warranted to assess the contribution of this pathway to enteric colonization and pathogenesis.

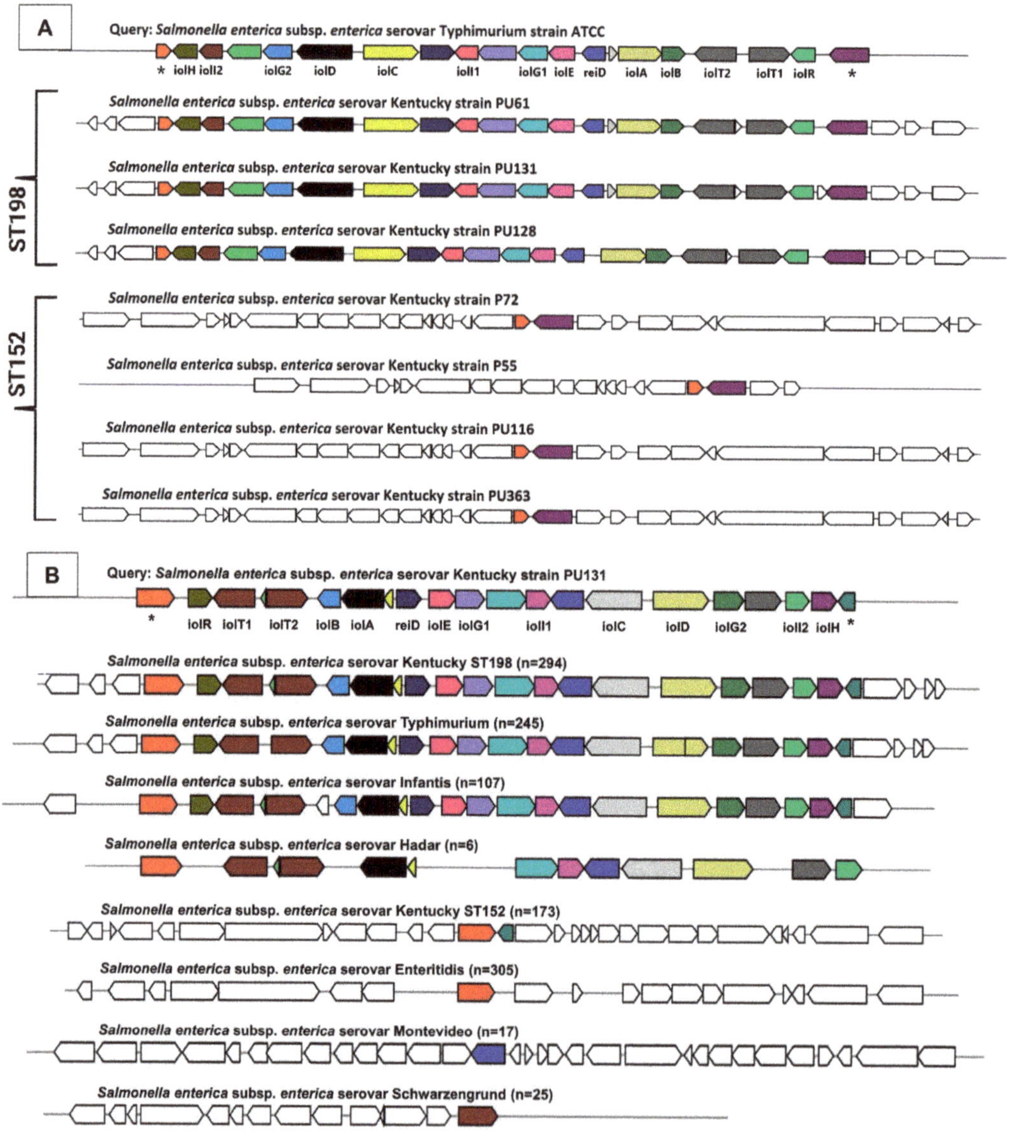

Figure 6. Distribution of *iolCatGC* involved in *myo*-inositol catabolism in *S.* Kentucky Flu^R ST198 (n = 3) and ST152 (n = 4) strains (**A**). Differential carriage of *iolCatGC* in different *Salmonella* serovars (**B**). * Flanking genes. Identical color shows identical genes.

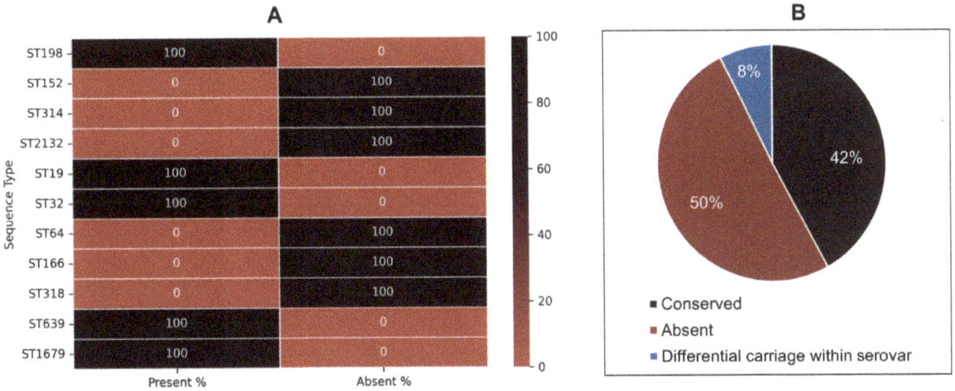

Figure 7. Differential carriage of inositol catabolic gene cluster *iolCatGC* across *S.* Kentucky ST198 ($n = 294$), ST152 ($n = 173$), ST314 ($n = 14$), ST2132 ($n = 3$), ST19 ($n = 2$), ST32, ST64, ST166, ST318, ST639, ST1679 ($n = 1$ each) (**A**), and 107 different NTS serovars including *S.* Kentucky (**B**).

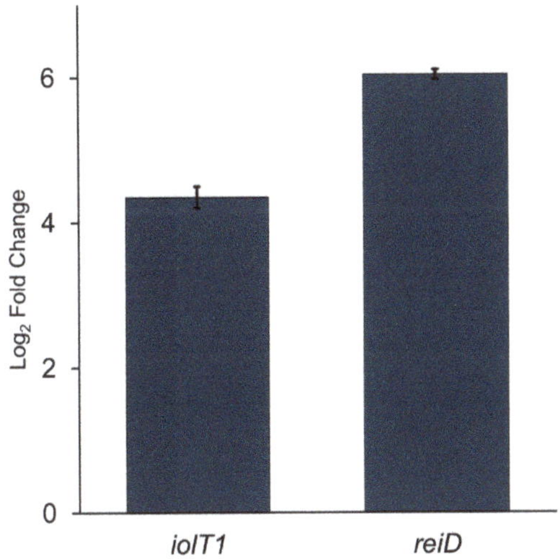

Figure 8. The expression of inositol catabolic genes (*iolT1* and *reiD*) was upregulated in response to *myo*-inositol when compared to glucose as a control condition.

3.4.2. Phenotype-to-Genotype Association for Lactulose Metabolism in *S.* Kentucky ST198 and ST152 Lineages and Across NTS

While lactulose metabolism by NTS has not been previously described, our data demonstrate that NTS, including *S.* Kentucky, can utilize lactulose as a sole carbon and energy source to support growth, despite lineage-specific variation in growth efficiency (Figures 4 and 9B). Growth assays confirmed that this capacity is not restricted to a single serovar; all five NTS serovars tested exhibited robust growth in media containing lactulose as the only carbon source (Figure 9B). These findings suggest that *S.* Kentucky and other NTS serovars possess a functional, yet uncharacterized, metabolic pathway for lactulose utilization.

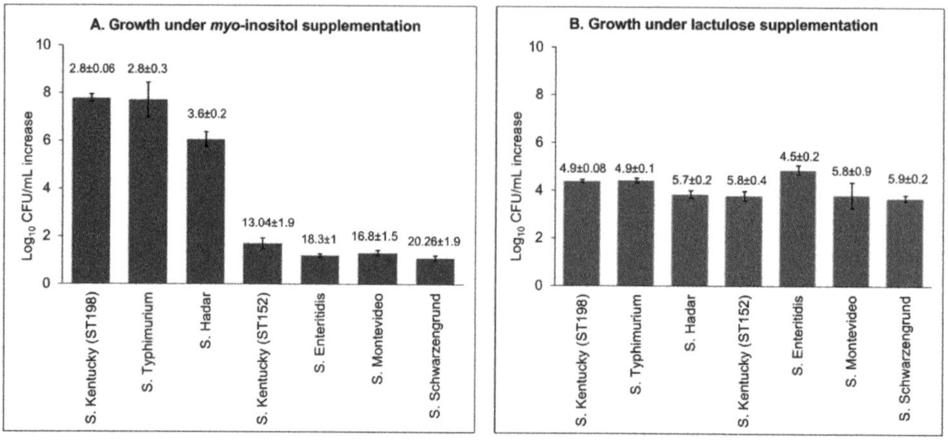

Figure 9. *S.* Kentucky ST198, *S.* Typhimurium, and *S.* Hadar exhibited significantly higher growth with lower Td under *myo*-inositol (20 mM) supplementation, whereas *S.* Kentucky ST152, *S.* Enteritidis, *S.* Montevideo, and *S.* Schwarzengrund had impaired growth, exhibited by lower $\log_{10}$ change with higher Td under the same conditions (**A**). All of these strains demonstrated serovar-independent growth when lactulose (20 mM) was supplemented as the sole energy source (**B**). The value above each column represents doubling time in hours at 72 h post-incubation.

The specific genes responsible for lactulose metabolism in NTS remain unidentified; however, glycoside hydrolase (GH) enzymes, which cleave disaccharides like lactulose into fermentable monosaccharides such as fructose and galactose, are widespread in gut microbes [51]. To identify candidate genes potentially involved in lactulose hydrolysis, we interrogated the genome of *S.* Kentucky strain PU131, a well-characterized Flu^R ST198 reference strain [33]. We identified six genes encoding putative GH family proteins (Table S3) and screened for their potential functions within the CAZy database. Notably, one gene (AUW48348.1) encoding a GH1 family protein with putative β-galactosidase activity was identified with the potential to hydrolyze β-1,4 linkages within disaccharides [30,31,34]. Additionally, we identified a gene (AUW48392.1) encoding a glycoside-pentoside-hexuronide (GPH) family transporter, a membrane protein family known to mediate disaccharide uptake.

Comparative genomics revealed that genes encoding the GH1 enzyme and the GPH transporter are widely conserved among *S.* Kentucky isolates, including ST198 (n = 294), ST152 (n = 173), and other sequence types such as ST314 (n = 14) and various minor STs (n = 11) (Table S3). To assess broader conservation across NTS, we screened 1792 genomes representing 107 NTS serovars. The GH1 gene was present in 102 (95.3%) serovars and absent in 5 (4.7%), while the GPH transporter was also conserved in 102 (95.3%) serovars, absent in 3 (2.8%), and variably present in 2 (1.9%) (Figure 10). These results indicate that genes encoding both GH1 and GPH are broadly conserved across NTS.

The roles of these proteins in lactulose hydrolysis or uptake remain unknown. These genes are likely transcriptionally responsive to lactulose or their activity may be regulated through post-transcriptional or post-translational mechanisms [52,53]. Additionally, the CAZy database highlights that GH enzymes display high substrate specificity and are not universally present across all microbes [51,54]. Thus, it is also likely that other glycoside hydroxylase enzymes may also contribute to lactulose hydrolysis in NTS. Collectively, these data support the hypothesis that NTS may rely on a distinct and currently uncharacterized lactulose metabolic pathway. Elucidating the genetic basis of lactulose metabolism in NTS will provide key insights into the metabolic adaptability of NTS.

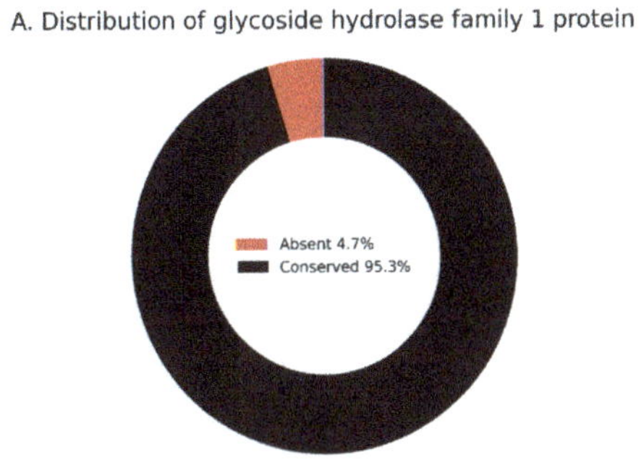

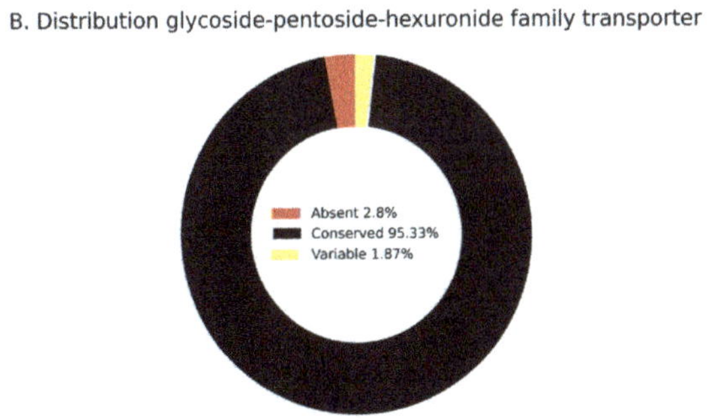

Figure 10. Differential carriage of Glycoside Hydrolase Family 1 (GH1) Protein (**A**) and Glycoside-pentoside-hexuronide (GPH) Family Transporter (**B**) across 107 different *Salmonella* serovars.

Lactulose is a synthetic disaccharide frequently used in clinical settings to manage hepatic encephalopathy and constipation. It is also present in heat-treated dairy products and commonly consumed as a dietary supplement at doses ranging from 10 to 60 g per day [55]. Historically, lactulose has been regarded as a prebiotic that is selectively fermented by beneficial gut bacteria, such as *Lactobacillus* and *Bifidobacterium*. However, several opportunistic pathogens, including *Escherichia*, *Enterococcus*, *Klebsiella*, *Pseudomonas*, *Streptococcus*, and *Cronobacter* species, have also been reported to metabolize lactulose [32]. Recent studies demonstrate that lactulose can enhance *Escherichia coli* (*E. coli*) colonization in both murine models and human subjects, raising concerns about a potential increased risk of systemic infection [56]. Similarly, lactulose supplementation has been shown to promote intestinal translocation of *Salmonella enterica* serovar Typhimurium [57]. In contrast, some human studies reported no significant effect of lactulose ingestion on *E. coli* abundance in stool samples of healthy volunteers [58], while others observed a reduction in potentially pathogenic *Clostridium* spp. following lactulose supplementation [59,60]. In animal models, lactulose reduced intestinal colonization by *Salmonella* and *E. coli* in chickens [61], but had no significant effect on *Salmonella* colonization in pigs [62]. Taken together, these findings suggest that the impact of lactulose on the metabolic fitness and colonization potential of enteric pathogens could be host-dependent and remains inconsistent across studies.

Our findings demonstrate that *S.* Kentucky and other NTS serovars can metabolize lactulose (Figures 4 and 9B). Given the widespread use of lactulose as a prebiotic in human and veterinary contexts, these results raise important questions regarding the potential for lactulose to modulate the metabolic fitness and colonization potential of pathogenic NTS in vivo. Further studies are warranted to identify genetic determinants of lactulose metabolism in NTS and assess the mechanisms underlying lactulose-mediated modulation of NTS metabolome and its impact on NTS colonization and pathogenesis in vivo.

4. Conclusions

S. Kentucky lineages ST152 and Flu^R ST198 have diverged both genotypically and phenotypically, following distinct evolutionary and epidemiological trajectories. This study provides the first direct evidence of lineage-specific metabolic specialization, revealing that ST198 exhibits broader physiological plasticity, greater metabolic versatility, and enhanced stress resilience compared to ST152. These traits are evident in their efficient utilization of substrates such as *myo*-inositol and lactulose at both avian and mammalian physiological temperatures.

Our data show that ST152 lacks the canonical *iol* gene cluster responsible for *myo*-inositol catabolism in enteric bacteria, as do several other NTS serovars. Nonetheless, ST152 exhibits weak but detectable growth on *myo*-inositol, suggesting the existence of a non-canonical or alternative metabolic pathway for inositol utilization in these strains. This finding challenges the current understanding of inositol metabolism in *Salmonella* and opens new avenues for discovering novel metabolic genes or regulatory networks.

We also report for the first time that multiple NTS serovars, including *S.* Kentucky, can metabolize lactulose, a synthetic disaccharide traditionally considered fermentable only by beneficial gut microbes. The conservation of genes encoding GH1 and GPH transporter in most NTS serovars points to the potential existence of mechanisms underlying lactulose metabolism in *Salmonella*.

Together, these findings underscore the metabolic adaptability of ST198 and reveal previously unrecognized metabolic capacity in NTS. Understanding how nutrient utilization pathways contribute to ecological fitness, host adaptation, and antimicrobial resistance in NTS, and, in particular, emerging lineages like Flu^R ST198, is critical for developing targeted, nutrition-based interventions. Such approaches could serve as alternatives to antibiotics for controlling MDR NTS and other enteric pathogens across human and animal health interfaces.

Supplementary Materials: The following supporting information can be downloaded at: https://www.mdpi.com/article/10.3390/genes16091051/s1, Figure S1. Hierarchical clustering showing distinct metabolic profiles of ST152 and Flu^R ST198 strains based on the differences in respiratory activity (RA) in different conditions. Table S1. Phenotype Microarray data. Table S2. Distribution of *iolCatGC* in 107 *Salmonella* genomes. Table S3. Distribution of GH1 and GPH in 107 *Salmonella* genomes. Table S4. Primers.

Author Contributions: Conceptualization, J.A. and D.H.S.; Methodology, J.A., R.C.S. and J.G.; Formal analysis, J.A., S.S. and D.H.S.; Investigation, J.A., R.C.S., J.G. and D.H.S.; Resources, D.H.S.; Data curation, J.A., S.S., J.G. and D.H.S.; Writing—original draft, J.A. and D.H.S.; Writing—review & editing, J.A., R.C.S., S.S., J.G., B.J.H. and D.H.S.; Visualization, J.A. and D.H.S.; Supervision, S.S. and D.H.S.; Project administration, D.H.S.; Funding acquisition, J.G. and D.H.S. All authors have read and agreed to the published version of the manuscript.

Funding: This study was funded in part by the Texas Tech University School of Veterinary Medicine (TTU-SVM) and the U.S. National Poultry Research Center, Agriculture Research Service (Research Agreement 6040-32000-011-05J), U.S. Department of Agriculture. Juned Ahmed was supported by the Distinguished Graduate Student Assistantship from Texas Tech University.

Data Availability Statement: All data are included in the manuscript or as Supplementary Files.

Acknowledgments: We thank Samuel Ajulo, Asmita Shrestha, and Vishakha Kulkarni for technical assistance and discussion, and Jonathan Thompson and Fernanda Rosa for technical discussions.

Conflicts of Interest: The authors declare no conflict of interest.

References

1. Majowicz, S.E.; Musto, J.; Scallan, E.; Angulo, F.J.; Kirk, M.; O'Brien, S.J.; Jones, T.F.; Fazil, A.; Hoekstra, R.M.; for the International Collaboration on Enteric Disease "Burden of Illness" Studies. The Global Burden of Nontyphoidal Salmonella Gastroenteritis. *Clin. Infect. Dis.* **2010**, *50*, 882–889. [CrossRef]
2. Haley, B.J.; Kim, S.W.; Haendiges, J.; Keller, E.; Torpey, D.; Kim, A.; Crocker, K.; Myers, R.A.; Van Kessel, J.A.S. Salmonella Enterica Serovar Kentucky Recovered from Human Clinical Cases in Maryland, USA (2011–2015). *Zoonoses Public Health* **2019**, *66*, 382–392. [CrossRef]
3. Richards, A.K.; Kue, S.; Norris, C.G.; Shariat, N.W. Genomic and Phenotypic Characterization of Salmonella Enterica Serovar Kentucky. *Microb. Genom.* **2023**, *9*, 001089. [CrossRef]
4. Tate, H.; Hsu, C.-H.; Chen, J.C.; Han, J.; Foley, S.L.; Folster, J.P.; Francois Watkins, L.K.; Reynolds, J.; Tillman, G.E.; Nyirabahizi, E.; et al. Genomic Diversity, Antimicrobial Resistance, and Virulence Gene Profiles of Salmonella Serovar Kentucky Isolated from Humans, Food, and Animal Ceca Content Sources in the United States. *Foodborne Pathog. Dis.* **2022**, *19*, 509–521. [CrossRef] [PubMed]
5. Biggel, M.; Horlbog, J.; Nüesch-Inderbinen, M.; Chattaway, M.A.; Stephan, R. Epidemiological Links and Antimicrobial Resistance of Clinical Salmonella Enterica ST198 Isolates: A Nationwide Microbial Population Genomic Study in Switzerland. *Microb. Genom.* **2022**, *8*, 000877. [CrossRef]
6. Hello, S.L.; Harrois, D.; Bouchrif, B.; Sontag, L.; Elhani, D.; Guibert, V.; Zerouali, K.; Weill, F.-X. Highly Drug-Resistant Salmonella Enterica Serotype Kentucky ST198-X1: A Microbiological Study. *Lancet Infect. Dis.* **2013**, *13*, 672–679. [CrossRef]
7. Mashe, T.; Thilliez, G.; Chaibva, B.V.; Leekitcharoenphon, P.; Bawn, M.; Nyanzunda, M.; Robertson, V.; Tarupiwa, A.; Al-Khanaq, H.; Baker, D.; et al. Highly Drug Resistant Clone of Salmonella Kentucky ST198 in Clinical Infections and Poultry in Zimbabwe. *Npj Antimicrob. Resist.* **2023**, *1*, 6. [CrossRef]
8. Slowey, R.; Kim, S.W.; Prendergast, D.; Madigan, G.; Van Kessel, J.A.S.; Haley, B.J. Genomic Diversity and Resistome Profiles of Salmonella Enterica Subsp. Enterica Serovar Kentucky Isolated from Food and Animal Sources in Ireland. *Zoonoses Public Health* **2022**, *69*, 1–12. [CrossRef] [PubMed]
9. Soltys, R.C.; Sakomoto, C.K.; Oltean, H.N.; Guard, J.; Haley, B.J.; Shah, D.H. High-Resolution Comparative Genomics of Salmonella Kentucky Aids Source Tracing and Detection of ST198 and ST152 Lineage-Specific Mutations. *Front. Sustain. Food Syst.* **2021**, *5*, 695368. [CrossRef]
10. Haley, B.J.; Kim, S.W.; Pettengill, J.; Luo, Y.; Karns, J.S.; Kessel, J.A.S.V. Genomic and Evolutionary Analysis of Two Salmonella Enterica Serovar Kentucky Sequence Types Isolated from Bovine and Poultry Sources in North America. *PLoS ONE* **2016**, *11*, e0161225. [CrossRef]
11. Wang, Z.; Jiang, Y.; Xu, H.; Jiao, X.; Wang, J.; Li, Q. Poultry Production as the Main Reservoir of Ciprofloxacin- and Tigecycline-Resistant Extended-Spectrum β-Lactamase (ESBL)-Producing Salmonella Enterica Serovar Kentucky ST198.2-2 Causing Human Infections in China. *Appl. Environ. Microbiol.* **2023**, *89*, e00944-23. [CrossRef]
12. Chen, H.; Song, J.; Zeng, X.; Chen, D.; Chen, R.; Qiu, C.; Zhou, K. National Prevalence of Salmonella Enterica Serotype Kentucky ST198 with High-Level Resistance to Ciprofloxacin and Extended-Spectrum Cephalosporins in China, 2013 to 2017. *mSystems* **2021**, *6*, e00935-20. [CrossRef]
13. Vázquez, X.; Fernández, J.; Bances, M.; Lumbreras, P.; Alkorta, M.; Hernáez, S.; Prieto, E.; de la Iglesia, P.; de Toro, M.; Rodicio, M.R.; et al. Genomic Analysis of Ciprofloxacin-Resistant Salmonella Enterica Serovar Kentucky ST198 From Spanish Hospitals. *Front. Microbiol.* **2021**, *12*, 720449. [CrossRef]
14. Wołkowicz, T.; Zacharczuk, K.; Gierczyński, R.; Nowakowska, M.; Piekarska, K. Antimicrobial Resistance and Whole-Genome Characterisation of High-Level Ciprofloxacin-Resistant Salmonella Enterica Serovar Kentucky ST 198 Strains Isolated from Human in Poland. *Int. J. Mol. Sci.* **2021**, *22*, 9381. [CrossRef] [PubMed]
15. Shah, D.H.; Paul, N.C.; Guard, J. Complete Genome Sequence of a Ciprofloxacin-Resistant Salmonella Enterica Subsp. Enterica Serovar Kentucky Sequence Type 198 Strain, PU131, Isolated from a Human Patient in Washington State. *Genome Announc.* **2018**, *6*, e00125-18. [CrossRef]
16. Cummings, K.J.; Siler, J.D.; Goodman, L.B.; Childs-Sanford, S.E. Ciprofloxacin-Resistant ST198 Salmonella Kentucky in a Hospitalized American Black Bear (*Ursus americanus*), with Evidence of Subsequent Nosocomial Transmission. *Zoonoses Public Health* **2023**, *70*, 657–664. [CrossRef]
17. Okoro, C.K.; Barquist, L.; Connor, T.R.; Harris, S.R.; Clare, S.; Stevens, M.P.; Arends, M.J.; Hale, C.; Kane, L.; Pickard, D.J.; et al. Signatures of Adaptation in Human Invasive Salmonella Typhimurium ST313 Populations from Sub-Saharan Africa. *PLoS Negl. Trop. Dis.* **2015**, *9*, e0003611. [CrossRef]
18. Serrano-Fujarte, I.; Calva, E.; García-Domínguez, J.; Ortiz-Jiménez, S.; Puente, J.L. Population Structure and Ongoing Microevolution of the Emerging Multidrug-Resistant Salmonella Typhimurium ST213. *Npj Antimicrob. Resist.* **2024**, *2*, 10. [CrossRef]
19. Tanner, J.R.; Kingsley, R.A. Evolution of *Salmonella* within Hosts. *Trends Microbiol.* **2018**, *26*, 986–998. [CrossRef] [PubMed]

20. Burin, R.; Shah, D.H. Phenelzine and Amoxapine Inhibit Tyramine and D-Glucuronic Acid Catabolism in Clinically Significant Salmonella in A Serotype-Independent Manner. *Pathogens* **2021**, *10*, 469. [CrossRef]
21. Chiok, K.L.; Addwebi, T.; Guard, J.; Shah, D.H. Dimethyl Adenosine Transferase (KsgA) Deficiency in Salmonella Enterica Serovar Enteritidis Confers Susceptibility to High Osmolarity and Virulence Attenuation in Chickens. *Appl. Environ. Microbiol.* **2013**, *79*, 7857–7866. [CrossRef] [PubMed]
22. Elder, J.R.; Paul, N.C.; Burin, R.; Guard, J.; Shah, D.H. Genomic Organization and Role of SPI-13 in Nutritional Fitness of Salmonella. *Int. J. Med. Microbiol. IJMM* **2018**, *308*, 1043–1052. [CrossRef] [PubMed]
23. Guard-Bouldin, J.; Morales, C.A.; Frye, J.G.; Gast, R.K.; Musgrove, M. Detection of Salmonella Enterica Subpopulations by Phenotype Microarray Antibiotic Resistance Patterns. *Appl. Environ. Microbiol.* **2007**, *73*, 7753–7756. [CrossRef]
24. Kolde, R. *Pheatmap: Pretty Heatmaps. R Package*, version 1.0.12. 2019. Available online: https://CRAN.R-project.org/package=pheatmap (accessed on 26 August 2025).
25. Dereeper, A.; Summo, M.; Meyer, D.F. PanExplorer: A Web-Based Tool for Exploratory Analysis and Visualization of Bacterial Pan-Genomes. *Bioinformatics* **2022**, *38*, 4412–4414. [CrossRef]
26. Kröger, C.; Rothhardt, J.E.; Brokatzky, D.; Felsl, A.; Kary, S.C.; Heermann, R.; Fuchs, T.M. The Small RNA RssR Regulates Myo-Inositol Degradation by Salmonella Enterica. *Sci. Rep.* **2018**, *8*, 17739. [CrossRef]
27. Weber, M.; Fuchs, T.M. Metabolism in the Niche: A Large-Scale Genome-Based Survey Reveals Inositol Utilization To Be Widespread among Soil, Commensal, and Pathogenic Bacteria. *Microbiol. Spectr.* **2022**, *10*, e02013-22. [CrossRef] [PubMed]
28. Medema, M.H.; Takano, E.; Breitling, R. Detecting Sequence Homology at the Gene Cluster Level with MultiGeneBlast. *Mol. Biol. Evol.* **2013**, *30*, 1218–1223. [CrossRef]
29. Olson, R.D.; Assaf, R.; Brettin, T.; Conrad, N.; Cucinell, C.; Davis, J.J.; Dempsey, D.M.; Dickerman, A.; Dietrich, E.M.; Kenyon, R.W.; et al. Introducing the Bacterial and Viral Bioinformatics Resource Center (BV-BRC): A Resource Combining PATRIC, IRD and ViPR. *Nucleic Acids Res.* **2023**, *51*, D678–D689. [CrossRef]
30. Baramee, S.; Siriatcharanon, A.-K.; Waeonukul, R.; Pason, P.; Tachaapaikoon, C.; Ratanakhanokchai, K. A Novel β-Galactosidase from Anaerobic Thermophilic Caldicellulosiruptor Saccharolyticus DSM 8903 with High Potential for Lactulose Production. *J. Sci. Food Agric.* **2025**, *105*, 6781–6789. [CrossRef]
31. Kumari, S.; Panesar, P.S.; Panwar, D.; Singla, G. Lactulose. In *Probiotics, Prebiotics and Synbiotics*; John Wiley & Sons, Ltd.: Hoboken, NJ, USA, 2022; pp. 338–360, ISBN 978-1-119-70216-0.
32. Mao, B.; Li, D.; Zhao, J.; Liu, X.; Gu, Z.; Chen, Y.Q.; Zhang, H.; Chen, W. In Vitro Fermentation of Lactulose by Human Gut Bacteria. *J. Agric. Food Chem.* **2014**, *62*, 10970–10977. [CrossRef]
33. Sahota, S.S.; Bramley, P.M.; Menzies, I.S. The Fermentation of Lactulose by Colonic Bacteria. *Microbiology* **1982**, *128*, 319–325. [CrossRef]
34. Vičič, V.; Pandel Mikuš, R.; Ferjančič, B. Review of History and Mechanisms of Action of Lactulose (4-O-β-D-Galactopyranosyl-β-D-Fructofuranose): Present and Future Applications in Food. *J. Food Sci. Technol.* **2024**, *61*, 2036–2045. [CrossRef]
35. Aung, K.T.; Khor, W.C.; Ong, K.H.; Tan, W.L.; Wong, Z.N.; Oh, J.Q.; Wong, W.K.; Tan, B.Z.Y.; Maiwald, M.; Tee, N.W.S.; et al. Characterisation of Salmonella Enteritidis ST11 and ST1925 Associated with Human Intestinal and Extra-Intestinal Infections in Singapore. *Int. J. Environ. Res. Public Health* **2022**, *19*, 5671. [CrossRef]
36. Pulford, C.V.; Perez-Sepulveda, B.M.; Canals, R.; Bevington, J.A.; Bengtsson, R.J.; Wenner, N.; Rodwell, E.V.; Kumwenda, B.; Zhu, X.; Bennett, R.J.; et al. Stepwise Evolution of Salmonella Typhimurium ST313 Causing Bloodstream Infection in Africa. *Nat. Microbiol.* **2021**, *6*, 327–338. [CrossRef] [PubMed]
37. Boas Lichty, K.E.; Loughran, R.M.; Ushijima, B.; Richards, G.P.; Boyd, E.F. Osmotic Stress Response of the Coral and Oyster Pathogen Vibrio Coralliilyticus: Acquisition of Catabolism Gene Clusters for the Compatible Solute and Signaling Molecule Myo-Inositol. *Appl. Environ. Microbiol.* **2024**, *90*, e00920-24. [CrossRef]
38. Manske, C.; Schell, U.; Hilbi, H. Metabolism of Myo-Inositol by Legionella Pneumophila Promotes Infection of Amoebae and Macrophages. *Appl. Environ. Microbiol.* **2016**, *82*, 5000–5014. [CrossRef]
39. Felsl, A.; Brokatzky, D.; Kröger, C.; Heermann, R.; Fuchs, T.M. Hierarchic Regulation of a Metabolic Pathway: H-NS, CRP, and SsrB Control Myo-Inositol Utilization by Salmonella Enterica. *Microbiol. Spectr.* **2023**, *12*, e02724-23. [CrossRef]
40. Rothhardt, J.E.; Kröger, C.; Broadley, S.P.; Fuchs, T.M. The Orphan Regulator ReiD of Salmonella Enterica Is Essential for Myo-Inositol Utilization. *Mol. Microbiol.* **2014**, *94*, 700–712. [CrossRef] [PubMed]
41. Kröger, C.; Stolz, J.; Fuchs, T.M. Myo-Inositol Transport by Salmonella Enterica Serovar Typhimurium. *Microbiology* **2010**, *156*, 128–138. [CrossRef] [PubMed]
42. Chhetri, D.R. Myo-Inositol and Its Derivatives: Their Emerging Role in the Treatment of Human Diseases. *Front. Pharmacol.* **2019**, *10*, 1172. [CrossRef]
43. Croze, M.L.; Soulage, C.O. Potential Role and Therapeutic Interests of *Myo*-Inositol in Metabolic Diseases. *Biochimie* **2013**, *95*, 1811–1827. [CrossRef]

44. Vitale, S.G.; Corrado, F.; Caruso, S.; Di Benedetto, A.; Giunta, L.; Cianci, A.; D'Anna, R. Myo-Inositol Supplementation to Prevent Gestational Diabetes in Overweight Non-Obese Women: Bioelectrical Impedance Analysis, Metabolic Aspects, Obstetric and Neonatal Outcomes—A Randomized and Open-Label, Placebo-Controlled Clinical Trial. *Int. J. Food Sci. Nutr.* **2021**, *72*, 670–679. [CrossRef]
45. *Food Phytates*, 1st ed.; Reddy, N.R., Sathe, S.K., Eds.; CRC Press: Boca Raton, FL, USA, 2001. [CrossRef]
46. Schlemmer, U.; Frølich, W.; Prieto, R.M.; Grases, F. Phytate in Foods and Significance for Humans: Food Sources, Intake, Processing, Bioavailability, Protective Role and Analysis. *Mol. Nutr. Food Res.* **2009**, *53*, S330–S375. [CrossRef] [PubMed]
47. Lam, S.; McWilliams, A.; LeRiche, J.; MacAulay, C.; Wattenberg, L.; Szabo, E. A Phase I Study of Myo-Inositol for Lung Cancer Chemoprevention. *Cancer Epidemiol. Biomark. Prev.* **2006**, *15*, 1526–1531. [CrossRef]
48. Regidor, P.-A.; Schindler, A.E. Myoinositol as a Safe and Alternative Approach in the Treatment of Infertile PCOS Women: A German Observational Study. *Int. J. Endocrinol.* **2016**, *2016*, 9537632. [CrossRef] [PubMed]
49. Norris, F.A.; Wilson, M.P.; Wallis, T.S.; Galyov, E.E.; Majerus, P.W. SopB, a Protein Required for Virulence of Salmonella Dublin, Is an Inositol Phosphate Phosphatase. *Proc. Natl. Acad. Sci. USA* **1998**, *95*, 14057–14059. [CrossRef]
50. Zhou, D.; Chen, L.M.; Hernandez, L.; Shears, S.B.; Galán, J.E. A Salmonella Inositol Polyphosphatase Acts in Conjunction with Other Bacterial Effectors to Promote Host Cell Actin Cytoskeleton Rearrangements and Bacterial Internalization. *Mol. Microbiol.* **2001**, *39*, 248–259. [CrossRef]
51. Berlemont, R.; Martiny, A.C. Glycoside Hydrolases across Environmental Microbial Communities. *PLoS Comput. Biol.* **2016**, *12*, e1005300. [CrossRef]
52. El Mouali, Y.; Gaviria-Cantin, T.; Sánchez-Romero, M.A.; Gibert, M.; Westermann, A.J.; Vogel, J.; Balsalobre, C. CRP-cAMP Mediates Silencing of Salmonella Virulence at the Post-Transcriptional Level. *PLoS Genet.* **2018**, *14*, e1007401. [CrossRef] [PubMed]
53. Hung, C.-C.; Garner, C.D.; Slauch, J.M.; Dwyer, Z.W.; Lawhon, S.D.; Frye, J.G.; McClelland, M.; Ahmer, B.M.M.; Altier, C. The Intestinal Fatty Acid Propionate Inhibits Salmonella Invasion through the Post-Translational Control of HilD. *Mol. Microbiol.* **2013**, *87*, 1045–1060. [CrossRef]
54. Lombard, V.; Golaconda Ramulu, H.; Drula, E.; Coutinho, P.M.; Henrissat, B. The Carbohydrate-Active Enzymes Database (CAZy) in 2013. *Nucleic Acids Res.* **2014**, *42*, D490–D495. [CrossRef] [PubMed]
55. Karakan, T.; Tuohy, K.M.; Janssen-van Solingen, G. Low-Dose Lactulose as a Prebiotic for Improved Gut Health and Enhanced Mineral Absorption. *Front. Nutr.* **2021**, *8*, 672925. [CrossRef] [PubMed]
56. Hecht, A.L.; Mahmud, N.; Chaudhry, S.; Cao, J.Y.; Branigan, G.P.; Lee, J.; Theiller, E.; Roggiani, M.; Friedman, E.S.; Herman, L.; et al. Carbohydrate Consumption Drives Adaptive Mutations in Escherichia Coli Associated with Increased Risk for Systemic Infection. *BioRxiv* **2025**. [CrossRef]
57. Bovee-Oudenhoven, I.M.J.; ten Bruggencate, S.J.M.; Lettink-Wissink, M.L.G.; Meer, R. van der Dietary Fructo-Oligosaccharides and Lactulose Inhibit Intestinal Colonisation but Stimulate Translocation of Salmonella in Rats. *Gut* **2003**, *52*, 1572–1578. [CrossRef]
58. Wotzka, S.Y.; Kreuzer, M.; Maier, L.; Zünd, M.; Schlumberger, M.; Nguyen, B.; Fox, M.; Pohl, D.; Heinrich, H.; Rogler, G.; et al. Microbiota Stability in Healthy Individuals after Single-Dose Lactulose Challenge—A Randomized Controlled Study. *PLoS ONE* **2018**, *13*, e0206214. [CrossRef]
59. Terada, A.; Hara, H.; Kataoka, M.; Mitsuoka, T. Effect of Lactulose on the Composition and Metabolic Activity of the Human Faecal Flora. *Microb. Ecol. Health Dis.* **1992**, *5*, 43–50. [CrossRef]
60. Tuohy, K.M.; Ziemer, C.J.; Klinder, A.; Knöbel, Y.; Pool-Zobel, B.L.; Gibson, G.R. A Human Volunteer Study to Determine the Prebiotic Effects of Lactulose Powder on Human Colonic Microbiota. *Microb. Ecol. Health Dis.* **2002**, *14*, 165–173. [CrossRef]
61. Mohammadi Gheisar, M.; Nyachoti, C.M.; Hancock, J.D.; Kim, I.H. Effects of Lactulose on Growth, Carcass Characteristics, Faecal Microbiota, and Blood Constituents in Broilers. *Veterinární Medicína* **2016**, *61*, 90–96. [CrossRef]
62. Martín-Peláez, S.; Costabile, A.; Hoyles, L.; Rastall, R.A.; Gibson, G.R.; La Ragione, R.M.; Woodward, M.J.; Mateu, E.; Martín-Orúe, S.M. Evaluation of the Inclusion of a Mixture of Organic Acids or Lactulose into the Feed of Pigs Experimentally Challenged with *Salmonella* Typhimurium. *Vet. Microbiol.* **2010**, *142*, 337–345. [CrossRef] [PubMed]

genes

MDPI

Article

Phenotypic and Whole-Genome Sequencing-Based Profiling of Antimicrobial Resistance and Virulence in *Pseudomonas aeruginosa* Isolated from Patients with Ventilator-Associated Pneumonia and Ventilator-Associated Tracheobronchitis in a Croatian Intensive Care Unit

Marija Cavka [1], Marija Kvesic Ivankovic [2], Ana Maravic [3,*], Mia Dzelalija [3], Jelena Marinovic [4], Ivana Goic-Barisic [4,5], Marija Tonkic [4,5] and Anita Novak [4,5,*]

[1] Department of Anesthesiology and Intensive Care, University Hospital of Split, Spinciceva 1, 21 000 Split, Croatia
[2] Institute of Oceanography and Fisheries, Setaliste Ivana Mestrovica 63, 21 000 Split, Croatia
[3] Department of Biology, Faculty of Science, University of Split, Rudera Boskovica 33, 21 000 Split, Croatia
[4] Department of Clinical Microbiology, University Hospital of Split, Spinciceva 1, 21 000 Split, Croatia
[5] School of Medicine, University of Split, Soltanska 2A, 21 000 Split, Croatia
* Correspondence: amaravic@pmfst.hr (A.M.); anitanovak1@net.hr or annovak@kbsplit.hr (A.N.)

Abstract

Background/Objectives: *Pseudomonas aeruginosa* is one of the leading causes of ventilator-associated pneumonia (VAP) and ventilator-associated tracheobronchitis (VAT), with a worldwide spread of difficult-to-treat high-risk clones. This study aimed to investigate the virulence potential and to characterize phenotypic and genotypic antimicrobial resistance (AMR) in *P. aeruginosa* causing VAP/VAT in the Intensive Care Unit (ICU), University Hospital of Split, Croatia. Methods: The study included *P. aeruginosa* isolates obtained from ICU patients who met the criteria for VAP or VAT, between January 2023 and January 2024. Isolates were identified using MALDI-TOF MS and tested for antimicrobial susceptibility (AST). A subset of phenotypically multidrug-resistant (MDR) isolates was further analyzed using whole-genome sequencing (WGS) and multilocus sequence typing. Results: A high rate of resistance was detected to ceftazidime (23.4%), imipenem (39.6%), and meropenem (43.8%). WGS confirmed the presence of multiple AMR genes, including the $bla_{\text{VIM-2}}$ gene, whose genetic environment highlights a complex MDR locus integrating multiple AMR determinants and mobile genetic elements. All tested isolates possessed genes for class C (bla_{PDC34}, bla_{PDC374} or bla_{PDC16}) and class D ($bla_{\text{OXA-2}}$, $bla_{\text{OXA-10}}$ or $bla_{\text{OXA-50}}$) β-lactamases, *fosA*, *aph(3′)-IIb* and *crpP* genes. Additionally, WGS analysis revealed the presence of numerous virulence genes including those for adherence (Type IV pili and Fap protein production), motility (such as *flgF*), biofilm formation (e.g., *algE* and *mucE*), quorum sensing (*lasI*, *lasR*, *rhlI* and *rhlR*), exotoxin (*toxA* and *plcH*) and exoenzyme activity (*exoU*, *exoT*, *exoS*, *exoY*, *pcrV*, *hcp1* and *lasA*). The isolates belonged to four different sequence types: ST235, ST446, the high-risk ST253 and the widely distributed ST395. Phylogenomic comparison demonstrated that the isolates from this study do not originate from a single clonal source, but instead represent multiple globally distributed high-risk *P. aeruginosa* lineages introduced into the clinical setting. Conclusions: Due to the emergence of high-risk clones with broad AMR and strong virulence potential, ineffectiveness of standard empirical therapy may be anticipated, highlighting the urgent need for new therapeutic approaches (including those targeting major virulence factors).

Genes **2026**, *17*, 130 https://doi.org/10.3390/genes17020130

Keywords: *Pseudomonas aeruginosa*; multidrug-resistance; hospital-acquired pneumonia; intensive care unit; whole-genome sequencing; VIM-2; genetic context

1. Introduction

Pseudomonas aeruginosa (*P. aeruginosa*) is a Gram-negative, non-fermenting bacillus, that can often be found in moist environments, soil and on plants. It is considered a common colonizer and an opportunistic pathogen in humans [1,2]. Numerous virulence factors, particularly the capacity for biofilm formation, as well as antimicrobial resistance (AMR), make this opportunistic microorganism one of the most important causes of hospital-acquired infections (HAI) with high morbidity and mortality rates [3].

There are many mechanisms, ranging from efflux pumps to the production of β-lactamases and gene mutations, that result in resistance not only to cephalosporins, which were traditionally the first-line therapy, but also to a broad range of antibiotics, posing a significant challenge for treatment [3–5]. Overproduction of its naturally occurring AmpC-type chromosomally-encoded cephalosporinase (so-called *Pseudomonas*-derived cephalosporinase, PDC) is one of the major mechanisms responsible for broad AMR (piperacillin, ceftazidime, cefepime, aztreonam) and even leads to reduced carbapenem susceptibility [6]. However, resistance to carbapenems is mainly linked to the production of metallo-β-lactamase (MBL) carbapenemases [7]. Carbapenem-resistant *P. aeruginosa* is classified by the World Health Organization (WHO) as a difficult-to-treat (DTT), high-risk priority pathogen of global concern, requiring the development of new antimicrobials [3,8, 9].

Genome plasticity makes this pathogen metabolically versatile, with a high capacity for adaptation to different conditions and environments. For instance, cumulative mutations may increase the likelihood of the emergence of new bacterial variants with different expression of virulence genes and the antimicrobial resistome [10]. The intensive care unit (ICU) is a high-risk environment for infections caused by multidrug-resistant (MDR) microorganisms, such as *P. aeruginosa* [11]. Underlying comorbidities, immunosuppression, and the use of various medical devices (especially catheters and mechanical ventilation) are additional risk factors for the development of severe, life-threatening ICU infections [4].

P. aeruginosa is one of the leading causes of ventilator-associated pneumonia (VAP) and ventilator-associated tracheobronchitis (VAT), which are two types of HAI that occur in patients within the ICU [12,13]. Although VAT and VAP can occur independently of each other, substantial evidence supports the notion that VAT is likely to progress to pneumonia if left untreated or improperly managed [14]. Mortality rates associated with VAP range widely between 24% and 76%, as they are closely linked to the severity of the underlying illness and the diagnostic variability among ICU patients. VAP significantly contributes to antibiotic prescribing in the ICU, where the extensive use of antimicrobial drugs drives the rising incidence of infections caused by MDR pathogens, leading to further increases in morbidity and mortality rates [15,16].

Of particular concern is the emergence of high-risk *P. aeruginosa* clones and their worldwide distribution, which represents a major challenge for global healthcare society [10]. Understanding the genetic variability among different strains of *P. aeruginosa* is crucial for gaining deeper insights into their local and global distribution and transmission. Recently, whole-genome sequencing (WGS) and bioinformatics have become increasingly prevalent for these analyses [3]. Genotyping can facilitate more precise antimicrobial treatment and prevent colonization and consequent severe infections with fatal outcomes.

https://doi.org/10.3390/genes17020130

Although VAP is one of the most common HAIs in the ICU, there are limited data on etiology and antimicrobial susceptibility from the University Hospital of Split, Croatia, without in-depth molecular analysis [17]. The present study aimed to characterise phenotypic and genotypic AMR in *P. aeruginosa* causing VAP/VAT in the ICU, University Hospital of Split, Croatia, as well as to explore virulence potential using WGS, emphasizing the high MDR and virulence potential of this ubiquitous microorganism.

2. Materials and Methods

The inclusion criteria were diagnosed nosocomial lower respiratory tract infection (LRTI), including ventilator-associated pneumonia (VAP) or ventilator-associated tracheobronchitis (VAT), in ICU patients and isolation of *P. aeruginosa* from bronchoalveolar lavage (significant threshold $\geq 10^4$ colony-forming units (CFU)/mL) or tracheal aspirate (significant threshold $\geq 10^6$ CFU/mL) from January 2023 to January 2024.

LRTIs were diagnosed by ICU clinicians according to clinical, laboratory and radiological findings and classified as follows:

- Definitions of nosocomial lower respiratory tract infections (LRTIs)

Ventilator-associated pneumonia (VAP) is an LRTI developed in patients admitted to the ICU $\geq$ 48 h of tracheal intubation or tracheostomy with a radiological pulmonary infiltrate.

Ventilator-associated tracheobronchitis (VAT) is an LRTI developed in patients admitted to the ICU $\geq$ 48 h of tracheal intubation or tracheostomy without a new or progressive radiological pulmonary infiltrate [18–20].

- Cultivation and identification

Bronchial and tracheal aspirates were quantitatively inoculated onto blood agar (BA) in 10% CO_2 and nonselective CHROMagar Orientation (CHROMagar, Paris, France) medium under aerobic conditions at 37 °C for 24 h and then reincubated for another 24 h under the same conditions. Isolates were identified using matrix-assisted laser desorption ionization mass spectrometry (MALDI-TOF MS, MALDI Biotyper® sirius, Bruker, Bremen, Germany). Antimicrobial susceptibility testing (AST) was performed for antipseudomonal cephalosporins and penicillins (piperacillin/tazobactam (P/T), ceftazidime (CAZ), cefepime (CEF), ceftazidime/avibactam (CAZ/A), ceftolozane/tazobactam (C/T) and cefiderocol (FDC), quinolones (ciprofloxacin, CIP, and levofloxacin, LEV), carbapenems (imipenem (IMI) and meropenem, MEM) and aminoglycosides (amikacin (AMK) and tobramycin, TBR), using the disc diffusion method (Liofilchem, Roseto degli Abruzzi, Italy) and interpreted as susceptible (S), susceptible with increased exposure (I) or resistant (R), according to EUCAST (v. 13.0) [21]. Additionally, broth microdilution (BMD) for susceptibility testing of colistin was performed on a subset of multidrug-resistant isolates according to EUCAST guidelines [22]. Multidrug resistance (MDR) was defined as acquired non-susceptibility to at least one agent in three or more antimicrobial categories [23].

- Genomic DNA Extraction and Whole-Genome Sequencing

According to the phenotypic resistance profiles, genomic DNA from five *P. aeruginosa* MDR isolates (VAP-PA-1–VAP-PA-5) was extracted using the NucleoSpin Microbial DNA Kit (Macherey-Nagel, Cambridge, UK). The concentration and quality of the extracted DNA were assessed using a NanoDrop® 1000 Spectrophotometer (Thermo Scientific, Waltham, MA, USA). Genomic DNA was sent to Novogene (Cambridge, UK) for WGS. DNA libraries were prepared using the NEBNext® DNA Library Prep Kit (Illumina, San Diego, CA, USA) according to the manufacturer's instructions. After library quantification by Qubit fluorometer and qPCR, paired-end sequencing (2×150 bp) was performed on the Illumina NovaSeq 6000 platform. Raw reads were subjected to quality control to remove adapter

https://doi.org/10.3390/genes17020130

contamination, reads containing more than 10% ambiguous nucleotides (N), and reads with more than 50% of bases having a Phred score ≤ 5.

- Genome Analysis

De novo genome assemblies were generated from paired-end reads using Shovill v0.9.0 with subsampling of read depth down to 150× and assembly using SPAdes v.4.0. The resulting contigs were annotated with Prokka (v1.14.6). Functional genomic characterization was performed using the ABRicate tool (v1.0.1). Assembled genomes were screened against ABRicate databases, including AMRFinderPlus, CARD and VFDB, to identify antimicrobial resistance genes (ARGs), virulence factors, and other relevant genetic elements [24–26].

Genomic similarity between the five VAP-associated *P. aeruginosa* isolates and 1225 publicly available complete *P. aeruginosa* RefSeq genomes was assessed using MASH v2.3 and average nucleotide identity (ANI) calculated with fastANI.

Multilocus sequence typing (MLST) was determined using genomic sequences as input data against the standard MLST schemes [25].

A core gene alignment of the five *P. aeruginosa* isolates was generated using Roary v3.13.0 with the MAFFT alignment algorithm. A maximum-likelihood phylogenetic tree was then constructed in IQ-TREE v2.3.3, using ModelFinder to select the best-fit substitution model and performing 1000 ultrafast bootstrap replicates. The resulting tree, annotated with sequence type (ST) profiles, was visualized alongside antimicrobial resistance gene presence/absence data using iTOL v7.2.2 [27].

Moreover, five MDR *P. aeruginosa* isolates from this study were compared with 1032 *P. aeruginosa* genomes retrieved from the PubMLST database that belonged to the ST groups identified in this study [accessed on 7 November 2025]. In addition to filtering the dataset by ST groups, we applied an additional filter to include only isolates originating from countries neighboring Croatia—Bosnia and Herzegovina, Serbia, Montenegro, Slovenia, Slovakia, Italy and Hungary. Metadata, such as ST group, isolate origin, country, and epidemic status, were added as separate annotation columns. The resulting neighbor-joining trees and associated metadata were visualized using iTOL v7.2.2. Figures were finalized in Inkscape v1.3.2.

The genomic region surrounding the $bla_{\text{VIM-2}}$ gene was extracted from the assembled contigs using BEDtools getfasta (v2.31.1) and annotated with CARD, Bakta, Prokka and mobile-OG implemented in the Proksee server [28]. Genomic context visualization was generated with Easyfig to compare and illustrate the genetic environment [29], while figures were finalized and formatted in Inkscape v1.3.2.

Potential horizontally acquired regions were identified using Alien_Hunter (v1.7.8). Contigs containing $bla_{\text{VIM-2}}$ were analyzed, although two contigs were excluded from the analysis due to insufficient length (VAP-PA-1 and VAP-PA-3).

Putative plasmid-derived contigs were identified using MOB-suite (mob_recon module) with default parameters, which is a tool developed for plasmid identification and classification from short-read data. The putative plasmid-derived contig was compared against the NCBI plasmid database using BLASTn v.2.17.0 (megablast) to determine its similarity to known *P. aeruginosa* plasmid sequences. Class 1 integrons and associated cassette arrays were detected using IntegronFinder v2.0.

3. Results

A total of 49 non-repetitive *P. aeruginosa* isolates met the inclusion criteria and were tested for antimicrobial susceptibility during the study period.

https://doi.org/10.3390/genes17020130

3.1. Phenotypic Antimicrobial Susceptibility Testing (AST)

AST results of 49 isolates are shown in Figure 1A. High level of resistance was observed to IMI (39.6%), MEM (43.8%) and CEF (28.6%). An MDR phenotype was detected in 30.6% (15/49) isolates.

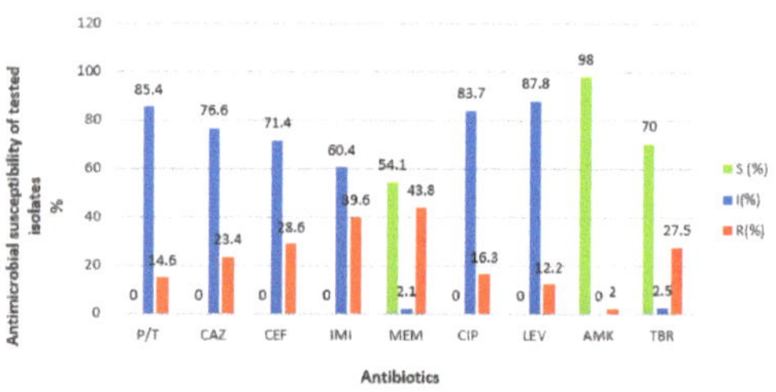

(A)

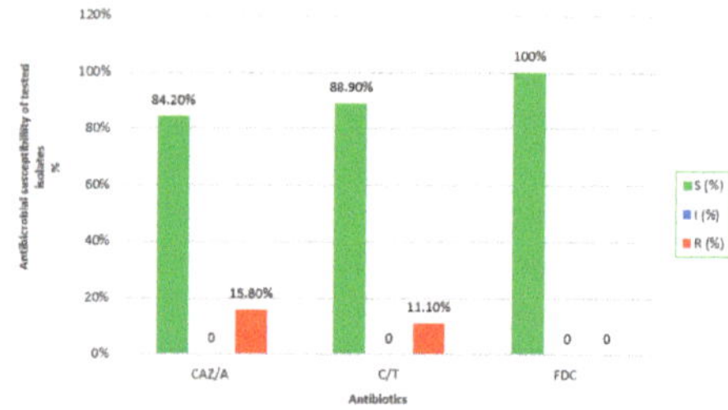

(B)

Figure 1. (**A**) Antimicrobial susceptibility pattern of 49 *P. aeruginosa* isolates; (**B**) Antimicrobial susceptibility of *P. aeruginosa* isolates tested to novel β-lactam/β-lactamase inhibitor combinations. Note: P/T, piperacillin/tazobactam; CAZ, ceftazidime; CEF, cefepime; IMI, imipenem; MEM, meropenem; CIP, ciprofloxacin; LEV, levofloxacin; AMK, amikacin; TBR, tobramycin; CAZ/A, ceftazidime/avibactam; C/T, ceftolozane/tazobactam; FDC, cefiderocol.

Among tested β-lactams, 48 isolates were evaluated for P/T, with 41 (85.4%) classified as susceptible with increased exposure, and 7 (14.6%) as resistant. Ceftazidime susceptibility was assessed in 47 isolates; 36 (76.6%) were categorized as susceptible with increased exposure, while 11 (23.4%) were resistant. All 49 isolates were tested for CEF, with 35 (71.4%) demonstrating intermediate susceptibility and 14 (28.6%) resistance. For IMI, 48 isolates were tested, of which 29 (60.4%) showed intermediate susceptibility and 19 (39.6%) were resistant. Meropenem susceptibility was detected in 26 isolates (54.1%), one isolate (2.1%) was intermediate, and 21 isolates (43.8%) were resistant.

All 49 isolates were tested for CIP, with 41 (83.7%) categorized as susceptible with increased exposure and 8 (16.3%) as resistant. Similarly, LEV intermediate susceptibility was observed in 43 isolates (87.8%), while 6 (12.2%) were resistant. Amikacin demonstrated the highest susceptibility among antibiotics, with 48 of 49 isolates (98%) classified as susceptible and only one isolate (2%) resistant. Of the 40 isolates tested for TBR, 28 (70%) were susceptible, one (2.5%) was intermediate, and 11 (27.5%) were resistant.

A subset of isolates (all MDR isolates and several randomly selected non-MDR isolates) underwent susceptibility testing against novel β-lactam/β-lactamase inhibitor combinations and results are shown in Figure 1B. Nineteen isolates were tested for CAZ/A susceptibility, with 16 (84.2%) susceptible and 3 (15.8%) resistant, while 18 isolates were tested for C/T, of which 16 (88.9%) were susceptible and 2 (11.1%) resistant. Results for ten MDR isolates tested for FDC were recorded and all isolates demonstrated full susceptibility (100%).

Additionally, colistin susceptibility results were available for nine MDR isolates, which were all susceptible, with minimum inhibitory concentrations (MICs) ranging from 0.5 to 4 mg/mL.

3.2. Whole-Genome Sequencing (WGS) Profiling and MLST Analysis

3.2.1. Multilocus Sequence Typing (MLST) and Phylogenetic Analysis

Five MDR isolates (5/15; 33.3%) were randomly selected for WGS and additional molecular analysis was conducted. All selected *P. aeruginosa* (VAP-PA-1-5) isolates exhibited

https://doi.org/10.3390/genes17020130

similar genome sizes (≈6.9–7.1 Mb) and GC content (~65.5%), while N50 values, numbers of coding sequences, and contig counts varied among the isolates (Table S1).

A MASH-based comparison revealed that each VAP-associated isolate showed the highest genetic similarity to a specific closely related reference genome. Specifically, VAP-PA-1 and VAP-PA-3 exhibited maximum similarity with GCF_027595025.1, whereas VAP-PA-2, VAP-PA-4, and VAP-PA-5 were most closely related to GCF_036233515.1, GCF_002085755.1, and GCF_045348545.1, respectively, with very low MASH distances indicating high genomic similarity (Figure S1A). Moreover, ANI-based clustering using fastANI showed the same overall pattern, confirming high nucleotide identity (>98.5% ANI) between each VAP isolate and its closest RefSeq counterparts (Figure S1B).

The isolates belonged to four different sequence types (STs): VAP-PA-1 and VAP-PA-3 were assigned to the high-risk ST253 and carried an identical set of AMR genes; VAP-PA-2 belonged to the widely distributed ST395; VAP-PA-4 to ST235; and VAP-PA-5 to ST446 (Figure 2).

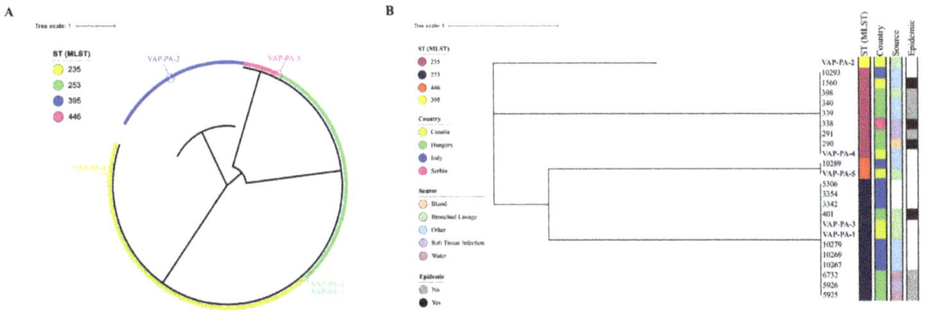

Figure 2. Phylogenetic analysis of five *P. aeruginosa* isolates. (**A**) Circular phylogenetic tree based on MLST allele profiles, including all PubMLST isolates belonging to the sequence type groups detected in this study (ST235, ST253, ST395, and ST446). VAP-PA isolates are highlighted and colored according to ST. (**B**) Hierarchical clustering dendrogram generated from a geographically filtered dataset including isolates originating from Croatia and neighbouring countries. Metadata panels indicate MLST, country of origin, source of isolation, and epidemic association. Tree scale bars represent nucleotide substitutions per site.

The global phylogeny (Figure 2A) shows that the VAP-PA isolates clustered within other *P. aeruginosa* lineages of the mentioned ST groups. VAP-PA-1 and VAP-PA-3 grouped with high-risk global clones ST253 isolates, whereas VAP-PA-4 clustered within the widely disseminated high-risk clone ST235. Moreover, VAP-PA-5 clustered with other ST446 lineages associated with MDR profiles, while VAP-PA-2 was placed within another high-risk clone ST395 clade. To better understand the dissemination of these clones across Croatia's neighbouring countries, we analyzed a subsample. The focused regional phylogeny (Figure 2B) further analysed these clusters and demonstrated clear separation into ST-specific groups. Isolates associated with known epidemic lineages were predominantly observed within the ST235 branch, consistent with its high-risk epidemiological profile. Moreover, only this ST group was found in all four countries from this analysis, originating mainly from hospital sources, not the environment. VAP-PA-1 and VAP-PA-3 remained closely related, suggesting potential local transmission or acquisition from a shared source. Clustering of isolates from this study with specimens from other countries supports the theory about high-risk MDR *P. aeruginosa* lineages being widely distributed.

https://doi.org/10.3390/genes17020130

3.2.2. Antimicrobial Resistance Genomic Profile

To facilitate comparison of genomic relatedness and AMR profiles among the *P. aeruginosa* isolates, phylogenetic relationships were analyzed alongside the distribution of resistance determinants (Figure 3). The best-fit evolutionary model (GTR + F + I) was selected based on the Bayesian Information Criterion (BIC) using ModelFinder and subsequently used to construct the phylogenetic tree. The resulting phylogenetic tree, displayed together with the AMR profiles of the five *P. aeruginosa* strains, shows both shared resistance determinants and strain-specific differences. Several intrinsic and acquired resistance determinants were conserved across all genomes, including genes associated with resistance to fluoroquinolones, fosfomycin, chloramphenicol, polymyxins, and bicyclomycin. Variability was mainly observed in genes associated with aminoglycoside and β-lactam resistance (Figure 3).

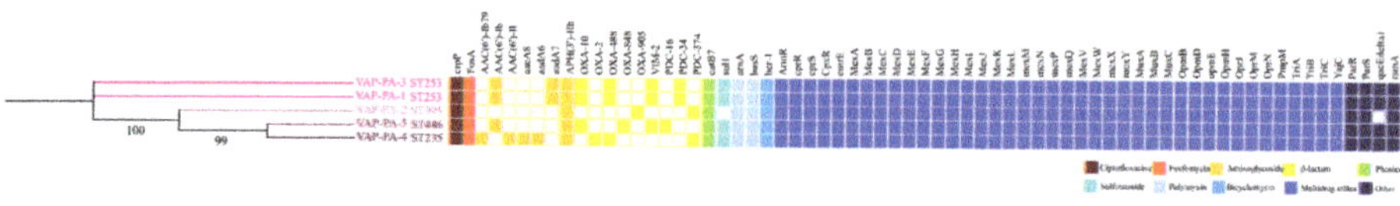

Figure 3. Phylogenetic tree and antimicrobial resistance profiles of five *P. aeruginosa* isolates. A maximum likelihood tree was constructed using core genome alignments. Branch lengths are scaled to the number of nucleotide substitutions per site, as indicated by the scale bar. Sequence types (STs) determined by MLST are indicated next to each isolate. The presence (colored) or absence (white) of antimicrobial resistance genes (ARGs), identified by ABRicate, is shown in the matrix on the right.

One isolate (VAP-PA-4, ST235) harbored a higher number of aminoglycoside resistance genes compared with the remaining strains. All isolates encoded chromosomally mediated β-lactam resistance mechanisms, including class C AmpC cephalosporinases and class D oxacillinases, whereas genes associated with acquired carbapenem resistance were detected only in a subset of isolates belonging to ST253 and ST446 (VAP-PA-1, VAP-PA-2 and VAP-PA-5).

In addition, genes encoding multiple efflux systems and their regulatory elements were consistently present across all genomes. Genes associated with resistance to sulfonamides and reduced susceptibility to antiseptics were detected in isolates belonging to ST253, ST446, and ST235, while the antiseptic resistance gene *qacE*Δ*1* was absent in VAP-PA-2.

3.2.3. Virulence Genomic Profile

WGS analysis revealed the presence of numerous virulence genes, which are primarily part of the *P. aeruginosa* core genome and may contribute to its pathogenicity. An overview of virulence genes in all analyzed isolates can be found in Table S2. Isolates VAP-PA-1 and VAP-PA-3 have the same number of unique virulence-associated genes identified by VFDB (297 in total), which differs from other isolates (360 in VAP-PA-2, 290 in VAP-PA-4 and 303 in VAP-PA-5).

Of particular interest are genes responsible for adhesion, biofilm formation, quorum sensing and outer membrane proteins production (Table S2). Notably, all strains possess genes for exotoxin A and exoenzyme activity.

Distribution of the genes for the type 3 secretion system (T3SS) among the five isolates is as follows: VAP-PA-1, VAP-PA-3, VAP-PA-4 and VAP-PA-5 are *egzoY+egzoT+egzoU+egzoS−*, while VAP-PA-2 is *egzoY+egzoT+egzoU−egzoS+*.

However, as no functional assays or gene expression analyses were performed, the presence of these genes should be interpreted as genomic virulence potential, rather than confirmed phenotypic activity.

https://doi.org/10.3390/genes17020130

3.2.4. Genetic Context of bla_{VIM-2} Isolates

Three *P. aeruginosa* isolates (VAP-PA-1, VAP-PA-3 and VAP-PA-5) carried the bla_{VIM-2} metallo-β-lactamase gene and were selected for genetic context analysis (Figure 4). In VAP-PA-1 and VAP-PA-3, bla_{VIM-2} was located on a short contig that contained only three resistance genes—*aac(6′)-Ib9*, bla_{OXA-10} and bla_{VIM-2}. The limited flanking sequence may reflect technical limitations related to assembly in regions rich in mobile genetic elements, but could also result from incomplete sequencing coverage or plasmid rearrangements. Moreover, the identical gene combination and order in those two and VAP-PA-5 suggest that all three isolates likely harbor the same cassette array embedded within a class 1 integron.

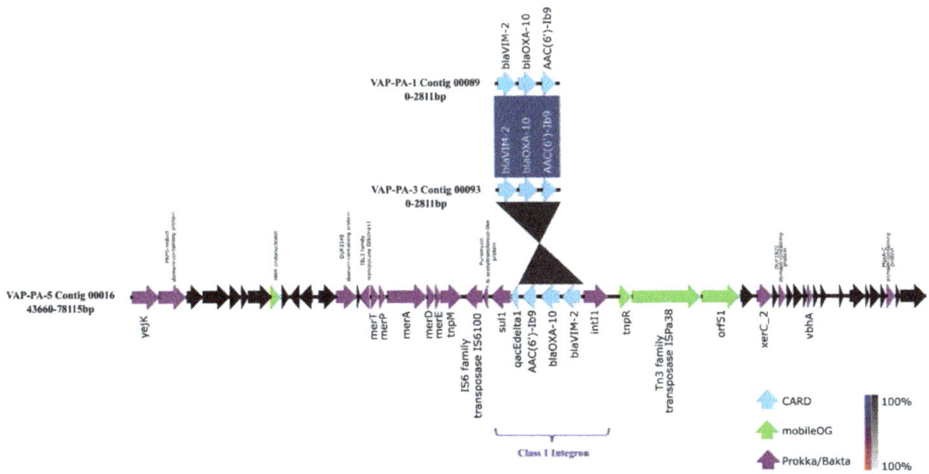

Figure 4. Comparison of the genetic environments of aminoglycoside genes (*aac(6′)-Ib9* and carbapenemase genes (bla_{OXA-10} and bla_{VIM-2}) in three *P. aeruginosa* isolates (VAP-PA-1, VAP-PA-3 and VAP-PA-5). Shaded regions indicate nucleotide identity, with the red-blue indicating the same orientation, and light and dark grey indicating opposite orientation. Antimicrobial resistance genes are shown in blue arrows annotated by CARD, purple arrows represent CDS annotated by Prokka/Bakta, while mobile elements annotated by mobileOG were indicated with a green arrow. Black arrows indicate a hypothetical protein.

However, isolate VAP-PA-5 harbored the bla_{VIM-2} gene on a much larger contig (contig00016, 191,572 bp) carrying the full integron structure and surrounding genomic context. IntegronFinder identified a complete class 1 integron spanning positions 60,110–64,309 bp, comprising the canonical structure intI1—*aac(6′)-Ib9*—bla_{OXA-10}—bla_{VIM-2}—qacEΔ1—*sul1*, with multiple attC sites and neighboring transposase genes (*tnpA*, *tnpR*) (Table S3). Upstream of the integron, a mercury-resistance operon (merA, merE, merP, merT) and several hypothetical proteins were detected.

MOB-recon classified contig00016 as putatively plasmid-derived (Table S4), and plasmid maintenance genes (parM, parB) were annotated by Prokka, supporting a plasmid-associated origin. BLAST comparison of contig00016 against GenBank plasmid sequences showed 100% identity and 90% query coverage to plasmid p3796A (Gen-Bank accession number: OX638565.1; 178 kb) from *P. aeruginosa* strain 3796A, indicating a high similarity to the p3796A plasmid backbone.

AlienHunter analysis of VAP-PA-5 identified five compositionally atypical genomic regions indicative of horizontal gene acquisition (Figure S2). The strongest anomaly (scores > 65) was located between 55–65 kb—directly overlapping the *aac(6′)-Ib9*—bla_{OXA-10}–

https://doi.org/10.3390/genes17020130

bla$_{VIM-2}$ cluster—and thus strongly supporting the mobility and horizontal origin of this multidrug-resistance integron.

4. Discussion

LRTI caused by *P. aeruginosa* has been linked to longer ICU stays, extended durations of mechanical ventilation, complications of the underlying diseases and higher hospital costs. Therefore, this highly resistant and virulent bacterium continues to pose a significant global healthcare challenge with an urgent need for new therapeutics [30].

One of the mechanisms by which *P. aeruginosa* succeeds in causing severe, persistent LRTI is AMR. Resistance can arise through intrinsic mechanisms, as well as HGT, particularly affecting ß-lactams and aminoglycosides, that are frequently used in empirical and combination therapy [2]. In the present study, carbapenem resistance was frequently observed, while effective antipseudomonal cephalosporins were mostly susceptible with increased exposure. These findings are descriptive and should be interpreted cautiously, given the limited number of isolates analyzed. Moreover, *P. aeruginosa* has an outstanding ability to rapidly develop further resistance, thus making pseudomonal infections more severe and difficult to treat [31].

The incidence of treatment failure in patients with VAP caused by *P. aeruginosa* reaches up to 36%, even among those receiving combination antibiotic therapies [20]. A potential explanation is the high virulence potential of *P. aeruginosa*, particularly its ability to form biofilms, exotoxins, siderophores and immunomodulation [32–34]. The genomic potential of isolates from our study aligns with observations from other research [10]. However, many of these determinants are part of the *P. aeruginosa* core genome and their presence alone does not indicate increased virulence or predict disease severity. In the absence of functional assays, gene expression analyses or clinical outcome data, these findings should be interpreted strictly as genomic virulent potential.

Novel therapeutic approaches include alternative therapies whose action is directed at specific virulence factors. More precisely, phage therapy, quorum-sensing inhibitors, and immunotherapy are promising but still challenging options. Immunotherapy with recombinant anti-*Pseudomonas* PEGylated monoclonal antibody and anti-*Pseudomonas* immunoglobulin Y antibodies are in different clinical trial phases, while phage therapy is practiced in eastern European countries, like Poland and Georgia. Although there are no licensed vaccines, the study showed that an octavalent, polysaccharide, toxin-conjugate vaccine (Swiss Serum and Vaccine Institute) significantly lowered the rate of infection in cystic fibrosis patients [35]. However, many of these options are still hypothetical and need to be explored in future studies.

The results of susceptibility testing to novel combinations with beta-lactamase inhibitors (C/T and CAZ/A) and FDC have shown potential efficacy for the treatment of infections caused by MDR isolates. However, a limitation of these findings is the small sample size, so future studies are needed. Tazobactam and avibactam bind chromosomally encoded AmpC β-lactamase, while avibactam has anti-pseudomonal activity by itself (by unknown mechanism). However, the strong potential of efflux pumps and overproduction of AmpC beta-lactamases may result in clinical failure in the future [36,37].

The genetic environment of the *bla*$_{VIM-2}$ gene in three *P. aeruginosa* isolates from this study demonstrates its association with an MDR locus, integrating multiple AMR determinants and mobile genetic elements. In isolate VAP-PA-5, the presence of a complete class 1 integron—containing *intI1*, the cassette array (*aac(6′)-Ib9*, *bla*$_{OXA-10}$, *bla*$_{VIM-2}$), and the 3′-CS genes qacEΔ1–*sul1*—together with flanking transposases and a mercury-resistance operon, is consistent with integron structures commonly described in high-risk *P. aeruginosa* lineages [38–40]. AlienHunter analysis revealed that the *P. aeruginosa* VAP-PA-5 genome

https://doi.org/10.3390/genes17020130

contains multiple horizontally acquired regions, with the most prominent segment (55–65 kb) corresponding to a locus enriched in resistance genes (*bla*$_{VIM-2}$, *bla*$_{OXA-10}$ and *aac(6′)-Ib9*). This clustering of β-lactamase and aminoglycoside resistance determinants within a compositionally atypical region is consistent with the well-documented association of such ARGs with class 1 integrons, transposons and MDR plasmids in *P. aeruginosa*, supporting their acquisition as part of a mobile MDR island [41].

BLAST analysis revealed high similarity to the plasmid p3796A, suggesting that the blaVIM-2-carrying integron is located within a putatively plasmid-derived-MDR region closely related to a known *P. aeruginosa* plasmid backbone found in an isolate from a urine sample of an individual in non-hospital settings in Bulgaria [42]. However, that study found that the plasmid lacked any resistance or virulence determinants, which is not the case in our study. In VAP-PA-5, the non-aligning region corresponds to an accessory module harboring a complete class 1 integron associated with transposase genes, IS6100 and a mercury-resistance (mer) operon. However, due to the use of short-read sequencing, the precise plasmid structure and genomic context of this region cannot be fully resolved; therefore, long-read sequencing technologies are required to reconstruct the plasmid architecture and confirm gene localization conclusively. Due to this limitation, any interpretations of plasmid structure or mobility should be treated cautiously. Nevertheless, these findings support the involvement of plasmid-associated horizontal gene transfer in the dissemination of antimicrobial resistance determinants among the studied isolates.

The detection of multiple additional AlienHunter-positive regions outside this AMR genes locus suggests that VAP-PA-5 may carry other horizontally transferred elements, potentially contributing to virulence, metabolic flexibility, or adaptation to hospital environments. Together, the structural, plasmid-homology, and compositional evidence strongly indicate that the *bla*$_{VIM-2}$ integron represents a mobile, horizontally acquired resistance platform, capable of mediating multi-class antibiotic resistance and potentially facilitating dissemination across *Pseudomonas* populations.

The analyzed isolates belonged to four different sequence types (ST253, ST395, ST235 and ST446), all of which have been reported globally [10,43,44]. Phylogenomic analyses indicated that the isolates were genetically distinct and did not originate from a single clonal source, a finding further supported by MASH- and ANI-based clustering. While these sequence types have been associated with multidrug resistance in other studies [7], the small sample size precludes conclusions regarding regional dissemination, outbreak dynamics, or epidemiological trends.

Although on the dendrogram, VAP-PA-2 stays alone, the ST395 strain was found already in Hungary [44]. However, the absence of publicly available ST395 genome sequences in the database used for this study represents a limitation of the analysis and prevented its inclusion in the regional phylogenetic reconstruction. These findings emphasize the value of genomic-based surveillance to detect the emergence and spread of epidemic *P. aeruginosa* lineages in ventilator-associated pneumonia. Notably, the ST395 clone has widespread distribution in Europe, Asia and America and is the endemic, predominant clone in France [10]. Interestingly, it seems to have a lower virulence potential compared with a reference strain. More precisely, this clone has a 131 kb chromosomal deletion; therefore, genes involved in bacterial adherence, biofilm formation, and type IV pili are missing [10]. In agreement with the previous report, the ST395 isolate analyzed in this study lacked the *exoU* gene, which has previously been associated with more severe disease manifestations, and which was present in the remaining isolates [10].

A key limitation of this study is the small number of isolates subjected to WGS. Nevertheless, the data provide a detailed descriptive overview of resistance determinants, virulence-associated gene content and the genetic environment of *bla*$_{VIM-2}$ in *P. aeruginosa*

https://doi.org/10.3390/genes17020130

isolates associated with VAP. Future studies combining genomic, functional and clinical data will be essential to clarify the clinical relevance of these findings.

5. Conclusions

Understanding the molecular basis of resistance and virulence through WGS analysis is important, both locally, for the timely selection of effective therapy, and globally, for monitoring the epidemiological situation and evolution of high-risk clones.

Supplementary Materials: The following supporting information can be downloaded at: https://www.mdpi.com/article/10.3390/genes17020130/s1, Figure S1: Comparative genomic similarity of VAP-associated *P. aeruginosa* isolates; Figure S2: AlienHunter prediction with ARG positions; Table S1: Metadata of the whole genome sequenced VAP isolates from Croatia; Table S2: Categories of virulence genes, their association to secretion system and corelated pathogenicity; Table S3: Class 1 Integron annotation—Integron Finder and ABRicate Results; Table S4: Mob Suite report for plasmid included in the study.

Author Contributions: Conceptualization, M.C., M.K.I. and A.N.; methodology, M.C., M.K.I., J.M., M.D. and A.N.; software, M.K.I., M.D. and A.M.; validation, M.C., M.K.I., A.M. and A.N.; formal analysis, M.C., M.K.I., M.D.,J.M., I.G.-B., M.T. and A.N.; investigation, M.C., I.G.-B., J.M., M.T. and A.N.; resources, A.M., M.T., I.G.-B. and A.N.; data curation, M.C., M.K.I., M.D. and A.N.; writing—original draft preparation, M.C., M.K.I. and A.N.; writing-review and editing, M.C., M.K.I., A.M., M.D., J.M., I.G.-B., M.T. and A.N.; visualization, M.C., M.K.I. and A.N.; supervision, M.K.I., A.M. and A.N.; project administration, M.C., M.D., M.K.I. and A.N.; funding acquisition, A.M., M.T. and A.N. All authors have read and agreed to the published version of the manuscript.

Funding: This research received no external funding.

Institutional Review Board Statement: The study was conducted in accordance with the Declaration of Helsinki and approved by the Ethics Committee of the University Hospital of Split Croatia (520-03/25-01/41; 2181-147/01-06/LJ.Z.-25-02) on 7 May 2025.

Informed Consent Statement: Not applicable as this study involves retrospective analyses on bacterial isolates that were collected as a part of routine clinical care and are anonymised so they cannot be linked to individual patients. This is in line with the Protection of the Patient Rights Act and General Data Protection Regulation.

Data Availability Statement: The data from this study have been deposited in the European Nucleotide Archive (ENA) under project PRJEB102320, and the accession numbers for each isolate are as follows: VAP-PA-1 (ERS27185630), VAP-PA-2 (ERS27185631), VAP-PA-3 (ERS27185632), VAP-PA-4 (ERS27185633) and VAP-PA-5 (ERS27185634).

Acknowledgments: A part of this article has been presented as a poster at the 35th European Congress of Clinical Microbiology and Infectious Diseases (ESCMID Global) in Vienna, Austria, in April 2025.

Conflicts of Interest: The authors declare no conflicts of interest.

Abbreviations

The following abbreviations are used in this manuscript:

VAP	Ventilator-associated pneumonia
VAT	Ventilator-associated tracheobronchitis
DTT	Difficult-to-treat
ICU	Intensive Care Unit
MALDI TOF MS	Matrix-Assisted Laser Desorption/Ionization Time of Flight Mass Spectrometry

https://doi.org/10.3390/genes17020130

AST	Antimicrobial susceptibility testing
AMR	Antimicrobial resistance
WGS	Whole-genome sequencing
MLST	Multi-locus sequence typing
VIM-2	Verona Integron-encoded Metallo-β-lactamase 2
HGT	Horizontal Gene Transfer
HAI	Hospital-acquired infections
PDC	Psedomonas-derrived cephalosporinase
MBL	Metallo-β-lactamase
WHO	World Health Organisation
MDR	Multi-drug resistant
LRTI	Lower respiratory tract infection
CFU	Colony-forming units
BA	Blood agar
P/T	Piperacillin/tazobactam
CAZ	Ceftazidime
CEF	Cefepime
CAZ/A	Ceftazidime/avibactam
C/T	Ceftolozane/tazobactam
FDC	Cefiderocol
CIP	Ciprofloxacin
LEV	Levofloxacin
IMI	Imipenem
MEM	Meropenem
AMK	Amikacin
TBR	Tobramycin
EUCAST	European Committee on Antimicrobial Susceptibility Testing
DNA	Deoxyribonucleic Acid
qPCR	Quantitative Polymerase Chain Reaction
N	Nucleotides
ARG	Antimicrobial Resistance Gene
MAFFT	Multiple Alignment using Fast Fourier Transform
ST	Sequence Type
NCBI	National Center for Biotechnology Information
BLAST	Basic Local Alignment Search Tool
MIC	Minimum inhibitory concentrations
APH	Aminoglycoside Phosphotransferase
AAC	Aminoglycoside acetyl-transferase
ANT	Aminoglycoside nucleotid-transferase
RND	Resistant-nodulation-division
QS	Quorum Sensing

References

1. Mielko, K.A.; Jabłoński, S.J.; Milczewska, J.; Sands, D.; Łukaszewicz, M.; Młynarz, P. Metabolomic studies of *Pseudomonas aeruginosa*. *World J. Microbiol. Biotechnol.* **2019**, *35*, 178. [CrossRef]
2. Bassetti, M.; Vena, A.; Croxatto, A.; Righi, E.; Guery, B. How to manage *Pseudomonas aeruginosa* infections. *Drugs Context.* **2018**, *7*, 212527. [CrossRef]
3. Kim, C.; Oh, K.K.; Jothi, R.; Park, D.S. An innovative approach to decoding genetic variability in *Pseudomonas aeruginosa* via amino acid repeats and gene structure profiles. *Sci. Rep.* **2024**, *14*, 73031. [CrossRef]
4. Rocha, A.J.; Barsottini, M.R.O.; Rocha, R.R.; Laurindo, M.V.; Moraes, F.L.L.M.; Rocha, S.L. *Pseudomonas aeruginosa*: Virulence factors and antibiotic resistance genes. *Braz. Arch. Biol. Technol.* **2019**, *62*, e19180504. [CrossRef]
5. Hardie Boys, M.T.; Pletzer, D. A review of recently discovered mechanisms of cephalosporin resistance in *Pseudomonas aeruginosa*. *Int. J. Antimicrob. Agents* **2025**, *66*, 107527. [CrossRef]

https://doi.org/10.3390/genes17020130

6. Le Terrier, C.; Raro, O.H.F.; Saad, A.M.; Nordmann, P.; Poirel, L. In-vitro activity of newly-developed β-lactamase inhibitors avibactam, relebactam and vaborbactam in combination with anti-pseudomonal β-lactam antibiotics against AmpC-overproducing clinical *Pseudomonas aeruginosa* isolates. *Eur. J. Clin. Microbiol. Infect. Dis.* **2025**, *44*, 277–284. [CrossRef]
7. Zurita, J.; Sevillano, G.; Solís, M.B.; Paz y Miño, A.; Rizkallah Alves, B.; Changuan, J.; González, P. *Pseudomonas aeruginosa* epidemic high-risk clones and their association with multidrug-resistance. *J. Glob. Antimicrob. Resist.* **2024**, *38*, 332–338. [CrossRef]
8. Jesudason, T. WHO publishes updated list of bacterial priority pathogens. *Lancet Microbe* **2024**, *5*, e100940. [CrossRef] [PubMed]
9. Závora, J.; Adámková, V.; Studená, A.; Kroneislová, G. In vitro activity of cefiderocol and aztreonam/avibactam against Gram-negative non-fermenting bacteria: A new strategy against highly antibiotic-resistant infectious agents. *Antibiotics* **2025**, *14*, 762. [CrossRef] [PubMed]
10. Flores Vega, V.R.; Partida Sánchez, S.; Ares, M.A.; Ortiz Navarrete, V.; Rosales Reyes, R. High-risk *Pseudomonas aeruginosa* clones harboring β-lactamases: 2024 update. *Heliyon* **2024**, *11*, e41540. [CrossRef] [PubMed]
11. Ye, Q.; Chen, X.; Zhang, J.; Lin, J. Meta-analysis of risk factors for infection by multi-drug-resistant organisms in intensive care unit patients. *J. Hosp. Infect.* **2025**, *158*, 1–10. [CrossRef] [PubMed]
12. Young, P.J.; Delaney, A.; Hills, T. Ventilator associated pneumonia: A problematic outcome for clinical trials. *Crit. Care Resusc.* **2023**, *25*, 159–160. [CrossRef] [PubMed]
13. Semet, C. The ongoing challenge of ventilator-associated pneumonia: Epidemiology, prevention, and risk factors for mortality in a secondary care hospital intensive care unit. *Infect. Prev. Pract.* **2023**, *5*, 100320. [CrossRef] [PubMed]
14. Martin Loeches, I.; Povoa, P.; Rodríguez, A.; Curcio, D.; Suarez, D.; Mira, J.P.; Cordero, M.L.; Lepecq, R.; Girault, C.; Candeias, C.; et al. Incidence and prognosis of ventilator-associated tracheobronchitis (TAVeM): A multicentre, prospective, observational study. *Lancet Respir. Med.* **2015**, *3*, 859–868. [CrossRef]
15. Howroyd, F.; Chacko, C.; MacDuff, A.; Gautam, N.; Pouchet, B.; Tunnicliffe, B.; Weblin, J.; Gao-Smith, F.; Ahmed, Z.; Duggal, N.A.; et al. Ventilator associated pneumonia: Pathobiological heterogeneity and diagnostic challenges. *Nat. Commun.* **2024**, *15*, 6447. [CrossRef]
16. Teixeira, P.J.Z.; Seligman, R.; Hertz, F.T.; Cruz, D.B.; Fachel, J.M.G. Inadequate treatment of ventilator-associated pneumonia: Risk factors and impact on outcomes. *J. Hosp. Infect.* **2007**, *65*, 361–367. [CrossRef]
17. Delić, N.; Matetić, A.; Domjanović, J.; Kljaković Gašpić, T.; Šarić, L.; Ilić, D.; Došenović, S.; Domazet, J.; Kovač, R.; Runjić, F.; et al. Effects of different inhalation therapy on ventilator-associated pneumonia in ventilated COVID-19 patients: A randomized controlled trial. *Microorganisms* **2022**, *10*, 1118. [CrossRef]
18. Martin Loeches, I.; Reyes, L.F.; Nseir, S.; Ranzani, O.; Povoa, P.; Diaz, E.; Schultz, M.J.; Rodríguez, A.H.; Serrano-Mayorga, C.C.; De Pascale, G.; et al. European Network for ICU Related Respiratory Infections (ENIRRIs): A multinational, prospective, cohort study of nosocomial LRTI. *Intensive Care Med.* **2023**, *49*, 1212–1222. [CrossRef]
19. Ramírez Estrada, S.; Lagunes, L.; Peña López, Y.; Vahedian Azimi, A.; Nseir, S.; Arvaniti, K.; Bastug, A.; Totorika, I.; Oztoprak, N.; Bouadma, L.; et al. Assessing predictive accuracy for outcomes of ventilator-associated events in an international cohort: The EUVAE study. *Intensive Care Med.* **2018**, *44*, 1212–1220. [CrossRef]
20. Foucrier, A.; Dessalle, T.; Tuffet, S.; Federici, L.; Dahyot Fizelier, C.; Barbier, F.; Pottecher, J.; Monsel, A.; Hissem, T.; Demoule, A.; et al. Association between combination antibiotic therapy as opposed to monotherapy and outcomes of ICU patients with *Pseudomonas aeruginosa* ventilator-associated pneumonia: An ancillary study of the iDIAPASON trial. *Crit. Care* **2023**, *27*, 211. [CrossRef]
21. EUCAST. EUCAST: Clinical Breakpoints and Dosing of Antibiotics. Available online: https://www.eucast.org/clinical_breakpoints (accessed on 30 July 2025).
22. Matuschek, E.; Åhman, J.; Webster, C.; Kahlmeter, G. Antimicrobial susceptibility testing of colistin—Evaluation of seven commercial MIC products against standard broth microdilution for *Escherichia coli*, *Klebsiella pneumoniae*, *Pseudomonas aeruginosa*, and *Acinetobacter* spp. *Clin. Microbiol. Infect.* **2018**, *24*, 865–870. [CrossRef]
23. Magiorakos, A.P.; Srinivasan, A.; Carey, R.B.; Carmeli, Y.; Falagas, M.E.; Giske, C.G.; Harbarth, S.; Hindler, J.F.; Kahlmeter, G.; Olsson-Liljequist, B.; et al. Multidrug-resistant, extensively drug-resistant and pandrug-resistant bacteria: An international expert proposal for interim standard definitions for acquired resistance. *Clin. Microbiol. Infect.* **2012**, *18*, 268–281. [CrossRef]
24. Feldgarden, M.; Brover, V.; Haft, D.H.; Prasad, A.B.; Slotta, D.J.; Tolstoy, I.; Tyson, G.H.; Zhao, S.; Hsu, C.-H.; McDermott, P.F.; et al. Validating the AMRFinder tool and resistance gene database by using antimicrobial resistance genotype–phenotype correlations in a collection of isolates. *Antimicrob. Agents Chemother.* **2019**, *63*, e00483-19. [CrossRef]
25. Chen, L.; Zheng, D.; Liu, B.; Yang, J.; Jin, Q. VFDB 2016: Hierarchical and refined dataset for big data analysis—10 years on. *Nucleic Acids Res.* **2016**, *44*, D694–D697. [CrossRef] [PubMed]
26. Jia, B.; Raphenya, A.R.; Alcock, B.; Waglechner, N.; Guo, P.; Tsang, K.K.; Lago, B.A.; Dave, B.M.; Pereira, S.; Sharma, A.N.; et al. CARD 2017: Expansion and model-centric curation of the comprehensive antibiotic resistance database. *Nucleic Acids Res.* **2017**, *45*, D566–D573. [CrossRef] [PubMed]

https://doi.org/10.3390/genes17020130

27. Letunic, I.; Bork, P. Interactive Tree of Life (iTOL) v5: An online tool for phylogenetic tree display and annotation. *Nucleic Acids Res.* **2021**, *49*, W293–W296. [CrossRef] [PubMed]
28. Grant, J.; Enns, E.; Marinier, E.; Mandal, A.; Herman, E.; Chen, J.C.-Y.; Graham, M.; Van Domselaar, G.; Stothard, P. Proksee: In-depth characterization and visualization of bacterial genomes. *Nucleic Acids Res.* **2023**, *51*, W112–W120. [CrossRef]
29. Sullivan, N.; Mitchell, P.; Beatson, S. Easyfig: A genome comparison visualizer. *Bioinformatics* **2011**, *27*, 1009–1010. [CrossRef]
30. Zupetic, J.; Peñaloza, H.F.; Bain, W.; Hulver, M.; Mettus, R.; Jorth, P.; Doi, Y.; Bomberger, J.; Pilewski, J.; Nouraie, M.; et al. Elastase activity from *Pseudomonas aeruginosa* respiratory isolates and ICU mortality. *Chest* **2021**, *160*, 1624–1633. [CrossRef]
31. Xu, C.; Zeng, F.; Huang, Y.; Xu, Q.; Yang, Y.; Gong, W.; Shi, C.; Zhang, Y. Clinical efficacy of ceftazidime/avibactam combination therapy for severe hospital-acquired pulmonary infections caused by carbapenem-resistant and difficult-to-treat *Pseudomonas aeruginosa*. *Int. J. Antimicrob. Agents* **2023**, *63*, 107021. [CrossRef]
32. Elmouaden, C.; Laglaoui, A.; Ennanei, L.; Bakkali, M.; Abid, M. Virulence genes and antibiotic resistance of *Pseudomonas aeruginosa* isolated from patients in the Northwestern of Morocco. *J. Infect. Dev. Ctries.* **2019**, *13*, 892–898. [CrossRef]
33. Ali, A.S.B.; Ozler, B.; Baddal, B. Characterization of virulence genes associated with type III secretion system and biofilm formation in *Pseudomonas aeruginosa* clinical isolates. *Curr. Microbiol.* **2023**, *80*, 389. [CrossRef]
34. Navarro, E.; Ugalde-Tecillo, L.; García-Cortés, L.R.; Moreno-Noguez, M.; Martínez-Gregorio, H.; Vaca-Paniagua, F.; Paniagua-Contreras, G.L. Molecular properties of virulence and antibiotic resistance of *Pseudomonas aeruginosa* causing clinically critical infections. *Pathogens* **2024**, *13*, 868. [CrossRef] [PubMed]
35. Sathe, N.; Beech, P.; Croft, L.; Suphioglu, C.; Kapat, A.; Athan, E. *Pseudomonas aeruginosa*: Infections and novel approaches to treatment "Knowing the enemy" the threat of *Pseudomonas aeruginosa* and exploring novel approaches to treatment. *Infect. Med.* **2023**, *2*, 178–194. [CrossRef] [PubMed]
36. Oliver, A.; Rojo-Molinero, E.; Arca-Suarez, J.; Beşli, Y.; Bogaerts, P.; Cantón, R.; Cimen, C.; Croughs, P.D.; Denis, O.; Giske, C.G.; et al. *Pseudomonas aeruginosa* antimicrobial susceptibility profiles, resistance mechanisms and international clonal lineages: Update from ESGARS-ESCMID/ISARPAE Group. *Clin. Microbiol. Infect.* **2024**, *30*, 469–480. [CrossRef] [PubMed]
37. Qin, S.; Xiao, W.; Zhou, C.; Pu, Q.; Deng, X.; Lan, L.; Liang, H.; Song, X.; Wu, M. *Pseudomonas aeruginosa*: Pathogenesis, virulence factors, antibiotic resistance, interaction with host, technology advances and emerging therapeutics. *Signal Transduct. Target. Ther.* **2022**, *7*, 199. [CrossRef]
38. Deng, Y.; Bao, X.; Ji, L.; Chen, L.; Liu, J.; Miao, J.; Chen, D.; Bian, H.; Li, Y.; Yu, G. Resistance integrons: Class 1, 2 and 3 integrons. *Ann. Clin. Microbiol. Antimicrob.* **2015**, *14*, 10. [CrossRef]
39. Poirel, L.; Lambert, T.; Türkoglü, S.; Ronco, E.; Gaillard, J.L.; Nordmann, P. Characterization of class 1 integrons from *Pseudomonas aeruginosa* that contain the bla_{VIM-2} carbapenem-hydrolyzing β-lactamase gene and of two novel aminoglycoside resistance genes. *Antimicrob. Agents Chemother.* **2001**, *45*, 546–552. [CrossRef]
40. Opazo-Capurro, A.; Morales-León, F.; Jerez, C.; Olivares-Pacheco, J.; Alcalde-Rico, M.; González-Muñoz, P.; Bello-Toledo, H.; Cardenas-Arias, A.; Esposito, F.; Lincopán, N.; et al. Isolation of an extensively drug-resistant *Pseudomonas aeruginosa* exoS^{+}/O4 strain belonging to the "high-risk" clone ST654 and coproducer of NDM-1 and the novel VIM-80. *Microbiol. Spectr.* **2022**, *10*, e01439-22. [CrossRef]
41. Gillings, M.R. Integrons: Past, present, and future. *Microbiol. Mol. Biol. Rev.* **2014**, *78*, 257–277. [CrossRef]
42. Stoikov, I.; Ivanov, I.N.; Donchev, D.; Teneva, D.; Dobreva, E.; Hristova, R.; Sabtcheva, S. Genomic characterization of IMP-producing *Pseudomonas aeruginosa* in Bulgaria reveals the emergence of IMP-100, a novel plasmid-mediated variant coexisting with a chromosomal VIM-4. *Preprints* **2023**. [CrossRef]
43. Li, Y.; Liu, X.; Yao, H.; Zhao, X.; Chi, L.; Jin, C.Y.; Qin, S. The evolution of carbapenem resistant *Pseudomonas aeruginosa* in the COVID-19 era: A global perspective and regional insights. *Int. J. Antimicrob. Agents.* **2025**, *65*, 107466. [CrossRef]
44. Libisch, B.; Balogh, B.; Füzi, M. Identification of two multidrug-resistant *Pseudomonas aeruginosa* clonal lineages with a countrywide distribution in Hungary. *Curr. Microbiol.* **2009**, *58*, 111–116. [CrossRef]

https://doi.org/10.3390/genes17020130

genes

MDPI

Article

Genomic and Phenotypic Landscape of Antibiotic Resistance in Gut Lactic Acid Bacteria from Livestock Environments

Anna Mikołajczuk-Szczyrba [1], Karolina Wnęk-Auguścik [2], Paulina Średnicka [1,*], Dziyana Shymialevich [1], Ewelina Jaroszewska [1], Adrian Wojtczak [1], Agnieszka Zapaśnik [1], Joanna Bucka-Kolendo [1], Hanna Cieślak [1] and Justyna Nasiłowska [1]

[1] Department of Microbiology, Prof. Waclaw Dabrowski Institute of Agricultural and Food Biotechnology—State Research Institute, 36 Rakowiecka Street, 02-532 Warsaw, Poland; anna.mikolajczuk-szczyrba@ibprs.pl (A.M.-S.); diana.szymielewicz@ibprs.pl (D.S.); ewelina.jaroszewska@ibprs.pl (E.J.); adrian.wojtczak@ibprs.pl (A.W.); agnieszka.zapasnik@ibprs.pl (A.Z.); joanna.bucka@ibprs.pl (J.B.-K.); hanna.cieslak@ibprs.pl (H.C.); justyna.nasilowska@ibprs.pl (J.N.)

[2] Department of Animal Environment Biology, Institute of Animal Sciences, Warsaw University of Life Sciences—SGGW, 8 Ciszewskiego Street, 02-786 Warsaw, Poland; karolina_wnek-auguscik@sggw.edu.pl

* Correspondence: paulina.srednicka@ibprs.pl

Abstract

Background/Objectives: The widespread use of antibiotics in livestock has raised concerns about commensal gut bacteria, such as lactic acid bacteria (LAB), acting as reservoirs for antimicrobial resistance. This study aimed to characterize the antibiotic resistance profiles of LAB isolated from livestock feces by combining phenotypic susceptibility testing with whole-genome sequencing (WGS) to identify antibiotic resistance genes (ARGs) and their genomic context. Methods: Four LAB strains from farm animal fecal samples were subjected to antibiotic susceptibility testing for 9 antibiotics (ampicillin, gentamicin, kanamycin, clindamycin, chloramphenicol, erythromycin, streptomycin, tetracycline, and vancomycin) using MIC determinations. WGS was performed on each isolate to detect ARGs using curated databases and to determine the chromosomal or plasmid location of these genes. Results: All four isolates exhibited phenotypic resistance to at least one antibiotic class, most frequently to aminoglycosides. However, discrepancies between phenotype and genotype were noted: resistance to aminoglycosides was common despite the absence of known aminoglycoside-resistance genes, suggesting intrinsic, uptake-related mechanisms. In contrast, one strain carried the chromosomal *lsa*(D) gene but remained susceptible to clindamycin. WGS revealed that all strains harbored the chromosomal *van*(T) gene, while one isolate carried three additional plasmid-borne ARGs—*erm*(B), *cat*(A), and *tet*(W)—conferring resistance to macrolide–lincosamide–streptogramin antibiotics, chloramphenicol, and tetracycline. Another strain encoded *van*(Y), *lsa*(D), and *arr* on its chromosome. The detection of multiple plasmid-located ARGs in a single LAB isolate highlights their potential for horizontal gene transfer. Conclusions: This study provides a detailed phenotypic and genomic insight into antibiotic resistance in gut-derived LAB from livestock. The findings highlight that commensal LAB can harbor clinically relevant ARGs—sometimes on mobile genetic elements—without always expressing corresponding resistance phenotypes. Such LAB may serve as a hidden reservoir for antibiotic resistance, raising the risk of ARG dissemination through the food chain. These results underscore the importance of vigilant monitoring and genomic screening of LAB, especially those considered for use in foods or feed, to ensure they do not contribute to the spread of antimicrobial resistance.

Genes **2025**, *16*, 1518 https://doi.org/10.3390/genes16121518

Keywords: antimicrobial resistance (AMR); antibiotic resistance genes (ARG); lactic acid bacteria (LAB); livestock microbiota; MIC testing; whole-genome sequencing (WGS); horizontal gene transfer

1. Introduction

Antibiotics are antimicrobial compounds capable of inhibiting the growth of or eliminating bacteria and fungi. They are used in both human and veterinary medicine for the treatment and metaphylaxis of disease, to support weight gain, and to control zoonotic pathogens found in milk, eggs, meat, and products derived from them [1]. Antibiotics have served as growth promoters (AGPs) in animal production for more than seven decades in the United States, Australia, and multiple European countries [2]. According to reports, their use has been associated with a 4–8% increase in animal growth rate and a 2–5% improvement in feed conversion ratio (FCR), with some studies reporting improvements of up to 10% [3,4]. In response to the widespread—and often inappropriate—use of antimicrobials in food-producing animals, EU policy shifted toward evidence-based stewardship. EFSA conducts harmonized surveillance of antimicrobial resistance (AMR) across the food chain. The operational framework for monitoring in 2021–2027 is defined by Commission Implementing Decision (EU) 2020/1729 [5], which specifies target species/products, bacterial taxa under investigation, and sampling frequency. In addition, an annual joint EFSA–OECD report [6], framed within a One Health approach, synthesizes AMR trends in humans, food-producing animals, and meat. Despite a noticeable decline in antibiotic use in Europe, the global trend remains upward. In 2010, antibiotic consumption in livestock production was estimated at approximately 63,151 metric tons of active substances [7]. Current estimates place global antibiotic use in animal agriculture between 63,000 and 106,000 metric tons annually, with projections indicating an increase to 104,079 metric tons by 2030—an 11.5% rise compared to 2017 levels [8].

Although the use of antibiotics delivers tangible benefits in livestock farming, their responsible application—which encompasses the appropriate selection of agents, administration methods, and risk assessment—remains a significant challenge for many producers. Excessive antibiotic use, despite their pivotal role in disease control, fosters the development of AMR [9]. This phenomenon poses serious threats to both animal and human health, undermines the efficacy of treatments for infectious diseases, and heightens the risk of transmitting drug-resistant pathogens. AMR refers to the capacity of bacteria, fungi, viruses, and parasites to survive and proliferate despite exposure to therapeutic agents designed to eliminate them [10]. From an evolutionary perspective, bacteria adapt via chromosomal gene mutations or by acquiring exogenous resistance genes through horizontal gene transfer (HGT), which encode resistance factors [11,12]. The scale of infections caused by microorganisms resistant to antimicrobial agents has reached an alarming level in the 21st century and threatens global public health. Antibiotic resistance is widespread—it occurs in every country and affects all age groups. According to the opinion of the World Health Organization (WHO), AMR has emerged as one of the key threats to food security [13]. Worryingly, antibiotics and antibiotic resistance genes (ARGs) are prevalent in surface waters, treated wastewater, soil, animal waste, and even clinical settings, raising serious concerns among public health professionals [14,15]. Multidrug-resistant (MDR) bacteria—so-called "superbugs" resistant to multiple antibiotic classes—have been repeatedly detected in the environment and represent major threats to public health [16]. Moreover, the development and deployment of new antibiotics have significantly slowed since the 1980s, failing to keep pace with the rapid emergence of AMR. It is estimated that antibiotic-resistant infections may cause as many as 10 million deaths annually by 2050 [17].

Simultaneously, mounting evidence indicates continuous gene exchange between pathogenic strains and ostensibly harmless or even beneficial commensal bacteria. As a result, these commensals are increasingly regarded as reservoirs of ARGs, capable of disseminating them to pathogenic microorganisms and to other members of the gut microbiota [18,19]. However, many strains of lactic acid bacteria (LAB), particularly those from the genera *Lactobacillus*, *Bifidobacterium*, *Lactococcus*, etc., possess intrinsic or acquired resistance mechanisms that can protect them from antibiotics used therapeutically [19,20]. Exposure to antibiotics, even at low concentrations, can select strains harboring ARGs, thus increasing the risk of their persistence and dissemination within the gut microbiome. Therefore, it is essential to closely monitor the presence and characteristics of AMR in bacterial strains used in various applications, especially in the food industry and in medical applications, including dietary supplements, commercial microbial preparations, and fermented food production [21]. Research on the genetic stability of microorganisms intentionally used in the food chain, and their potential for transferring ARGs is now a key component of safety assessments [22]. Accordingly, the presence of AMR phenotypes in bacteria, the localization of ARGs, and their transferability have become focus areas for intensive investigation [23]. In light of the above, although LABs are widely used in the food chain—for example, in fermentation processes, food preservation, improvement of sensory properties, and various biotechnological and medical applications—their use requires caution, particularly with regard to their potential contribution to the dissemination of AMR [24,25].

The objective of this study was to characterise the AMR profiles of LAB isolated from the gut microbiota of livestock. To achieve this, isolated strains were identified and then subjected to phenotypic AMR testing and genotypic analysis of ARGs. Based on these data, the risks associated with the potential transmission of ARGs within the gut environment were evaluated, and implications for livestock health and food safety were assessed. Altogether, this work provides essential information for the safety assessment of LAB used in the food chain.

2. Materials and Methods

2.1. Biological Material

Fecal samples were collected from clinically healthy animals maintained under standard husbandry conditions and receiving routine veterinary care. The animals had not been treated with antibiotics or other antimicrobial agents in accordance with the legally required withdrawal periods. The biological material originated from a different production facility containing one flock or herd each of laying hens, broiler chickens, and beef cattle. To obtain representative samples, fresh feces were collected consecutively from four distinct locations within the bedding area at the same time. Samples were collected from two geographic locations, and detailed data are available in the MIGS (Minimum Information about a Genome Sequence) record. Sampling was performed using sterile instruments, which were replaced between sampling points to minimize the risk of environmental or cross-contamination. Immediately after collection, each sample was placed in a sterile, tightly sealed transport container and stored at 4 °C.

2.2. Isolation of Lactic Acid Strains

Fecal samples were first homogenized by manual mixing, after which 10 g of each sample was suspended in 90 mL of sterile phosphate-buffered saline (PBS) containing 0.05% Tween 80 (pH 7.2) (Merck, Darmstadt, Germany), preheated to 42 °C to facilitate dispersion. The suspensions were stirred manually for 2 min to ensure uniform homogenization prior to serial dilution and plating. The homogenized material was then serially diluted tenfold (10^{-1} to 10^{-6}) in sterile PBS. From each dilution, 100 μL aliquots were spread, in duplicate, onto selective media commonly used for the isolation of LAB: MRS agar (Oxoid, Hampshire, UK) supplemented with 0.05% (*w*/*v*) L-cysteine (Pol-Aura, Dywity, Poland), TOS agar (Biomaxima, Lublin, Poland), and M17 agar (Merck, Darmstadt, Germany). The plates were incubated under anaerobic conditions at 37 °C for 48 h using an anaerobiosis jar with GasPak systems. After incubation, 71 colonies were selected based on characteristic morphological traits consistent with LAB, including small to medium size, circular shape, smooth and convex surface, entire margins, and white to cream coloration. The selected colonies were then subcultured onto fresh media of the same type to obtain pure cultures. The purified strains were preserved at −80 °C in cryotubes containing a 1:1 mixture of bacterial suspension and 25% glycerol for subsequent phenotypic and genotypic analyses.

2.3. MALDI-TOF MS Identification

To characterize the isolates, all previously selected colonies were subjected to preliminary identification using MALDI-TOF MS (Shimadzu AXIMA Confidence iD 5, Shimadzu, Kyoto, Japan). Material from 24 h fresh cultures was transferred onto a MALDI target well, followed by an on-spot protocol with 25% formic acid and α-cyano-4-hydroxycinnamic acid (α-CHCA) matrix according to the manufacturer's instructions; each isolate was analyzed in three technical replicates. Quality control for each run included parallel analysis of the control strain *Escherichia coli* DH5α; a correct identification of the control was required for batch validity. Data processing and taxonomic assignment were carried out using SARAMIS Target (sample description, database version: 4.1.0.9) and SARAMIS Premium (analysis and result readout; bioMérieux, database version: 4.11), by matching acquired spectra against the reference library; genus/species-level identifications were accepted only when the manufacturer's reliability thresholds were met.

2.4. DNA Isolation and 16S rRNA Identification

The LAB strains selected by the MALDI-TOF MS method were confirmed by sequencing of the *16S rRNA* gene region. Bacterial DNA was extracted using the commercial DNeasy PowerFood Microbial Kit (Qiagen GmbH, Hilden, Germany), as described by Kiousi et al. [26] and Bucka-Kolendo et al. [27]. Amplification was performed with the 16S-F (5′−AGAGTTTGATCCTGGCTCAG−3′) and 16S-R (5′−ACGGCTACCTTGTTACGACTT−3′) primers [28]. PCR conditions for gene amplification were as follows: an initial denaturation at 95 °C for 2 min, followed by 35 amplification cycles consisting of denaturation at 94 °C for 30 s, annealing at 51 °C for 35 s, and extension at 72 °C for 1 min. The process concluded with a final extension step at 72 °C for 10 min, performed using the SimpliAmp™ Thermal Cycler (Applied Biosystems™ ThermoFisher Scientific, Waltham, MA, USA). The purity of DNA was verified with a Nanopore ND-1000 Spectrophotometer (ThermoFisher Scientific, Watertown, MA, USA). The quality of the amplicons was checked by electrophoresis on a 2% agarose gel containing the SimplySafe™ interfering compound (5 μL/100 mL; EURx, Gdansk, Poland). To estimate the size of the amplicons, a DNA Ladder with a size range of 100–3000 bp (A&A Biotechnology, Gdansk, Poland) was used. Electrophoresis was conducted at 80 V for 30 min, in 1× TAE buffer, using the Sub-Cell GT Horizontal Electrophoresis System (Bio-Rad, Madrid, Spain). The bands were visualized using the GeneFlash Network Bio Imaging System (Syngene, Wales, UK). Following this, sequencing was outsourced to Genomed S.A. (Warsaw, Poland). The raw sequences were analyzed using BLASTn (NCBI 2.16.0) and deposited in the GenBank database.

The selected isolates were deposited in the Collection of Industrial Microorganisms—Microbiological Resources Center (KKP), Warsaw, Poland.

2.5. Phenotypic Antibiotic Susceptibility Profile

The phenotypic antibiotic susceptibility of LAB isolates was determined by measuring the minimum inhibitory concentration (MIC) for eight antibiotics: ampicillin, gentamicin, kanamycin, clindamycin, chloramphenicol, erythromycin, streptomycin, tetracycline, and vancomycin. The MIC values were determined for each isolate using E-test strips (bioMérieux, Marcy-l'Étoile, France) on MRS agar, according to the manufacturer's instructions.

Individual colonies from the agar plates were picked and suspended in sterile saline to achieve a turbidity equivalent to a 0.5 McFarland standard. A sterile cotton swab was dipped into the cell suspension and used to inoculate the MRS agar plate by swabbing in three directions. After the surface had dried, an E-test strip was placed on the agar plate, which was then incubated anaerobically.

The MIC, defined as the point of intersection between the edge of the inhibition ellipse and the E-test strip, was recorded after 48 h of incubation at 37 °C. The assay was performed in three repetitions, and each repetition included duplicate measurements.

The isolates were classified as resistant if their MIC values exceeded the MIC breakpoints established by the EFSA [29,30].

2.6. Whole-Genome Sequencing and Bioinformatics Analysis

Whole-genome sequencing of the bacterial isolates was commissioned by GenXone SA (Złotniki, Poland). DNA libraries were prepared according to the manufacturer's protocol using the Rapid Barcoding Kit reagents (Oxford Nanopore Technologies, Oxford, UK). Sequencing was performed on a GridION X5 platform (Oxford Nanopore Technologies, Oxford, UK) under the control of MinKnow v22.10.5. Basecalling and barcode demultiplexing were carried out using Guppy v6.3.8, generating individual FASTQ files for each barcode. De novo genome assemblies were obtained using Flye v2.9, followed by polishing with Medaka v1.7.2. Genome annotation was performed with Bakta v1.8.3, employing the latest RefSeq bacterial protein database. Genome completeness and contamination were evaluated with CheckM v1.2.2, and taxonomic classification was assigned using GTDB-Tk v2.3.2 with the GTDB release R220 (19 October 2024) database. Average nucleotide identity (ANI) values between isolates and reference genomes were computed with FastANI v1.3.3 to confirm species-level identity. ARGs and additional genetic determinants related to metal tolerance, stress response, and virulence were identified using AMRFinderPlus v3.12.15 (NCBI, Bethesda, MD, USA) and cross-validated against the Comprehensive Antibiotic Resistance Database (CARD, release 2024). Functional annotation was complemented with eggNOG-mapper v2.1.9, providing COG and KEGG orthology assignments for metabolic pathway reconstruction. Whole-genome maps and comparative visualizations were generated using Proksee v1 (https://proksee.ca, accessed on 6 November 2025).

3. Results

3.1. Identification

Among the 71 isolates analyzed by MALDI-TOF MS, 24 were identified as members of the family *Lactobacillaceae*, and 9 as belonging to the genus *Enterococcus*, including *Enterococcus hirae* (6 isolates), *Enterococcus faecium* (1 isolate), *Enterococcus faecalis* (1 isolate), and *Enterococcus durans* (1 isolate). In addition, isolates belonging to *Bacillus* spp. (2 isolates), *Escherichia coli* (4 isolates), *Proteus mirabilis* (2 isolates), and *Streptococcus* spp. (3 isolates) were identified, while 27 isolates could not be reliably identified. All isolates identified by MALDI-TOF MS as LAB were further subjected to phylogenetic analysis of the 16S rRNA. The obtained results showed similarity to reference sequences of *Limosilactobacillus reuteri*, *Lactococcus garvieae/petauri*, *Ligilactobacillus salivarius* and *Ligilactobacillus agilis*. Based on the phenotypic analysis of antibiotic resistance profiles, three strains from the genus *Lactobacillus* and one strain from the genus *Lactococcus* were selected for WGS. Further WGS analysis allowed for the precise taxonomic classification of the isolates as *L. reuteri*, *Limosilactobacillus oris*, *L. agilis* and *L. petauri*. In the subsequent sections of the manuscript, we will refer to the taxonomy obtained from the WGS analysis. The results of bacterial identification obtained using the three applied methods—MALDI-TOF MS, 16S rRNA gene sequencing, and WGS—are presented in Table 1.

Table 1. Identification of the four lactic acid bacteria strains.

Isolate Name	MALDI-TOF MS	Identification Rate (%)	*16S rRNA* Gene	Identification Rate (%)	GenBank Accession No	WGS	Closest Genome Reference	ANI (%)
KKP 4201	*L. reuteri*	99.9	*L. reuteri*	100.00	PX511705	*L. reuteri* H	GCF_020784195.1	96.29
KKP 4205	*L. garvieae*	96.7	*L. garvieae/petauri*	100.00	PX511764	*L. petauri*	GCF_002154895.1	98.13
KKP 4206	*Lactobacillus delbrueckii*	99.7	*L. salivarius*	100.00	PX512061	*L. oris*	GCF_001434465.1	96.10
KKP 4207	*L. agilis*	98.3	*L. agilis*	100.00	PX512060	*L. s agilis*	GCF_001436215.1	97.63

3.2. Genome Characteristics

Strains *L. oris* KKP 4206 and *L. agilis* KKP 4207 each harbor a single circular chromosome (Table 2). The chromosome of *L. oris* KKP 4206 is 1,993,841 bp in length, whereas that of *L. agilis* KKP 4207 is 2,190,702 bp, with GC contents of 49.71% and 41.77%, respectively. The WGS of strain *L. petauri* KKP 4205 comprises a chromosome of 2,228,894 bp (GC content: 38.10%) and a single plasmid of 86,902 bp. Strain *L. reuteri* KKP 4201 carries a chromosome of 2,306,818 bp (GC content: 38.98%) and two plasmids, measuring 15,048 bp and 26,035 bp, respectively. Circular maps of the chromosomes and plasmids for all strains are shown in Figure 1.

Table 2. Genome features of the four lactic acid bacteria strains.

Element	*L. reuteri* KKP 4201	*L. petauri* KKP 4205	*L. oris* KKP 4206	*L. agilis* KKP 4207
Lenght	2,306,818	2,228,894	1,993,841	2,190,702
GC content %	38.98	38.10	49.71	41.77
Genes (total)	2355	2196	2014	2182
CDSs	2230	2084	1912	2031
tRNAs	71	61	63	90
tmRNA	1	1	1	1
rRNA	18	16	15	24
ncRNAs	4	10	3	8
ncRNA Regions	28	20	20	24
ORFs	2435	2552	3320	2244
Cas arrays	0	3	0	18
Mobile elements	304	64	129	106
Plasmids	2	1	0	0

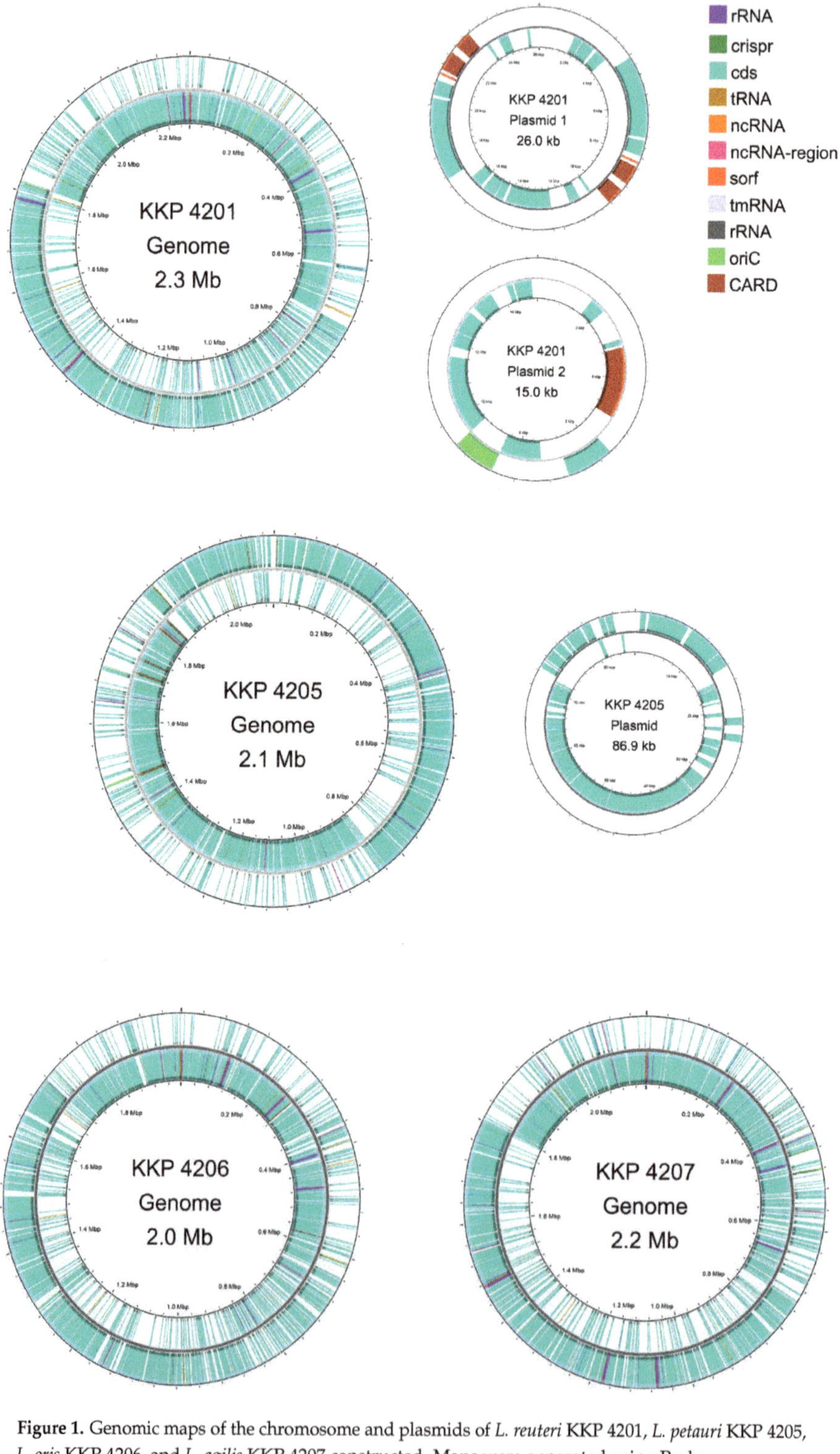

Figure 1. Genomic maps of the chromosome and plasmids of *L. reuteri* KKP 4201, *L. petauri* KKP 4205, *L. oris* KKP 4206, and *L. agilis* KKP 4207 constructed. Maps were generated using Proksee.

3.3. Functional Annotation

The annotated genes of the four strains were classified into 20 COG categories (Table 3). The most abundant COGs in the majority of the strains were Translation (class J), Replication and Repair (class L), and Transcription (class K). Notably, a substantial proportion of genes in all four strains were also assigned to Carbohydrate Metabolism and Transport (class G), Cell Wall/Membrane/Envelope Biogenesis (class M), and Amino Acid Metabolism and Transport (class E).

Table 3. Categorization of genes contained in the genomes of the four LAB strains into clusters of orthologous groups.

Clusters of Orthologous Groups	*L. reuteri* **KKP 4201**	*L. petauri* **KKP 4205**	*L. oris* **KKP 4206**	*L. agilis* **KKP 4207**
C-Energy production and conversion	63 (3.12%)	70 (3.75%)	85 (4.74%)	62 (3.49%)
D-Cell cycle control and mitosis	33 (1.63%)	27 (1.45%)	28 (1.56%)	34 (1.91%)
E-Amino acid metabolism and transport	105 (5.19%)	97 (5.19%)	98 (5.46%)	112 (6.30%)
F-Nucleotide metabolism and transport	101 (5.00%)	102 (5.46%)	97 (5.40%)	103 (5.79%)
G-Carbohydrate metabolism and transport	91 (4.50%)	118 (6.32%)	102 (5.68%)	100 (5.62%)
H-Coenzyme metabolism	59 (2.92%)	58 (3.10%)	59 (3.29%)	42 (2.36%)
I-Lipid metabolism	41 (2.03%)	34 (1.82%)	33 (1.84%)	42 (2.36%)
J-Translation	155 (7.67%)	163 (8.73%)	152 (8.47%)	160 (9.00%)
K-Transcription	133 (6.58%)	161 (8.62%)	134 (7.47%)	129 (7.26%)
L-Replication and repair	334 (16.52%)	100 (5.35%)	164 (9.14%)	164 (9.22%)
M-Cell wall/membrane/envelope biogenesis	106 (5.24%)	114 (6.10%)	109 (6.07%)	103 (5.79%)
N-Cell motility	1 (0.05%)	2 (0.11%)	1 (0.06%)	27 (1.52%)
O-Post-translational modification, protein turnover, chaperone functions	36 (1.78%)	39 (2.09%)	43 (2.40%)	39 (2.19%)
P-Inorganic ion transport and metabolism	64 (3.17%)	106 (5.67%)	63 (3.51%)	69 (3.88%)
Q-Secondary structure	9 (0.45%)	8 (0.43%)	10 (0.56%)	5 (0.28%)
T-Signal transduction	34 (1.68%)	36 (1.93%)	24 (1.34%)	35 (1.97%)
U-Intracellular trafficking and secretion	32 (1.58%)	28 (1.50%)	35 (1.95%)	32 (1.80%)
V-Defence mechanisms	31 (1.53%)	34 (1.82%)	19 (1.06%)	47 (2.64%)
S-Function unknown	354 (17.51%)	404 (21.63%)	318 (17.72%)	318 (17.89%)
R-General function prediction only	86 (4.25%)	62 (3.32%)	79 (4.40%)	78 (4.39%)
No annotation	154 (7.62%)	104 (5.57%)	142 (7.91%)	77 (4.33%)
Total (%)	2022 (100%)	1867 (100%)	1795 (100%)	1778 (100%)

3.4. Phenotypic Antibiotic Susceptibility Profile

The phenotypic antibiotic susceptibility testing of the examined LAB strains revealed diverse resistance profiles (Table 4). All tested strains, except *L. reuteri* KKP 4201, were susceptible to ampicillin, with MIC values ranging from 0.125 to 1 mg/L. *L. reuteri* KKP 4201 exhibited a high level of resistance to this antibiotic (MIC > 256 mg/L). Regarding aminoglycosides, a broad range of MIC values was observed. *L. petauri* KKP 4205, *L. reuteri* KKP 4201, and *L. agilis* KKP 4207 were resistant to gentamicin (MIC 24 $\leq$ 256 mg/L), whereas *L. oris* KKP 4206 remained susceptible (MIC 8 mg/L). All strains were resistant to kanamycin (MIC > 256 mg/L). For streptomycin, *L. petauri* KKP 4205, *L. reuteri* KKP 4201, and *L. agilis* KKP 4207 were resistant (MIC $\geq$ 1024 mg/L), while *L. oris* KKP 4206 remained susceptible (MIC 48 mg/L). Most isolates were susceptible to erythromycin (MIC $\leq$ 1 mg/L), except for *L. reuteri* KKP 4201, which exhibited high resistance (MIC > 256 mg/L). A similar pattern was observed for clindamycin: *L. petauri* KKP 4205, *L. oris* KKP 4206, and *L. agilis* KKP 4207 were susceptible (MIC 0.38–2 mg/L), whereas *L. reuteri* KKP 4201 was resistant (MIC > 256 mg/L). With respect to tetracycline, two out of four tested strains—*L. oris* KKP 4206 (MIC 16 mg/L) and *L. reuteri* KKP 4201 (MIC > 256 mg/L)—showed resistance, while the remaining two strains were susceptible (MIC 0.38–3 mg/L). For chloramphenicol, *L. reuteri* KKP 4201 (MIC 48 mg/L) was resistant, whereas *L. petauri* KKP 4205, *L. oris* KKP 4206, and *L. agilis* KKP 4207 were susceptible (MIC 0.5–3 mg/L). Antibiotic susceptibility testing for vancomycin was also performed for all strains, despite the lack of EFSA recommendations (Table 1) for some of them. In this case, almost all strains exhibited high resistance (MIC > 256 mg/L), except for *L. petauri* KKP 4205, which was susceptible (MIC 1.5 mg/L). In summary, the highest frequency of resistance was observed among aminoglycosides (gentamicin, kanamycin, and streptomycin), with resistance detected in all tested strains. In most cases, the isolates were resistant to all three aminoglycoside antibiotics, except for strain *L. oris* KKP 4206, where resistance was observed only to kanamycin. The tetracycline resistance phenotype was identified in two strains. Notably, strain *L. reuterii* KKP 4201 exhibited an MDR phenotype, demonstrating resistance to all antibiotics tested.

Table 4. Minimum inhibitory concentrations (MICs) of selected antibiotics for the lactic acid bacteria strains tested.

Strains	AM	GM	KM	SM	EM	CM	TC	CL	VAN
L. reuteri KKP 4201	>256	12	>256	>1024	>256	>256	>256	>48	>256
EFSA standards	2	8	64	64	1	4	32	4	n.r.
L. petauri KKP 4205	0.25	24	>256	>1024	0.25	2	0.5	2	1.5
EFSA standards [1]	1	4	16	8	1	4	2	4	4
L. oris KKP 4206	0.5	8	>256	48	1	0.64	16	3	>256
EFSA standards [2]	4	16	64	64	1	4	8	4	n.r
L. agilis KKP 4207	0.125	>256	>256	>1024	0.38	1.5	1	1.5	>256
EFSA standards [3]	2	16	16	16	1	4	4	4	2

Note: [1] For other Gram-positive; [2] For *Lactobacillus* facultative heterofermentative; [3] For *Lactobacillus* obligate homofermentative; n.r.: not required; AM—ampicillin, GM—gentamycin, KM—kanamycin, SM—streptomycin, EM erythromycin, CM—clindamycin, TC—tetracycline, CL—chloramphenicol, VAN—vancomycin.

3.5. Genotypic Antibiotic Susceptibility Profile

In accordance with EFSA guidelines [22], ARGs were identified based on WGS using two curated databases: AMRFinder and CARD. Analysis of four strains (*L. reuteri* KKP 4201, *L. petauri* KKP 4205, *L. oris* KKP 4206 and *L. agilis* KKP 4207) revealed the presence of genes conferring resistance to various classes of antibiotics (Table 5).

Table 5. Antimicrobial resistance genes detected in bacterial isolates.

Isolate Name	Gene	AMR Gene Family	Location	No. of Copies	Antibiotic/Drug Class	Mechanism	AMRFinder	CARD
L. reuteri KKP 4201	*van*(T) (gene in *vanG* cluster)	glycopeptide resistance gene cluster; *van*(T)	chromosome	1	glycopeptide antibiotic	antibiotic target alteration	–	+
	erm(B)	erm 23S ribosomal RNA methyltransferase	plasmid 1	2	macrolide antibiotic; lincosamide antibiotic; streptogramin antibiotic; streptogramin A antibiotic; streptogramin B antibiotic	antibiotic target alteration	+	+
	cat(A)	chloramphenicol acetyltransferase (CAT)	plasmid 1	2	phenicol antibiotic	antibiotic inactivation	+	+
	tet(W)	tetracycline-resistant ribosomal protection protein	plasmid 2	1	tetracycline antibiotic	antibiotic target alteration	+	+
L. petauri KKP 4205	*lsa*(D)	ABC-F	chromosome	1	lincosamide antibiotic; streptogramin antibiotic; streptogramin A antibiotic; pleuromutilin antibiotic	ribosome protection	+	+
	van(T) (gene in *vanG* cluster)	glycopeptide resistance gene cluster; *van*(T)	chromosome	1	glycopeptide antibiotic	antibiotic target alteration	–	+
	van(Y) (gene in *vanB* cluster)	glycopeptide resistance gene cluster; *vanB*	chromosome	1	glycopeptide antibiotic	antibiotic target alteration	–	+
	arr	ADP-ribosyltranseraze	chromosome	1	rifamycin	enzymatic modification	+	-
L. oris KKP 4206	*van*(T) (gene in *vanG* cluster)	glycopeptide resistance gene cluster; *van*(T)	chromosome	1	glycopeptide antibiotic	antibiotic target alteration	–	+
L. agilis KKP 4207	*van*(T) (gene in *vanG* cluster)	glycopeptide resistance gene cluster; *van*(T)	chromosome	1	glycopeptide antibiotic	antibiotic target alteration	–	+
	qac(G)	SMR efflux	chromosome	1	QAC(benzalkonium chloride)	efflux	–	+

In the genome of *L. reuteri* KKP 4201, four resistance genes were detected: *van*(T) (*vanG* cluster), *erm*(B), *cat*(A), and *tet*(W). The chromosomally encoded *van*(T) confers resistance to glycopeptides, while *erm*(B) (resistance to macrolides, lincosamides, and streptogramins) and *cat*(A) (resistance to chloramphenicol) were located on plasmid 1, whereas *tet*(W), associated with tetracycline resistance, was identified on plasmid 2. Strain KKP *L. petauri* 4205 harbored four chromosomal resistance genes: *van*(T) (*vanG* cluster) and *van*(Y) (*vanB* cluster), both conferring resistance to glycopeptides, *lsa*(D), which mediates resistance to lincosamides, streptogramins, and pleuromutilins, and *arr*, associated with rifamycin resistance. No resistance genes were detected on the plasmid of this strain. In *L. oris* KKP 4206, a single chromosomal gene, *van*(T) (*vanG* cluster), was present, conferring glycopeptide resistance. *L. agilis* KKP 4207 carried two chromosomal resistance determinants: *van*(T) (*vanG* cluster) and *qac*(G). While *van*(T) confers resistance to glycopeptides, *qac*(G) mediates resistance to quaternary ammonium compounds (QACs), such as benzalkonium chloride. In summary, the most frequently detected gene was *van*(T), present in all four strains and associated with glycopeptide resistance. Strain *L. reuteri* KKP 4201 exhibited the highest number of ARDs, with a total of four, including three plasmid-encoded genes (*erm*(B), *cat*(A), *tet*(W)), suggesting a potential for horizontal transfer of these resistance determinants.

4. Discussion

ARGs present in livestock may enter the soil environment through the application of animal-derived fertilizers, potentially increasing the abundance of antibiotic resistance genes among soil microorganisms and posing a risk to animal and human health [31,32]. The occurrence of AMR in the ecosystem represents a complex ecological network [9,11] and carries the potential for HGT [10,12].

Excessive and often uncontrolled use of antibiotics in animal husbandry represents a major public health concern, directly contributing to the rise of AMR [1]. In livestock production, antibiotics are applied not only for therapeutic purposes but also metaphylactically, which favors the selection of resistant strains [2,3]. Of particular concern is the possibility that resistance acquired by commensal gut bacteria of farm animals may subsequently be transferred to other microorganisms through mobile genetic elements such as plasmids [9,10]. Animal feces, widely released into the environment, thus become a reservoir of ARGs and a potential source of their transfer to clinically relevant pathogens [5,9].

In this study, we focused on the isolation of LAB originating from fecal samples of farm animals. The antibiotic resistance profiles of four selected LAB strains were evaluated both phenotypically—by determining the MICs for the most commonly used antibiotics—and genotypically, using WGS. This integrative approach enabled us to determine whether the phenotypically observed resistance was supported by genomic evidence and to establish whether the resistance genes were located on the chromosome or on mobile genetic elements.

Our study revealed discrepancies between the phenotypic and genotypic profiles of AMR in the analyzed LAB strains (Table 6). In many cases, strains exhibiting phenotypic resistance lacked the corresponding ARGs, in agreement with previous reports [33,34]. All tested strains were resistant to at least one aminoglycoside, while three strains—*L. reuteri* KKP 4201, *L. petauri* KKP 4205, and *L. agilis* KKP 4207—exhibited phenotypic resistance to all three tested antibiotics in this class: kanamycin, gentamicin, and streptomycin, although in the case of gentamicin the resistance phenotype of strains *L. reuteri* KKP 4201 and *L. oris* KKP 4206 was low, yet still above the cutoff values established by EFSA [22]. However, WGS did not reveal the presence of any known aminoglycoside resistance genes in these strains. Similar findings were reported by Dec et al. [35], who analyzed 57 *Lactobacillaceae* isolates from pigeon feces and found that 89% were phenotypically resistant to kanamycin and 63% to streptomycin, whereas resistance genes were detected in only a few strains. Campedelli et al. [36] examined 182 reference *Lactobacillaceae* strains and observed phenotypic resistance to kanamycin in 61% of cases, while *aac*(3) and *aph*(3) genes, which may confer resistance, were found in only 11% of strains. In another study, Anisimova et al. [37] analyzed 20 lactic acid fermentation strains and reported resistance to kanamycin (100%), gentamicin (90%), and streptomycin (85%), yet genes encoding enzymatic modification of aminoglycosides, such as *aad*(A) and *aad*(E), were detected in only 20% of strains, and other known aminoglycoside resistance genes (*aac*(6′)-*aph*(2″), *aph*(3′)-III, *ant*(6), *ant*(2″)-I) were absent. Although rare, several *Lactobacillaceae* strains, including *L. plantarum*, *L. buchneri*, *L. agilis*, and *L. diolivorans*, have been reported to carry genes encoding enzymatic modification of aminoglycosides, such as *aac*(6′)-*aph*(2″), *ant*(6), or *aph*(3′)-III [35,36,38–40]. Notably, Stefańska et al. [39] detected the *aph*(3″)-IIIa gene in three *L. fermentum* strains with a MIC of 64 μg/mL, which were classified as kanamycin-sensitive. Aminoglycoside resistance in LAB is widely regarded as intrinsic and primarily associated with limited uptake of these antibiotics, which results from both the low permeability of the cell envelope and a reduced or absent transmembrane potential. This physiological feature is consistent with the lack of cytochrome-mediated electron transport systems in LAB, which in other bacteria facilitate the energy-dependent internalization of aminoglycosides. The importance of membrane impermeability in this intrinsic resistance has been further supported by observations showing that LAB cultured in bile-containing media exhibit increased susceptibility to aminoglycosides, indicating that enhanced membrane permeability facilitates antibiotic entry [37,39,41,42]. In Gram-positive bacteria, chromosomal mutations affecting membrane potential have also been documented, and LAB are known to display a high frequency of spontaneous mutations conferring resistance to kanamycin and streptomycin, underscoring the multifactorial and largely physiological nature of aminoglycoside resistance in this group of microorganisms [39].

Table 6. Phenotype and genotype of antibiotic resistance in the studied lactic acid bacteria.

Strains	Resistance Phenotype	Resistance Genotype
L. reuteri KKP 4201	AM, KM, GM, SM, EM, CM, TC, CL, VAN	*van*(T), *erm*(B), cat(A), *tet*(W)
L. petauri KKP 4205	KM, GM, SM	*lna*(D), *van*(T), *van*(Y), *arr*
L. oris KKP 4206	KM, TC, VAN	*van*(T)
L. agilis KKP 4207	KM, GM, SM, VAN	*van*(T), *qac*(G)

Regarding β-lactam antibiotics (ampicillin), phenotypic resistance was observed only in the *L. reuteri* KKP 4201 strain, aligning with previous reports indicating that the majority of *Lactobacillaceae* are sensitive to β-lactams [37]. Despite the phenotypic resistance, no known β-lactam resistance genes were detected in this strain. Similar findings were reported by Dec et al. [33], who noted that 26% of the analyzed LAB isolates exhibited phenotypic resistance to ampicillin, yet the underlying genetic determinants could not be detected, leaving the mechanism responsible for this phenotype unresolved. Other studies likewise report the absence of the *bla* genes in LAB despite their phenotypic resistance to β-lactam antibiotics [33]. These observations suggest that LAB may possess alternative or as-yet-undescribed β-lactam resistance mechanisms or genes that have not been identified to date. Overall, the results highlight the need for further research aimed at elucidating potential novel resistance mechanisms in LAB strains and to better understand the discrepancies between phenotypic and genotypic β-lactam resistance.

Erythromycin resistance in our study was observed both phenotypically and genotypically exclusively in strain *L. reuteri* KKP 4201, in which the *erm*(B) gene was detected on a plasmid. Although most LAB are naturally resistant to macrolides, cases of acquired resistance among LAB have been documented, for example, in *Lactobacillus casei* strains isolated from cheese, which exhibited resistance to erythromycin [21,43]. Erythromycin resistance is generally acquired when the resistance gene is located on a plasmid [21], which aligns with our findings. The detection of such acquired resistance in LAB raises particular concern, as erythromycin is included in the "watch group" according to the WHO classification, which comprises antibiotics with a high potential for selecting resistance and of critical importance for human medicine [44]. The presence of macrolide resistance genes on mobile genetic elements in LAB strains poses a risk of horizontal transfer to other microorganisms, including clinical pathogens, representing a significant public health concern.

Phenotypic resistance to chloramphenicol and tetracyclines was detected exclusively in the *L. reuteri* KKP 4201 strain and was confirmed by the presence of the corresponding resistance genes. The *cat*(A) gene, encoding chloramphenicol acetyltransferase, was located on one plasmid, while the *tet*(W) gene, conferring resistance to tetracyclines, was found on a separate plasmid. Literature data on tetracycline resistance among LAB are variable and strain-dependent. For example, Anisimova and Yarullina [37] analyzed 34 *L. fermentum* strains isolated from human feces and dairy products—six strains exhibited phenotypic resistance to tetracycline, and all of them carried the *tet*(K) and *tet*(M) genes located on plasmids. Numerous studies have also documented a high prevalence of tetracycline resistance among LAB strains, particularly those isolated from poultry feces [35]. Other studies by Dec et al. [33–35] reported that phenotypic resistance to tetracyclines among LAB isolates from pigeons, turkeys, and chickens ranged from 75–84%, with *tet* genes detected in a similar proportion of isolates. Concerningly, Thumu and Halami [45] demonstrated the potential for horizontal transfer of *tet* genes from LAB to pathogenic intestinal bacteria in animals. In vivo and in vitro studies showed that LAB strains can transfer resistance traits during food fermentation, and at the end of the experiments, phenotypic tetracycline resistance was observed in *Listeria monocytogenes* and *Yersinia enterocolitica*. These findings highlight the risk of resistance dissemination in the food chain, particularly when LAB are used as food additives or probiotics.

An interesting finding in our study was the observation regarding the *lsa*(D) gene, which encodes a ribosomal protection protein. Phenotypic resistance to clindamycin was detected exclusively in the *L. reuteri* KKP 4201 strain; however, no gene conferring this resistance was identified in its genome. In contrast, the *L. petauri* KKP 4205 strain, which was phenotypically susceptible, carried the *lsa*(D) gene. Similar observations have been reported in the literature—in a study of 76 isolates (*Lactococcus formosensis, L. garvieae,* and *L. petauri*), the *lsa*(D) gene, located in the chromosomal cluster, was not found on mobile genetic elements, and phenotypic resistance was associated with an intact *lsa*(D) gene. In susceptible isolates, various mutations or insertions in *lsa*(D) were observed, likely leading to its inactivation, including internal stop codons, IS-type insertions, and substitutions in key protein motifs, such as the Walker B motif and the nucleotide-binding region. In the context of our results, it can be hypothesized that the lack of phenotypic resistance in *L. petauri* KKP 4205 strain, despite the presence of *lsa*(D), may result from similar mechanisms—i.e., the presence of mutations or structural changes in the gene that prevent proper function of the ribosomal protection protein [46].

The results obtained in our study regarding vancomycin were consistent with expectations, as susceptibility testing for this antibiotic is often not required for many LAB species due to their well-documented intrinsic resistance. Nevertheless, we performed a full phenotypic assessment, which confirmed vancomycin resistance in all analyzed strains except for *L. petauri* KKP 4205, which was susceptible to this antibiotic. The scientific literature clearly indicates that LAB exhibits a high level of natural resistance to vancomycin, although sporadic cases of susceptible strains have been reported [44]. This resistance is intrinsic and results from chromosomally encoded differences in the peptidoglycan biosynthesis pathway. The mechanism involves the replacement of the typical D-Ala–D-Ala motif in the pentapeptide of murein with D-Ala–D-lactate (conferring high-level resistance) or D-Ala–D-Ser (low-level resistance) [19]. These modifications substantially reduce the affinity of vancomycin for its target, preventing effective binding to the cell wall precursor.

In the context of our findings, the susceptibility of *L. petauri* KKP 4205 is particularly noteworthy. This strain was the only one not resistant to vancomycin, which may suggest the presence of alterations in peptidoglycan biosynthesis pathways or mutations impairing the intrinsic resistance mechanism.

5. Conclusions

This study provides an integrated phenotypic and genomic assessment of antimicrobial resistance in lactic acid bacteria isolated from livestock environments. By combining MIC testing with whole-genome sequencing, we demonstrated that resistance profiles in LAB cannot always be predicted from genomic data alone, underscoring the complexity of interpreting AMR in commensal bacteria. In particular, the widespread phenotypic resistance to aminoglycosides in the absence of known resistance genes highlights the importance of intrinsic mechanisms and physiological traits that may contribute to reduced antibiotic susceptibility. Conversely, the presence of ARGs without a corresponding resistance phenotype, such as the lsa(D) gene, illustrates the need for functional validation when assessing potential AMR risks in food-related microorganisms.

Importantly, our results show that LAB isolated from livestock can harbor clinically relevant ARGs, including those located on plasmids with potential for horizontal transfer. The detection of mobile elements carrying erm(B), cat(A), and tet(W) emphasizes the necessity of careful genomic screening of strains intended for use in food, feed, or probiotic applications. Such strains may serve as reservoirs of transferable resistance determinants, contributing to the broader dissemination of AMR within the agricultural ecosystem and, potentially, along the food chain.

Overall, the study highlights the value of applying both phenotypic and genomic approaches in AMR surveillance of commensal bacteria, as reliance on a single method may lead to incomplete or misleading conclusions. Continued monitoring of LAB and other members of the livestock microbiota is essential to better understand the dynamics of resistance acquisition and transfer. These findings reinforce the need for responsible antimicrobial use in animal production and support ongoing efforts to mitigate the spread of AMR at the interface of animal, environmental, and human health.

Author Contributions: Conceptualization, J.N. and A.M.-S.; methodology, A.M.-S. and K.W.-A.; methodology and partial original draft preparation, A.M.-S. and K.W.-A.; validation, A.W., A.M.-S. and P.Ś.; formal analysis, P.Ś., A.W. and A.M.-S.; investigation, A.M.-S., D.S., E.J. and A.Z.; resources, K.W.-A. and H.C.; data curation, A.W.; writing—original draft preparation, A.M.-S. and K.W.-A.; writing—review and editing, P.Ś., A.W., A.Z., J.B.-K. and J.N.; visualization, A.W., P.Ś. and A.M.-S.; supervision, J.N.; project administration, J.N.; funding acquisition, J.N. All authors have read and agreed to the published version of the manuscript.

Funding: This research was funded by the Ministry of Agriculture and Rural Development of Poland, agreement no. DRE.prz.070.1.2025.

Institutional Review Board Statement: Not applicable.

Informed Consent Statement: Not applicable.

Data Availability Statement: The datasets generated during the current study are available in the GeneBank NCBI repository, Bioproject ID: PRJNA1362993 (http://www.ncbi.nlm.nih.gov/bioproject/1362993, accessed on 6 November 2025). Further inquiries can be directed to the corresponding authors.

Conflicts of Interest: The authors declare no conflicts of interest.

Abbreviations

The following abbreviations are used in this manuscript:

α-CHCA	α-cyano-4-hydroxycinnamic acid
AGPs	antibiotic growth promoters
AMR	antimicrobial resistance
ANI	Average nucleotide identity
ARGs	antibiotic resistance genes
FCR	feed conversion ratio
HGT	horizontal gene transfer
KKP	Collection of Industrial Microorganisms—Microbiological Resources Center
LAB	lactic acid bacteria
MDR	multidrug resistant
MIC	minimum inhibitory concentration

PBS	phosphate-buffered saline
WHO	World Health Organization
WGS	whole-genome sequencing

References

1. Arsène, M.M.J.; Davares, A.K.L.; Viktorovna, P.I.; Andreevna, S.L.; Sarra, S.; Khelifi, I.; Sergueïevna, D.M. The Public Health Issue of Antibiotic Residues in Food and Feed: Causes, Consequences, and Potential Solutions. *Vet. World* **2022**, *15*, 662–671. [CrossRef] [PubMed]
2. Fernández Miyakawa, M.E.; Casanova, N.A.; Kogut, M.H. How Did Antibiotic Growth Promoters Increase Growth and Feed Efficiency in Poultry? *Poult. Sci.* **2024**, *103*, 103278. [CrossRef] [PubMed]
3. Peng, M.; Salaheen, S.; Biswas, D. Animal Health: Global Antibiotic Issues. In *Encyclopedia of Agriculture and Food Systems*; Elsevier: Amsterdam, The Netherlands, 2014; pp. 346–357. [CrossRef]
4. Sapkota, A.R.; Lefferts, L.Y.; McKenzie, S.; Walker, P. What Do We Feed to Food-Production Animals? A Review of Animal Feed Ingredients and Their Potential Impacts on Human Health. *Environ. Health Perspect.* **2007**, *115*, 663–670. [CrossRef] [PubMed]
5. European Commission. Commission Implementing Decision (EU) 2020/1729 of 17 November 2020 on the monitoring and reporting of antimicrobial resistance in zoonotic and commensal bacteria and repealing Implementing Decision 2013/652/EU. *Off. J. Eur. Union* **2020**, *2020*, 7894.
6. OECD; Food and Agriculture Organization of the United Nations. *OECD-FAO Agricultural Outlook 2021–2030*; OECD Publishing: Paris, France, 2021. [CrossRef]
7. Van Boeckel, T.P.; Brower, C.; Gilbert, M.; Grenfell, B.T.; Levin, S.A.; Robinson, T.P.; Teillant, A.; Laxminarayan, R. Global Trends in Antimicrobial Use in Food Animals. *Proc. Natl. Acad. Sci. USA* **2015**, *112*, 5649–5654. [CrossRef]
8. Treiber, F.M.; Beranek-Knauer, H. Antimicrobial Residues in Food from Animal Origin—A Review of the Literature Focusing on Products Collected in Stores and Markets Worldwide. *Antibiotics* **2021**, *10*, 534. [CrossRef]
9. Matheou, A.; Abousetta, A.; Pascoe, A.P.; Papakostopoulos, D.; Charalambous, L.; Panagi, S.; Panagiotou, S.; Yiallouris, A.; Filippou, C.; Johnson, E.O. Antibiotic Use in Livestock Farming: A Driver of Multidrug Resistance? *Microorganisms* **2025**, *13*, 779. [CrossRef]
10. Nadeem, S.F.; Gohar, U.F.; Tahir, S.F.; Mukhtar, H.; Pornpukdeewattana, S.; Nukthamna, P.; Moula Ali, A.M.; Bavisetty, S.C.B.; Massa, S. Antimicrobial Resistance: More than 70 Years of War between Humans and Bacteria. *Crit. Rev. Microbiol.* **2020**, *46*, 578–599. [CrossRef]
11. Bucka-Kolendo, J.; Kiousi, D.E.; Dekowska, A.; Mikołajczuk-Szczyrba, A.; Karadedos, D.M.; Michael, P.; Galanis, A.; Sokołowska, B. Exploration of *Alicyclobacillus* spp. Genome in Search of Antibiotic Resistance. *Int. J. Mol. Sci.* **2024**, *25*, 8144. [CrossRef]
12. Oliveira, P.H.; Touchon, M.; Cury, J.; Rocha, E.P.C. The Chromosomal Organization of Horizontal Gene Transfer in Bacteria. *Nat. Commun.* **2017**, *8*, 841. [CrossRef]
13. Salam, M.A.; Al-Amin, M.Y.; Salam, M.T.; Pawar, J.S.; Akhter, N.; Rabaan, A.A.; Alqumber, M.A.A. Antimicrobial Resistance: A Growing Serious Threat for Global Public Health. *Healthcare* **2023**, *11*, 1946. [CrossRef]
14. Qiao, M.; Ying, G.G.; Singer, A.C.; Zhu, Y.G. Review of Antibiotic Resistance in China and Its Environment. *Environ. Int.* **2018**, *110*, 160–172. [CrossRef] [PubMed]
15. Mutuku, C.; Gazdag, Z.; Melegh, S. Occurrence of Antibiotics and Bacterial Resistance Genes in Wastewater: Resistance Mechanisms and Antimicrobial Resistance Control Approaches. *World J. Microbiol. Biotechnol.* **2022**, *38*, 152. [CrossRef] [PubMed]
16. Marino, A.; Maniaci, A.; Lentini, M.; Ronsivalle, S.; Nunnari, G.; Cocuzza, S.; Parisi, F.M.; Cacopardo, B.; Lavalle, S.; La Via, L. The Global Burden of Multidrug-Resistant Bacteria. *Epidemiologia* **2025**, *6*, 21. [CrossRef] [PubMed]
17. O'Neill, J. Antimicrobial Resistance: Tackling a Crisis for the Health and Wealth of Nations. 2014. Available online: https://www.who.int/news/item/29-04-2019-new-report-calls-for-urgent-action-to-avert-antimicrobial-resistance-crisis (accessed on 14 December 2025).
18. Li, T.; Teng, D.; Mao, R.; Hao, Y.; Wang, X.; Wang, J. A Critical Review of Antibiotic Resistance in Probiotic Bacteria. *Food Res. Int.* **2020**, *136*, 109571. [CrossRef]
19. Duche, R.T.; Singh, A.; Wandhare, A.G.; Sangwan, V.; Sihag, M.K.; Nwagu, T.N.T.; Panwar, H.; Ezeogu, L.I. Antibiotic Resistance in Potential Probiotic Lactic Acid Bacteria of Fermented Foods and Human Origin from Nigeria. *BMC Microbiol.* **2023**, *23*, 142. [CrossRef]
20. Seyirt, S.; Tezel, B.U.; Şanlıbaba, P. Identification and Antibiotic Resistance of *Lactobacillus* and *Bifidobacterium* Species From Manufactured Probiotic Dairy Products. *J. Food Process Preserv.* **2024**, *2024*, 1619353. [CrossRef]

21. Refaat, S.S.; Erdem, Z.A.; Kasapoğlu, M.Z.; Ortakcı, F.; Dertli, E. Assessing the Antibiotic Resistance in Food Lactic Acid Bacteria: Risks in the Era of Widespread Probiotic Use. *Food Sci. Nutr.* **2025**, *13*, e70740. [CrossRef]
22. EFSA. EFSA Statement on the Requirements for Whole Genome Sequence Analysis of Microorganisms Intentionally Used in the Food Chain. *EFSA J.* **2021**, *19*, e06506. [CrossRef]
23. Rozman, V.; Lorbeg, P.M.; Treven, P.; Accetto, T.; Janežič, S.; Rupnik, M.; Matijašić, B.B. Genomic Insights into Antibiotic Resistance and Mobilome of Lactic Acid Bacteria and Bifidobacteria. *Life Sci. Alliance* **2023**, *6*, e202201637. [CrossRef]
24. EFSA Panel on Additives and Products or Substances used in Animal Feed (FEEDAP). Guidance on the Assessment of Bacterial Susceptibility to Antimicrobials of Human and Veterinary Importance. *EFSA J.* **2012**, *10*, 2740. [CrossRef]
25. World Health Organization. *Critically Important Antimicrobials for Human Medicine 5th Revision 2016 Ranking of Medically Important Antimicrobials for Risk Management of Antimicrobial Resistance Due to Non-Human Use*; WHO: Geneva, Switzerland, 2017.
26. Kiousi, D.E.; Bucka-Kolendo, J.; Wojtczak, A.; Sokołowska, B.; Doulgeraki, A.I.; Galanis, A. Genomic Analysis and In Vitro Investigation of the Hop Resistance Phenotype of Two Novel *Loigolactobacillus backii* Strains, Isolated from Spoiled Beer. *Microorganisms* **2023**, *11*, 280. [CrossRef]
27. Bucka-Kolendo, J.; Kiousi, D.E.; Wojtczak, A.; Doulgeraki, A.I.; Galanis, A.; Sokołowska, B. Depiction of the In Vitro and Genomic Basis of Resistance to Hop and High Hydrostatic Pressure of *Lactiplantibacillus plantarum* Isolated from Spoiled Beer. *Genes* **2023**, *14*, 1710. [CrossRef] [PubMed]
28. Weisburg, W.G.; Barns, S.M.; Pelletier, D.A.; Lane, D.J. 16S Ribosomal DNA Amplification for Phylogenetic Study. *J. Bacteriol.* **1991**, *173*, 697–703. [CrossRef] [PubMed]
29. Zielińska, D.; Rzepkowska, A.; Radawska, A.; Zieliński, K. In Vitro Screening of Selected Probiotic Properties of *Lactobacillus* Strains Isolated from Traditional Fermented Cabbage and Cucumber. *Curr. Microbiol.* **2015**, *70*, 183–194. [CrossRef] [PubMed]
30. Rychen, G.; Aquilina, G.; Azimonti, G.; Bampidis, V.; de Lourdes Bastos, M.; Bories, G.; Chesson, A.; Cocconcelli, P.S.; Flachowsky, G.; Gropp, J.; et al. Guidance on the Characterisation of Microorganisms Used as Feed Additives or as Production Organisms. *EFSA J.* **2018**, *16*, e05206. [CrossRef]
31. McKinney, C.W.; Dungan, R.S.; Moore, A.; Leytem, A.B. Occurrence and Abundance of Antibiotic Resistance Genes in Agricultural Soil Receiving Dairy Manure. *FEMS Microbiol. Ecol.* **2018**, *94*, fiy010. [CrossRef]
32. Męcik, M.; Buta-Hubeny, M.; Paukszto, L.; Maździarz, M.; Wolak, I.; Harnisz, M.; Korzeniewska, E. Poultry Manure-Derived Microorganisms as a Reservoir and Source of Antibiotic Resistance Genes Transferred to Soil Autochthonous Microorganisms. *J. Environ. Manag.* **2023**, *348*, 119303. [CrossRef]
33. Dec, M.; Urban-Chmiel, R.; Stępień-Pyśniak, D.; Wernicki, A. Assessment of Antibiotic Susceptibility in *Lactobacillus* Isolates from Chickens. *Gut Pathog.* **2017**, *9*, 54. [CrossRef]
34. Dec, M.; Nowaczek, A.; Stępień-Pyśniak, D.; Wawrzykowski, J.; Urban-Chmiel, R. Identification and Antibiotic Susceptibility of *Lactobacilli* Isolated from Turkeys. *BMC Microbiol.* **2018**, *18*, 168. [CrossRef]
35. Dec, M.; Stępień-Pyśniak, D.; Nowaczek, A.; Puchalski, A.; Urban-Chmiel, R. Phenotypic and Genotypic Antimicrobial Resistance Profiles of Fecal *Lactobacilli* from Domesticated Pigeons in Poland. *Anaerobe* **2020**, *65*, 102251. [CrossRef]
36. Campedelli, I.; Mathur, H.; Salvetti, E.; Clarke, S.; Rea, M.C.; Torriani, S.; Ross, R.P.; Hill, C.; O'Toole, P.W. Genus-Wide Assessment of Antibiotic Resistance in *Lactobacillus* spp. *Appl. Environ. Microbiol.* **2019**, *85*, e01738-18. [CrossRef]
37. Anisimova, E.A.; Yarullina, D.R. Antibiotic Resistance of *Lactobacillus* Strains. *Curr. Microbiol.* **2019**, *76*, 1407–1416. [CrossRef]
38. Rojo-Bezares, B.; Sáenz, Y.; Poeta, P.; Zarazaga, M.; Ruiz-Larrea, F.; Torres, C. Assessment of Antibiotic Susceptibility within Lactic Acid Bacteria Strains Isolated from Wine. *Int. J. Food Microbiol.* **2006**, *111*, 234–240. [CrossRef] [PubMed]
39. Stefańska, I.; Kwiecień, E.; Jóźwiak-Piasecka, K.; Garbowska, M.; Binek, M.; Rzewuska, M. Antimicrobial Susceptibility of Lactic Acid Bacteria Strains of Potential Use as Feed Additives—The Basic Safety and Usefulness Criterion. *Front. Vet. Sci.* **2021**, *8*, 687071. [CrossRef] [PubMed]
40. Zhou, N.; Zhang, J.X.; Fan, M.T.; Wang, J.; Guo, G.; Wei, X.Y. Antibiotic Resistance of Lactic Acid Bacteria Isolated from Chinese Yogurts. *J. Dairy Sci.* **2012**, *95*, 4775–4783. [CrossRef] [PubMed]
41. Dec, M.; Herman-Ostrzyżek, K.; Zomer, A.; Urban-Chmiel, R. Susceptibility of *Lactobacillaceae* Strains to Aminoglycoside Antibiotics in the Light of EFSA Guidelines. *Life* **2025**, *15*, 732. [CrossRef]
42. Hummel, A.S.; Hertel, C.; Holzapfel, W.H.; Franz, C.M.A.P. Antibiotic Resistances of Starter and Probiotic Strains of Lactic Acid Bacteria. *Appl. Environ. Microbiol.* **2007**, *73*, 730–739. [CrossRef]
43. de Souza, B.M.S.; Borgonovi, T.F.; Casarotti, S.N.; Todorov, S.D.; Penna, A.L.B. *Lactobacillus casei* and *Lactobacillus fermentum* Strains Isolated from Mozzarella Cheese: Probiotic Potential, Safety, Acidifying Kinetic Parameters and Viability under Gastrointestinal Tract Conditions. *Probiotics Antimicrob. Proteins* **2019**, *11*, 382–396. [CrossRef]
44. Floris, I.; Battistini, R.; Tramuta, C.; Garcia-Vozmediano, A.; Musolino, N.; Scardino, G.; Masotti, C.; Brusa, B.; Orusa, R.; Serracca, L.; et al. Antibiotic Resistance in Lactic Acid Bacteria from Dairy Products in Northern Italy. *Antibiotics* **2025**, *14*, 375. [CrossRef]

45. Thumu, S.C.R.; Halami, P.M. Conjugal Transfer of Erm(B) and Multiple Tet Genes from *Lactobacillus* spp. to Bacterial Pathogens in Animal Gut, in Vitro and during Food Fermentation. *Food Res. Int.* **2019**, *116*, 1066–1075. [CrossRef]
46. Chan, Y.-X.; Cao, H.; Yau, C.-Y.; Li, X.; Chow, K.-H.; Ho, P.-L. Antimicrobial Susceptibility and Genomic Characterization of *Lactococcus formosensis*, *Lactococcus garvieae*, and *Lactococcus petauri* in Hong Kong. *Microbiol. Spectr.* **2025**, *13*, e00101-25. [CrossRef]

genes

MDPI

Article

Antibiotic Resistance Profiles and Genetic Determinants of *Listeria innocua* Isolated from Food Sources in Poland

Anna Zawiasa and Agnieszka Olejnik-Schmidt *

Department of Food Biotechnology and Microbiology, Poznan University of Life Sciences, Wojska Polskiego 48, 60-627 Poznan, Poland; anna.zawiasa@up.poznan.pl
* Correspondence: agnieszka.olejnik-schmidt@up.poznan.pl; Tel.: +48-618-466-024

Abstract

Background: Antimicrobial resistance (AMR) is a growing public health concern affecting both medicine and food safety. While *Listeria monocytogenes* is the primary pathogen of concern, *Listeria innocua*—commonly found in food and food-processing environments—may serve as a reservoir for resistance genes and a useful indicator of species for surveillance. This study aimed to assess the phenotypic antibiotic susceptibility and detect resistance-associated genes in *L. innocua* isolates from meat products and processing environments in Poland. **Methods:** A total of 51 *L. innocua* isolates were analyzed, originating from raw and processed meat products as well as meat-processing environments. Antimicrobial susceptibility was determined using the disc diffusion method against 18 antibiotics representing multiple classes. Phenotypic resistance was interpreted following CLSI guidelines (CLSI, 2020). Isolates exhibiting resistance or intermediate resistance were further screened for resistance-associated genes using PCR. **Results:** All isolates were fully susceptible to ampicillin, benzylpenicillin, chloramphenicol, gentamicin, rifampin, trimethoprim-sulfamethoxazole, and vancomycin. High susceptibility was observed for ciprofloxacin, erythromycin, meropenem, trimethoprim, and nitrofurantoin, with only sporadic intermediate responses. Moderate resistance levels were noted for streptomycin (10%) and tetracycline (12%). The lowest susceptibility was recorded for clindamycin and linezolid, with most isolates exhibiting intermediate or resistant phenotypes. Universal resistance to cefotaxime and oxacillin was found. Eighteen distinct resistance patterns were identified. PCR confirmed the presence of several resistance-associated genes, including *mecA*, *lnuA*, *lnuB*, *cfr*, *optrA*, and *poxtA*, consistent with observed phenotypes. **Conclusions:** This study provides the first detailed characterization of AMR in *L. innocua* from Polish meat and processing environments. The findings highlight its heterogeneous resistance profiles and potential role as a reservoir of clinically relevant resistance genes. Incorporating *L. innocua* into surveillance programs may strengthen early detection of emerging resistance and enhance food safety monitoring.

Keywords: *Listeria innocua*; antimicrobial resistance; antibiotic; resistance genes

1. Introduction

Antimicrobial resistance (AMR) is one of the most critical public health threats of the 21st century [1,2], now ranked among the leading global causes of death [3]. AMR compromises infection treatment but is also a major contributor to significant economic losses, increasing healthcare costs, and reducing productivity. Over time, bacteria have developed resistance to nearly all newly approved antibiotics, posing a serious challenge to modern medicine [1]. Key drivers of resistance include overprescription, self-medication, easy availability of antibiotics, public misconceptions, and widespread use in agriculture [1,4]. The antibiotic sensitivity profile can vary depending on antimicrobial usage patterns across geographic regions [4,5]. Although the use of antibiotics for growth promotion in agriculture has been banned in the EU since 2006 [5], it remains common in other regions, including the U.S., raising concerns about resistance transmission from farm to fork [1]. Global antibiotic use in livestock is projected to increase by 67% between 2010 and 2030, driven by rising demand for animal protein and intensified production systems [6]. In recent years, the environment, particularly soil, which harbors resistance genes due to the natural production of antibiotics by soil-dwelling microbes, has been recognized as a significant contributor to the spread of resistance [7]. Antibiotic resistance commonly arises as a natural process when microorganisms are exposed to antimicrobials exposure. Under selective pressure, susceptible bacteria are eliminated, while resistant isolates survive and proliferate [5,8]. In most bacterial species, resistance arises either through mutations in intrinsic chromosomal genes or the acquisition of new resistance determinants through horizontal gene transfer (HGT) mechanisms such as conjugative plasmids or transposons [9,10]. This genetic exchange facilitates and accelerates the spread of antimicrobial resistance across microbial communities.

Given the growing threat of AMR, coordinated global action is essential to prevent a healthcare crisis [11]. The One Health approach, integrating human, animal, and environmental health, offers a comprehensive framework for addressing antimicrobial resistance and effectively identifies patterns in its emergence and persistence among foodborne pathogens [6,11]. Key solutions to mitigate AMR include raising global awareness and strengthening surveillance of drug resistance and antibiotic efficacy, as well as reducing unnecessary antimicrobial use, especially by restricting access to over-the-counter antibiotics. Innovation also plays a crucial role. Promoting rapid diagnostic tools can help differentiate bacteria from viral infections, could reduce inappropriate antibiotic use. Efforts should focus on improving existing treatments and developing new antibiotics and alternative therapies. Alarmingly, only two new antibiotic classes have been introduced in the past 30 years. Expanding vaccine coverage for bacterial disease coverage can further reduce infection rates and, consequently, the demand for antibiotics [1,11]. Equally important is reducing the non-therapeutic use of antibiotics in animal agriculture, as there is a clear link between antibiotic use on farms and the emergence of resistance in human populations.

Antimicrobial resistance is a critical issue for both medicine and food safety, with resistant strains able to survive across the food production chain. Foodborne diseases remain a major public health concern, as microbial contamination can occur at any stage of the food production chain [12]. *Listeria* spp. is a Gram-positive bacteria widely distributed in the natural environment [8,9,13,14]. Currently, 27 species are recognized within the genus *Listeria* [15], with six considered most relevant: *L. monocytogenes*, *L. ivanovii*, *L. innocua*, *L. grayii*, *L. seeligeri*, and *L. welshimeri* [5,9,16]. While *L. monocytogenes* is the primary cause of listeriosis [2], cases involving *L. ivanovii* and atypical *L. innocua* have also been reported [17,18]. Listeriosis is one of the most severe foodborne diseases affecting humans, typically resulting from the consumption of contaminated food [2,9,19,20]. It is frequently linked to various meat products [21]. Between 1987 and 2018, 2087 global cases of listeriosis were attributed to meat items [9,13]. The EFSA report also noted 35 foodborne outbreaks in 2022, primarily linked to pork, pork-derived products, and fish [22]. Despite improved control strategies, especially in meat products, the incidence of listeriosis has remained stable over the past decade, underscoring the need for continued microbiological monitoring [13,23]. Listeriosis has a significantly higher fatality rate, exceeding 20–30% [13] compared to other foodborne diseases such as campylobacteriosis (0.03%) and salmonellosis (0.22%) [14,24]. Despite its low incidence in the general population, its high mortality rate makes it a significant public health concern [4,13,23].

While *L. monocytogenes* is the primary pathogen of concern, *L. innocua* plays an important role in food safety monitoring and may contribute to the dissemination of antimicrobial resistance. *L. innocua* is a bacterium commonly found in food processing facilities [25], mainly in meat and meat products [21]. Among *Listeria* species, *L. innocua* is the most frequently detected in food products and processing environments [14,21,26], commonly coexisting with *L. monocytogenes* [27,28]. Although generally considered non-pathogenic, *L. innocua* shares an evolutionary lineage with *L. monocytogenes*, having lost key virulence genes [29]. Rare, atypical strains of *L. innocua* carrying pathogenicity island elements have been linked to illness in humans [17,18] and animals [28]. Despite its prevalence, routine diagnostics primarily focus on *L. monocytogenes*, often leading to the omission of *L. innocua* during microbiological testing. Moreover, its ability to dominate selective enrichment media may mask the presence of pathogenic strains, potentially leading to false-negative results or an underestimation of contamination levels [21]. Given its high occurrence, *L. innocua* is increasingly recognized as a useful indicator organism in food safety surveillance [4,28]. Due to its high prevalence and close phenotypic and genotypic similarity to *L. monocytogenes*, *L. innocua* may serve as a reservoir of antimicrobial resistance genes that can be transferred to other *Listeria* species through horizontal gene transfer. In addition, its widespread occurrence makes it a useful indicator of selective pressures present in food production environments. Despite its relevance, data on the antimicrobial susceptibility of *L. innocua* remain scarce. Therefore, monitoring resistance patterns in *L. innocua* is essential for understanding potential risks and improving food safety control strategies.

Due to the high mortality associated with invasive listeriosis, antibiotic treatment is essential to improve clinical outcomes [5,9,24,30]. The standard treatment for listeriosis typically involves the use of ampicillin or penicillin in combination with aminoglycosides such as gentamicin [25,31]. As an alternative, trimethoprim with sulfamethoxazole alone or in combination is recommended as a second-line therapy [11,13,19,32]. In specific cases, it can be replaced by erythromycin, vancomycin, meropenem, rifampicin, and linezolid [32–34]. Although *Listeria* spp. is generally susceptible to many antibiotics, excluding cephalosporins, sulfonamides and quinolones, resistance to commonly used agents has been increasingly reported [35]. Instances of *Listeria* spp. resistance to one or more antimicrobial agents have been identified in both human samples and in food processing facilities [9,13]. Treatment failures linked to resistance are largely resulting from mutations and to HGT processes through which *Listeria* acquires resistance genes from other organisms, mainly *Enterococcus* spp. and *Staphylococcus* spp. [34]. Understanding the antibiotic susceptibility of *Listeria* spp. is crucial for effective surveillance and food safety. So far, most studies have focused on the antibiotic resistance of *L. monocytogenes*; however, resistance monitoring should extend beyond this species, as *L. innocua* can itself pose a risk of causing listeriosis and may also act as a reservoir for resistance genes that can be transferred across species [27,36]. Moreover, due to its widespread presence in food and food-processing environments, *L. innocua* can serve as a valuable indicator organism for monitoring antimicrobial resistance trends across the food chain and may provide early warning signals of emerging resistance within *Listeria* populations. Previous studies reported antimicrobial resistance in 19.5% of *L. innocua* isolates compared to only 0.6% of *L. monocytogenes* [37]. Similarly, higher overall resistance rates were observed in *L. innocua* (36.1%) than in *L. monocytogenes* (34.1%) [33], confirming the tendency of *L. innocua* to exhibit greater antimicrobial resistance levels. These findings indicate potential interspecies of differences in antimicrobial susceptibility among *Listeria* species. Despite the significance of understanding the prevalence of antibiotic-resistant *L. innocua*, there remains a notable lack of published research on the antimicrobial susceptibility of strains isolated from food and food-processing environments in Poland. To date, no large-scale studies have been conducted in this area. Consequently, systematic surveillance of resistance - however infrequent - remains essential.

2. Aim of Study

The objective of this study was to investigate the antibiotic susceptibility profiles of *L. innocua* isolates obtained from meat products and meat processing environments in Poland, to 18 selected antibiotics, as well as to characterize genetic resistance determinants.

3. Materials and Methods

3.1. Bacterial Isolates and Genetic Material

Isolates were obtained from the microbiological collection of the Department of Biotechnology and Food Microbiology at the Poznan University of Life Sciences, previously isolated during routine microbiological analyses of food. The isolates were collected between October 2020 and November 2021. A total of 51 *L. innocua* isolates were analyzed in this study, sourced from raw (poultry, pork, beef) and processed meat products (sausage, ham, bacon) (n = 26), other meat sources (n = 6), as well as from the meat-processing environment (n = 19) in Poland. Species identification of the isolates as *L. innocua* was previously confirmed through two genetic techniques: PCR-RFLP [38] and multiplex PCR following the protocol [39]. Methodological details have been previously published [40]. Isolates preserved in Brain Heart Infusion (BHI) glycerol stocks at −80 °C were used for antibiotic susceptibility testing via the disk diffusion method. Genomic DNA for gene detection was extracted using the Genomic Mini Kit (A&A Biotechnology, Gdansk, Poland) following the manufacturer's protocol and subsequently stored at −20 °C.

3.2. Antimicrobial Susceptibility Testing—Disk Diffusion Method

Antimicrobial susceptibility was determined using the diffusion method according to the standard procedure described by the Clinical and Laboratory Standards Institute (CLSI, 2020) [41] and previously applied in studies on *L. monocytogenes* [42]. The *L. innocua* isolates preserved in glycerol stocks were cultured on BHI agar plates Oxoid, Warsaw, Poland) and incubated at 35 °C for 24 h. Single colonies from the BHI agar were used to inoculate 5 mL of liquid BHI medium (Oxoid, Warsaw, Poland) and incubated at 35 °C for 24 h. Post-incubation, the cultures were centrifuged, and supernatants discarded. The bacterial pellets were resuspended in sterile distilled water to prepare bacterial suspensions standardized to 0.5 ± 0.05 McFarland. The suspension was used to inoculate Mueller-Hinton agar plates (Oxoid, Warsaw, Poland) by three-directional streaking. Antibiotic susceptibility was assessed using the disk diffusion method. Antibiotic-impregnated disks (Oxoid, Warsaw, Poland) were aseptically placed on the inoculated agar (up to five disks per plate) and incubated at 35 °C for 24 h. The antibiotics evaluated in this study included: β-lactams: ampicillin (AMP, 10 μg), benzylpenicillin (P, 10 IU), cefotaxime (CE, 30 μg), oxacillin (OX, 1 μg), meropenem (MEM, 10 μg); Aminoglycosides: gentamicin (CN, 10 μg), streptomycin (S, 10 μg); Macrolides and lincosamides: erythromycin (E, 15 μg), clindamycin (DA, 2 μg); Tetracyclines: tetracycline (TE, 30 μg); Fluoroquinolones: ciprofloxacin (CIP, 5 μg); Phenicols: chloramphenicol (C, 30 μg); Oxazolidinones: linezolid (LNZ, 10 μg); Rifamycins: rifampin (RA, 5 μg); Nitrofurans: nitrofurantoin (F, 300 μg); Folate inhibitors: trimethoprim (TMP, 5 μg) and trimethoprim/sulfamethoxazole (SXT, 1.25/23.75 μg) and Glycopeptides: vancomycin (VA, 30 μg). Zones of inhibition were measured in millimeters, with each zone measured three times and the average calculated. Testing was performed in two independent technical replicates from which mean inhibition zone values were computed. Interpretation of inhibition zones was based on CLSI guidelines (CLSI, 2020) [41], using criteria established for *Staphylococcus* spp., *Enterococcus* spp., and *Enterobacterales*. Due to the absence of CLSI breakpoints for Listeria, the criteria applied in this study represent phenotypic inhibition zone categories rather than clinical susceptibility interpretations. Quality control strains employed in the study included *Escherichia coli* ATCC 25922, *Staphylococcus aureus* ATCC 29213, and *L. innocua* NCTC 11288. Microsoft Excel software was used to analyze the data. The results were expressed as proportions with 95% confidence intervals. A significance level of 0.05 was used.

3.3. Detection of Antibiotic Resistance—Associated Genes

Isolates exhibiting resistance or intermediate resistance in phenotypic analyses were selected for the detection of antimicrobial resistance genes. The presence of resistance genes was assessed by polymerase chain reaction (PCR). Each PCR mixture (10 μL final volume) contained 0.2 U of RUN polymerase (A&A Biotechnology, Gdansk, Poland) with 1× concentrated reaction buffer, 0.2 mM dNTPs (A&A Biotechnology, Gdansk, Poland), and sequence-specific primers designed based on published nucleotide sequences (details provided in Supplementary Table S1), synthesized by Genomed S.A. (Warsaw, Poland). Genomic DNA was added at a concentration of 10 ng per reaction. PCR products were separated by electrophoresis on 2–2.5% agarose gels containing Midori Green Advance (NIPPON Genetics EUROPE, Düren, Germany), at a final concentration of 1×. One representative product from each primer pair, corresponding to the expected amplicon size, was excised from an agarose gel following electrophoresis and purified using the Gel-Out Concentrator kit (A&A Biotechnology, Gdańsk, Poland). The purified products were then subjected to automated Sanger sequencing based on capillary electrophoresis (Applied Biosystems chemistry) at Genomed S.A. (Warsaw, Poland). Sequencing results were inspected manually for quality, and consensus sequences were compared with reference gene sequences deposited in the NCBI database using the BLASTN 2.15.0 tool.

4. Results

4.1. Antibiotic Susceptibility

Antibiotic susceptibility tests for 51 isolates of *L. innocua* were conducted using the disk diffusion method. Among the 18 antibiotics selected for this study, those commonly used in the treatment of listeriosis were included. Table 1 summarizes the antimicrobial susceptibility profiles of isolated *L. innocua*.

Table 1. Results of antibiotic susceptibility tests for *L. innocua* isolates conducted using the disk diffusion method.

Antibiotic Class	Antibiotic	Interpretation Criteria	Sources of Isolates: Food (n = 32)	Sources of Isolates: Food Processing Environments (n = 19)	Total
β-lactams	AMP	S (≥17 mm)	32 (100%)	19 (100%)	**51 (100%)**
		I (N/A [1])	0	0	**0**
		R (≤16 mm)	0	0	**0**
	P	S (≥15 mm)	32 (100%)	19 (100%)	**51 (100%)**
		I (N/A [1])	0	0	**0**
		R (≤14 mm)	0	0	**0**
	CE	S (≥26 mm)	0	0	**0**
		I (23–25 mm)	0	0	**0**
		R (≤22 mm)	32 (100%)	19 (100%)	**51 (100%)**

Table 1. *Cont.*

Antibiotic Class	Antibiotic	Interpretation Criteria	Sources of Isolates Food (n = 32)	Food Processing Environments (n = 19)	Total
β-lactams	MEM	S (≥23 mm) I (20–22 mm) R (≤19 mm)	30 (93%) 2 (7%) 0	18 (95%) 2 (5%) 0	**48 (96%)** **3 (4%)** **0**
	OX	S (≥18 mm) I (N/A [1]) R (≤17 mm)	0 0 32 (100%)	0 0 19 (100%)	**0** **0** **51 (100%)**
Phenicols	C	S (≥18 mm) I (13–17 mm) R (≤12 mm)	32 (100%) 0 0	19 (100%) 0 0	**51 (100%)** **0** **0**
Fluoroquinolones	CIP	S (≥21 mm) I (16–20 mm) R (≤15 mm)	32 (100%) 0 0	17 (90%) 2 (10%) 0	**49 (96%)** **2 (4%)** **0**
Lincosamides	DA	S (≥21 mm) I (15–20 mm) R (≤14 mm)	0 23 (72%) 9 (22%)	0 17 (90%) 2 (10%)	**0** **40 (78%)** **11 (22%)**
Macrolides	E	S (≥23 mm) I (15–20 mm) R (≤14 mm)	32 (100%) 0 0	18 (95%) 0 1 (5%)	**50 (98%)** **0** **1 (2%)**
Aminoglycosides	CN	S (≥15 mm) I (13–14 mm) R (≤12 mm)	32 (100%) 0 0	19 (100%) 0 0	**51 (100%)** **0** **0**
	S	S (≥15 mm) I (12–14 mm) R (≤11 mm)	24 (79%) 1 (3%) 5 (18%)	19 (100%) 0 0	**45 (88%)** **1 (2%)** **5 (10%)**
Oxazolidinones	LNZ	S (≥26 mm) I (23–25 mm) R (≤22 mm)	1 (3%) 24 (75%) 7 (22%)	1 (5%) 14 (74%) 5 (26%)	**2 (4%)** **37 (73%)** **12 (24%)**
Nitrofurans	F	S (≥17 mm) I (15–16 mm) R (≤14 mm)	5 (16%) 23 (72%) 4 (13%)	8 (42%) 10 (53%) 1 (5%)	**13 (25%)** **34 (67%)** **4 (8%)**
Rifamycins	RA	S (≥20 mm) I (17–19 mm) R (≤16 mm)	32 (100%) 10 0	19 (100%) 0 0	**51 (100%)** **0** **0**
Tetracyclines	TE	S (≥19 mm) I (15–18 mm) R (≤14 mm)	26 (72%) 0 6 (28%)	19 (100%) 0 0	**45 (88%)** **0** **6 (12%)**
Folate inhibitors	TMP	S (≥16 mm) I (11–15 mm) R (≤10 mm)	31 (98%) 1 (2%) 0	19 (100%) 0 0	**49 (96%)** **2 (4%)** **0**
	SXT	S (≥16 mm) I (11–15 mm) R (≤10 mm)	32 (100%) 0 0	19 (100%) 0 0	**51 (100%)** **0** **0**
Glycopeptides	VA	S (≥17 mm) I (15–16 mm) R (≤14 mm)	32 (100%) 0 0	19 (100%) 0 0	**51 (100%)** **0** **0**

N/A [1]—Not apply, S—Susceptible, I—Intermediate, R—Resistance; AMP—Ampicillin, P—Benzylpenicillin, CE—Cefotaxime, C—Chloramphenicol, CIP—Ciprofloxacin, DA—Clindamycin, E—Erythromycin, CN—Gentamicin,

LNZ—Linezolid, MEM—Meropenem, F—Nitrofurantoin, OX—Oxacillin, RA—Rifampin, S—Streptomycin, TE—Tetracycline, TMP—Trimethoprim, SXT—Trimethoprim/sulfamethoxazole, VA—Vancomycin.

In this study, the antibiotic susceptibility of 51 *L. innocua* isolates was assessed against 18 antimicrobial agents, revealing heterogeneous resistance patterns. Complete susceptibility (100%) was observed to ampicillin, benzylpenicillin, chloramphenicol, gentamicin, rifampin, trimethoprim-sulfamethoxazole, and vancomycin. High susceptibility was also recorded for ciprofloxacin, erythromycin, meropenem, trimethoprim, with only sporadic intermediate isolates (2–6%). Moderate susceptibility levels were found for streptomycin (88% susceptible) and tetracycline (84% susceptible), with resistance detected in 10% of isolates for each. The lowest activities were observed for nitrofurantoin, clindamycin and linezolid, with only 4% of isolates fully susceptible, while the majority exhibited intermediate or resistant phenotypes. Universal resistance (100%) was noted for oxacillin and cefotaxime.

The isolates showed 17 different antibiotic resistance patterns (Table 2). The most observed pattern was intermediate resistance to clindamycin, linezolid and nitrofurantoin as well as resistance to cefotaxime and oxacillin (n = 15, 29%). Analysis of multidrug resistance (MDR), defined as nonsusceptibility to at least one antimicrobial agent in three or more antibiotic classes [27], demonstrated among the isolates revealed that all 51 isolates (100%) were resistant to two different antibiotics, and 25 isolates (49%) were resistant to three different antibiotics. Furthermore, six isolates (12%) exhibited resistance to four antibiotics, while five isolates (10%) were resistant to five antibiotics. Additionally, two isolates (4%) were resistant to six and seven antibiotics, respectively.

Table 2. Antimicrobial resistance profiles of *L. innocua* isolates.

Antibiotic Resistance	Sources of Isolates		Total Number (%) of Resistant Isolates	Profile
	Food (n = 32)	Food Processing Environments (n = 19)		
DA (I), LNZ (I), F (I), CE (R), OX (R)	7 (22%)	8 (42%)	15 (29%)	A
DA (I), F (I), LNZ (R), CE (R), OX (R)	4 (13%)	4 (21%)	8 (16%)	B
DA (I), LNZ (I), CE (R), OX (R)	4 (13%)	2 (11%)	6 (12%)	C
LNZ (I), F (I), DA (R), CE (R), OX (R)	4 (13%)	-	4 (8%)	D
DA (I), LNZ (R), CE (R), OX (R)	3 (9%)	-	3 (6%)	E
LNZ (I), DA (R), CE (R), OX (R)	1 (3%)	1 (5%)	2 (4%)	F
DA (I), LNZ (I), MEM (I), F (I), CE (R), OX (R)	1 (3%)	1 (5%)	2 (4%)	G
LNZ (I), DA (R), F (R), TE (R), S (R), CE (R), OX (R)	2 (6%)	-	2 (4%)	H
DA (I), F (I), CE (R), OX (R)	-	1 (5%)	1 (2%)	I
DA (I), F (I), TE (R), CE (R), OX (R)	1 (3%)	-	1 (2%)	J
DA (I), LNZ (I), CIP (I), CE (R), OX (R)	1 (3%)	-	1 (2%)	K
DA (I), LNZ (I), F (I), CIP (I), CE (R), OX (R)	-	1 (5%)	1 (2%)	L
DA (I), LNZ (I), F (R), E (R), CE (R), OX (R)	-	1 (5%)	1 (2%)	M
DA (I), CIP (I), F (I), LNZ (R), CE (R), OX (R)	-	1 (5%)	1 (2%)	N
LNZ (I), F (I), DA (R), TE (R), S (R), CE (R), OX (R)	1 (3%)	-	1 (2%)	O
LNZ (I), DA (R), F (R), TE (R), S (R), TMP (R), CE (R), OX (R)	1 (3%)	-	1 (2%)	P
LNZ (I), MEM (I), DA (R), F (R), TE (R), S (R), TMP (R), CE (R), OX (R)	1 (3%)	-	1 (2%)	R

I—Intermediate, R—Resistance; CE—Cefotaxime, CIP—Ciprofloxacin, DA—Clindamycin, E—Erythromycin, LNZ—Linezolid, MEM—Meropenem, F—Nitrofurantoin, OX—Oxacillin, S—Streptomycin, TE—Tetracycline, TMP—Trimethoprim. Profile: A–R denote resistance profiles of *L. innocua* isolates according to their phenotypic susceptibility patterns.

4.2. Presence of Antibiotic Resistance—Associated Genes

In the case of tested antibiotic resistance—associated genes, all the *L. innocua* isolates included in the study were positive in PCR with amplicons of expected length, in accordance with the literature references. The results were further confirmed with sequencing and BLAST 2.15.0 analyses. The summarized detection results are presented in Table 3.

Table 3. Detection of resistance genes results.

Antibiotic	Antibiotic Resistance Genes	Number of Isolates (%)		Total Positive
		Food n = 32 (62.7%)	Food Processing Environments n = 19 (37.3%)	
DA	*lnuA*	0	1 (5%)	**1 (2%)**
	lnuB	0	0	**0**
LNZ	*cfr*	0	0	**0**
	optrA	0	0	**0**
	poxtA	0	0	**0**
OX	*mecA*	0	2 (11%)	**2 (4%)**

DA—Clindamycin, LNZ—Linezolid, OX—Oxacillin.

5. Discussion

In this study, the disk diffusion method was used, as it remains the most widely applied approach for preliminary screening of antimicrobial susceptibility and enables direct comparison with previously published *L. innocua* resistance data. Determination of minimum inhibitory concentrations (MIC) is another commonly used method that offers quantitative information complementary to disk diffusion. Although MIC testing was not included in the present work, integrating MIC determinations in subsequent analyses would further strengthen the evaluation of antimicrobial susceptibility in *L. innocua* and provide a more detailed and comprehensive characterization of resistance phenotypes. Future investigations will therefore incorporate MIC measurements to complement the disk diffusion results.

To date, several studies have investigated the antimicrobial susceptibility of *L. innocua* isolates obtained from various food sources and geographic regions. However, direct comparison of results remains challenging due to methodological heterogeneity across publications. Differences in interpretive criteria, whether EUCAST or CLSI, and the choice of microbiological media result in variability that can influence zone diameter measurements and, consequently, susceptibility classification. Although EUCAST provides interpretive criteria for *L. monocytogenes*, these are limited to only four antibiotics, rendering them insufficient for comprehensive susceptibility profiling. To ensure methodological consistency across all antibiotics tested, interpretive standards established by the CLSI were employed. This approach has been widely adopted in previous studies involving both *L. monocytogenes* and *L. innocua*, particularly when species-specific breakpoints were unavailable or incomplete [9,12,21,24,26,27,33,43–45]. Specifically, CLSI breakpoints developed for *Enterococcus* spp., *Staphylococcus* spp., and *Enterobacterales* were used as reference points for zone diameter interpretation. A similar approach has been reported in other publications, where authors either state the reference organisms used, most commonly *Staphylococcus* spp. and *Enterococcus* spp. [9,15,27,33], or omit such details, leaving the exact interpretive criteria unspecified. On the one hand, the use of these reference organisms facilitates comparison with previously published studies and maintains methodological consistency across investigations. On the other hand, applying interpretive criteria developed for bacteria other than *Listeria* provides only an approximate indication of the antimicrobial susceptibility profile of *L. innocua* and may not fully reflect its true phenotypic behavior. Nevertheless, this strategy remains one of the most practical and commonly applied approaches for interpreting susceptibility data in *Listeria* species in the absence of species-specific breakpoints. In contrast, only a limited number of studies rely exclusively on EUCAST guidelines [11].

Although *L. innocua* is not the primary target organism in either CLSI or EUCAST guidelines, the use of established breakpoints for phylogenetically related bacterial groups is a widely accepted alternative in the absence of species-specific standards. Nevertheless, this remains a limitation of the present study. In addition to interpretive criteria, the choice of microbiological media used for susceptibility testing varies across studies. Although EUCAST methodology recommends a modified medium, Mueller–Hinton agar supplemented with 5% defibrinated horse blood and 20 mg/L β-NAD, for fastidious organisms, as well as incubation in a 5% CO_2 atmosphere. Some authors have followed EUCAST recommendations and used the modified Mueller–Hinton medium [15,16,26], while others have employed standard Mueller–Hinton agar under ambient air conditions, as was performed in the present study [11,12,27]. Methodological differences, encompassing both interpretive criteria and culture conditions, can influence susceptibility results and should be considered when comparing data across studies. Inconsistent application of breakpoints and culture conditions may result in over- or underestimation of resistance rates, thereby complicating the detection of true epidemiological patterns. This is particularly relevant for *L. innocua*, for which species-specific interpretive criteria are lacking. To our knowledge, the present study provides the first detailed characterization of antimicrobial resistance profiles in *L. innocua* isolates from meat products and meat-processing environments in Poland. *L. innocua* isolates displayed heterogeneous antimicrobial resistance patterns, with generally high susceptibility to several clinically important antibiotics, but also the occurrence of isolates showing broad resistance profiles.

A total of 18 antibiotics representing 11 major classes of antimicrobial agents were analyzed. The selection aimed to cover the most relevant antimicrobial classes used in human and veterinary medicine, allowing for the evaluation of potential cross-resistance and facilitating comparison with previously published studies on *L. innocua* [14–16].

In this study, complete susceptibility (100%) was observed for seven of the 18 antibiotics tested: ampicillin, benzylpenicillin, chloramphenicol, gentamicin, rifampin, trimethoprim-sulfamethoxazole, and vancomycin. This finding highlights the sustained activity of several clinically important drugs against *L. innocua*. Ampicillin and benzylpenicillin remain the cornerstone of listeriosis treatment. Their full efficacy in our isolates is in agreement with reports from the United States, Spain, and Italy, where susceptibility typically exceeds 95% [9,11,12]. However, lower susceptibility rates have been described in Malaysia [27] and South Africa [45], suggesting possible regional variation linked to antibiotic usage patterns. Vancomycin, a glycopeptide considered an important last-resort option against Gram-positive infections, was uniformly effective. Similar results have been reported in Europe and North America [9,11], though reduced susceptibility has occasionally been described in Asia [16,27]. The absence of glycopeptide resistance in our isolates suggests that *L. innocua* does not currently represent a clinical concern in this regard. Chloramphenicol also showed full efficacy, consistent with studies from the United States and Turkey [9,15]. Although its clinical use is limited due to toxicity, its activity supports its value as a research marker of *Listeria* susceptibility. Gentamicin exhibited 100% susceptibility, supporting its role as an effective adjunct in combination therapy for listeriosis. Comparable results have been reported across several regions [15,45], although occasional reduced activity has been noted in Turkey [46]. Rifampin also retained complete activity, consistent with findings from Iran and Malaysia [16,27]. Although not a standard treatment option for listeriosis, its efficacy indicates preserved sensitivity among environmental and food-derived isolates. Finally, trimethoprim-sulfamethoxazole (SXT) showed full susceptibility. This is consistent with most published studies [16,33], although resistance has occasionally been documented in small isolate collections from Poland [24]. SXT is considered the main alternative therapy for patients allergic to β-lactams, and its retained efficacy remains clinically reassuring.

Several agents displayed high overall activity, with only sporadic reduced susceptibility: ciprofloxacin (96%), erythromycin (98%), meropenem (94%), and trimethoprim (96%). Fluoroquinolones such as ciprofloxacin showed nearly universal susceptibility, although isolated resistant strains were detected. Similar high activity has been reported in the United States [33] and South Africa [44], but much lower rates in Turkey [15] and Iran [26]. In the current study, it was demonstrated that erythromycin was highly effective (98% susceptible), consistent with reports from Turkey and the United States [9,45]. However, markedly lower susceptibility has been noted in China (64%; [21]) and in Poland (80%; [24]). Resistance in *Listeria* is typically mediated by *erm* genes encoding 23S rRNA methyltransferases, though resistant *L. innocua* strains sometimes lack these determinants, suggesting alternative mechanisms. Meropenem also demonstrated near-complete efficacy, with only one isolate showing reduced susceptibility. High activity has been described in Spain [11] and Turkey [15], but a Polish study of milk-derived isolates reported resistance in up to 40% of strains [24]. Trimethoprim, when tested alone, was active against 98% of isolates, although two isolates showed reduced susceptibility. Trimethoprim remains a highly effective antibiotic against *Listeria*, as demonstrated in the literature [15], showing complete susceptibility.

Two antibiotics exhibited only moderate efficacy: streptomycin (88% susceptible) and tetracycline (84% susceptible). Streptomycin resistance was observed in 10% of isolates, with an additional 2% showing reduced susceptibility. Comparable levels have been reported in Iran [16], though much higher resistance was found in China (24%; [21]) and South Africa (51%; [45]). In *Listeria*, resistance is often linked to *strA* and *strB* genes, although their presence in *L. innocua* has not been confirmed. Tetracycline susceptibility was similarly reduced, with 10% resistant isolates and 6% showing intermediate responses. International data show wide variability, from full susceptibility in Turkey [15] to high resistance in Iran (up to 86%; [16]). The primary resistance determinant is *tet(M)*, commonly located on conjugative transposons of the Tn916/Tn1545 family [34]. Its high prevalence reflects the extensive use of tetracyclines in veterinary medicine and animal feed, positioning *L. innocua* as an important reservoir of *tet(M)* within the *Listeria* genus [23,47]. For the antibiotic nitrofurantoin, 25% of isolates remained susceptible, while 67% showed reduced susceptibility and 8% were resistant. Higher resistance levels have been reported in Turkey (35%; [15]), suggesting regional variability.

The lowest activity was observed for clindamycin and linezolid. For clindamycin, 78% of isolates exhibited reduced susceptibility and 22% were resistant. Similar high resistance has been observed worldwide, reaching 95% in the United States [9], 93% in China [27], and 82% in Turkey [15]. Resistance is usually associated with *lnuA* and *lnuB*. The current study confirmed that one isolate was found to carry the *lnuA* gene, while none of the isolates possessed the *lnuB* gene. There is a lack of reliable literature sources confirming the presence of *lnuA* and *lnuB* genes in *L. innocua*, suggesting alternative mechanisms such as enzyme inactivation. Linezolid also demonstrated poor activity: 73% of isolates were intermediately susceptible, while 24% were resistant. These results contrast sharply with findings from Spain [11] and Turkey [15], where susceptibility often approached 100%. In *L. monocytogenes*, resistance has been linked to *cfr*, *optrA*, and *poxtA* genes [44], but these determinants have not yet been reported in *L. innocua*. Similar results were obtained in this study, as none of the isolates carried out the mentioned genes.

Finally, the findings of this study indicate that all isolates were resistant to oxacillin and cefotaxime. This observation is consistent with reports from Spain and Turkey [11,15], though some variability exists, with higher susceptibility observed in Italy (37%; [12]) and complete susceptibility in one Turkish study [46]. The literature has indicated that resistance to oxacillin may be associated with the presence of the *mecA* gene. The current study confirmed that, two isolates carried this gene, while the remaining ones did not; therefore, the exact mechanism of resistance to this antibiotic cannot be determined.

The antimicrobial susceptibility profiles of *L. innocua* isolates obtained in this study generally align with findings from other European countries. Uniform susceptibility to clinically important antibiotics like ampicillin, gentamicin, erythromycin, trimethoprim/sulfamethoxazole, and vancomycin, was observed both in this study and in reports from Spain and Italy, indicating a stable sensitivity pattern across these regions. However, comparable data from countries geographically closer to Poland are not available, this gap in the literature highlights the uniqueness and importance of the current study, as it provides essential baseline information on antimicrobial resistance in *L. innocua* in this part of Europe. On the contrary, other global studies from Asia, Africa, and North America demonstrated greater variability in resistance profiles, particularly tetracycline. These differences may be attributed to regional antibiotic usage practices, environmental factors, and strain differences. Importantly, the antibiotics commonly used in the treatment of listeriosis showed high phenotypic susceptibility among the *L. innocua* isolates, indicating that these agents remain effective against this species based on the obtained laboratory results. However, the fact that bacteria can develop resistance mechanisms or acquire resistance through horizontal gene transfer from other bacterial species has contributed to the global rise in antimicrobial resistance and may significantly complicate efforts to control bacterial infections. In addition to the presence of specific resistance genes, other mechanisms may also contribute to antimicrobial resistance, such as alterations in cell membrane permeability, modification of antibiotic target sites, or enzymatic degradation of the antimicrobial agent. Therefore, to fully elucidate the mechanisms underlying resistance, it is essential to investigate not only genetic determinants, but also potential physiological and biochemical processes involved in resistance expression [1]. The presence of multidrug-resistant strains and regional differences in resistance patterns highlights the need for ongoing monitoring of antimicrobial usage. In this study, resistance genes associated with clindamycin (*lnuA*, *lnuB*), linezolid (*cfr*, *optrA*, *poxtA*), and oxacillin (*mecA*) were analyzed, as these antibiotics showed the most variable and interesting phenotypic results. Cefotaxime was not included because *Listeria* spp. are known to exhibit natural resistance to this antibiotic. Although only a limited number of resistance genes were examined, this represents a limitation of the present study. Future investigations should include a broader range of resistance genes, covering additional antibiotic classes and alternative variants of the genes already tested, to provide a more comprehensive understanding of resistance mechanisms in *L. innocua*.

6. Conclusions

The findings of this study provide valuable insights into the antimicrobial resistance patterns of *L. innocua*, a species often overlooked in resistance surveillance compared to *L. monocytogenes*. Consistent with previous reports, our results indicate that resistance profiles in *L. innocua* are influenced by the type of food product and by geographical factors reported globally. However, due to the lack of detailed regional data, this aspect could not be specifically evaluated in the present study. This underscores the importance of localized monitoring, as resistance patterns vary considerably between isolates from different sources. The results obtained expand the knowledge of the pathogen's prevalence in the study region and highlight a significant public health concern that poses a threat to consumer health. Although data on *L. innocua* from this region are currently limited, this emphasizes the importance of the present study in providing valuable baseline information and helping to fill a critical knowledge gap. Each set of isolates contributes uniquely to the global understanding of antimicrobial resistance. Although *L. innocua* has traditionally been considered non-pathogenic, it has demonstrated the capacity to harbor clinically relevant resistance genes and may serve as a reservoir for horizontal gene transfer to other bacteria, including pathogenic *Listeria* species. The coexistence of *L. innocua* with other *Listeria* spp. in the same food sample or processing environment could further facilitate the dissemination of antimicrobial resistance between these species. The potential for resistance acquisition—even in environments where resistance is not yet apparent—emphasizes the need for proactive surveillance. Future research should focus on whole-genome sequencing (WGS) to identify novel resistance genes and assess the potential for horizontal gene transfer. While WGS offers a comprehensive view of genetic mechanisms, PCR-based screening remains a rapid, affordable, and accessible tool for detecting known resistance genes. Together, these complementary approaches can enhance the understanding of resistance dynamics in *Listeria* species. Therefore, antimicrobial resistance monitoring should not be limited to *L. monocytogenes* but should also include *L. innocua* as part of a comprehensive strategy to combat antibiotic resistance. Incorporating *L. innocua* into routine resistance surveillance programs may function as an early warning system and a critical component in preserving the efficacy of antimicrobial agents in the context of food safety and public health.

Supplementary Materials: The following supporting information can be downloaded at https://www.mdpi.com/article/10.3390/genes16121455/s1. Table S1. Sequences of primers used for detection of antibiotic resistance genes.

Author Contributions: Conceptualization, A.O.-S. and A.Z.; formal analysis, A.O.-S.; data curation, A.Z.; writing—original draft preparation, A.Z.; writing—review and editing, A.Z. and A.O.-S.; supervision, A.O.-S.; project administration, A.O.-S.; funding acquisition, A.O.-S. All authors have read and agreed to the published version of the manuscript.

Funding: The publication was financed by the Polish Minister of Science and Higher Education as part of the Strategy of the Poznan University of Life Sciences for 2024–2026 in the field of improving scientific research and development work in priority research areas.

Institutional Review Board Statement: Not applicable.

Informed Consent Statement: Not applicable.

Data Availability Statement: The original contributions presented in the study are included in the article (and Supplementary Material); further inquiries can be directed to the corresponding authors.

Acknowledgments: We sincerely thank Iwona Kawacka (Department of Food Biotechnology and Microbiology, Poznan University of Life Sciences) for establishing the collection of *L. innocua* isolates used in this study.

Conflicts of Interest: The authors declare no conflicts of interest.

References

1. Prestinaci, F.; Pezzotti, P.; Pantosti, A. Antimicrobial Resistance: A Global Multifaceted Phenomenon. *Pathogens Glob. Health* **2015**, *109*, 309–318. [CrossRef] [PubMed]
2. Aksoy, A.; Sezer, Ç.; Vatansever, L.; Gülbaz, G. Presence and Antibiotic Resistance of *Listeria monocytogenes* in Raw Milk and Dairy Products. *Kafkas Univ. Vet. Fak. Derg.* **2018**, *24*, 415–421.

3. Hanes, R.M.; Huang, Z. Investigation of Antimicrobial Resistance Genes in *Listeria monocytogenes* from 2010 through 2021. *Int. J. Environ. Res. Public Health* **2022**, *19*, 5506. [CrossRef]
4. Menon, K.V.; Sunil, B.; Latha, C. Prevalence and Antibiotic Resistance Profile of *Listeria* spp. Associated with Seafoods from Fish Catchment Areas in Kerala, India. *Vet. World* **2021**, *14*, 777–783. [CrossRef]
5. Korsak, D.; Borek, A.; Daniluk, S.; Grabowska, A.; Pappelbaum, K. Antimicrobial Susceptibilities of *Listeria monocytogenes* Strains Isolated from Food and Food Processing Environment in Poland. *Int. J. Food Microbiol.* **2012**, *158*, 203–208. [CrossRef]
6. Myers, J.; Hennessey, M.; Arnold, J.C.; McCubbin, K.D.; Lembo, T.; Mateus, A.; Kitutu, F.E.; Samanta, I.; Hutchinson, E.; Davis, A.; et al. Crossover-Use of Human Antibiotics in Livestock in Agricultural Communities: A Qualitative Cross-Country Comparison between Uganda, Tanzania and India. *Antibiotics* **2022**, *11*, 1342. [CrossRef]
7. Goh, Y.X.; Anupoju, S.M.B.; Nguyen, A.; Zhang, H.; Ponder, M.; Krometis, L.A.; Pruden, A.; Liao, J. Evidence of Horizontal Gene Transfer and Environmental Selection Impacting Antibiotic Resistance Evolution in Soil-Dwelling *Listeria*. *Nat. Commun.* **2024**, *15*, 10034. [CrossRef]
8. Maung, A.T.; Mohammadi, T.N.; Nakashima, S.; Liu, P.; Masuda, Y.; Honjoh, K.I.; Miyamoto, T. Antimicrobial Resistance Profiles of *Listeria monocytogenes* Isolated from Chicken Meat in Fukuoka, Japan. *Int. J. Food Microbiol.* **2019**, *304*, 49–57. [CrossRef]
9. Chen, B.Y.; Pyla, R.; Kim, T.J.; Silva, J.L.; Jung, Y.S. Antibiotic Resistance in *Listeria* Species Isolated from Catfish Fillets and Processing Environment. *Lett. Appl. Microbiol.* **2010**, *50*, 626–632. [CrossRef] [PubMed]
10. Li, L.; Olsen, R.H.; Ye, L.; Wang, W.; Shi, L.; Yan, H.; Meng, H. Characterization of Antimicrobial Resistance of *Listeria monocytogenes* Strains Isolated from a Pork Processing Plant and Its Respective Meat Markets in Southern China. *Foodborne Pathog. Dis.* **2016**, *13*, 262–268. [CrossRef] [PubMed]
11. Gómez, D.; Azón, E.; Marco, N.; Carramiñana, J.J.; Rota, C.; Ariño, A.; Yangüela, J. Antimicrobial Resistance of *Listeria monocytogenes* and *Listeria innocua* from Meat Products and Meat-Processing Environment. *Food Microbiol.* **2014**, *42*, 61–65. [CrossRef]
12. Pesavento, G.; Ducci, B.; Nieri, D.; Comodo, N.; Lo Nostro, A. Prevalence and Antibiotic Susceptibility of *Listeria* spp. Isolated from Raw Meat and Retail Foods. *Food Control* **2009**, *21*, 708–713. [CrossRef]
13. Arslan, S.; Özdemir, F. Prevalence and Antimicrobial Resistance of *Listeria* Species and Molecular Characterization of *Listeria monocytogenes* Isolated from Retail Ready-to-Eat Foods. *FEMS Microbiol. Lett.* **2020**, *367*, fnaa006. [CrossRef]
14. Erol, Z.; Polat, Z.; Soyuçok, A.; Yalçın, H.; Taşçı, F. Antimicrobial Resistance and Prevalence of *Listeria* Species from Raw Milk and Dairy Products in Burdur, Turkey. *Vet. Med. Sci.* **2024**, *10*, e1551. [CrossRef]
15. Cebeci, T.; Otlu, B. Prevalence, Virulence Potential, Antibiotic Resistance Profile, Heavy Metal Resistance Genes of *Listeria innocua*: A First Study in Consumed Foods for Assessment of Human Health Risk in Northern Turkey. *Environ. Sci. Pollut. Res.* **2024**, *31*, 65078–65091. [CrossRef] [PubMed]
16. Akrami-Mohajeri, F.; Derakhshan, Z.; Ferrante, M.; Hamidiyan, N.; Soleymani, M.; Conti, G.O.; Tafti, R.D. The Prevalence and Antimicrobial Resistance of *Listeria* spp. in Raw Milk and Traditional Dairy Products Delivered in Yazd, Central Iran (2016). *Food Chem. Toxicol.* **2018**, *114*, 141–144. [CrossRef] [PubMed]
17. Perrin, M.; Bemer, M.; Delamare, C. Fatal Case of *Listeria innocua* Bacteremia. *J. Clin. Microbiol.* **2003**, *41*, 5308–5309. [CrossRef]
18. Favaro, M.; Sarmati, L.; Sancesario, G.; Fontana, C. First Case of *Listeria innocua* Meningitis in a Patient on Steroids and Eternecept. *JMM Case Rep.* **2014**, *1*, e003103. [CrossRef]
19. Bilung, L.M.; Chai, L.S.; Tahar, A.S.; Ted, C.K.; Apun, K. Prevalence, Genetic Heterogeneity, and Antibiotic Resistance Profile of *Listeria* spp. and *Listeria monocytogenes* at Farm Level: A Highlight of ERIC- and BOX-PCR to Reveal Genetic Diversity. *Biomed Res. Int.* **2018**, *2018*, 3067494.
20. Rippa, A.; Bilei, S.; Peruzy, M.F.; Marrocco, M.G.; Leggeri, P.; Bossù, T.; Murru, N. Antimicrobial Resistance of *Listeria monocytogenes* Strains Isolated in Food and Food-Processing Environments in Italy. *Antibiotics* **2024**, *13*, 525. [CrossRef]
21. Dong, Z.; Sun, Y.; Cao, Q.; Liu, H.; Liu, Y.; Cao, Q.; Wei, H.; Song, C.; Gou, H.; Xue, H. Prevalence and Biological Characteristics of *Listeria* Species Isolated from Livestock and Poultry Meat in Gansu Province, China. *Pol. J. Microbiol.* **2023**, *72*, 11–20. [CrossRef]
22. EFSA (European Food Safety Authority). The European Union One Health 2021 Zoonoses Report. *EFSA J.* **2022**, *20*, e07666. [CrossRef]
23. Li, Q.; Sherwood, J.S.; Logue, C.M. Antimicrobial Resistance of *Listeria* spp. Recovered from Processed Bison. *Lett. Appl. Microbiol.* **2007**, *44*, 86–91. [CrossRef]
24. Sołtysiuk, M.; Wiszniewska, A.; Wojtacka, J.; Wysok, B. Prevalence and Antimicrobial Susceptibility of *Listeria* spp. Isolated from Bulk Raw Milk in North-Eastern Poland. *Medycyna Wet.* **2022**, *78*, 6628. [CrossRef]
25. Manqele, A.; Adesiyun, A.; Mafuna, T.; Pierneef, R.; Moerane, R.; Gcebe, N. Virulence Potential and Antimicrobial Resistance of *Listeria monocytogenes* Isolates Obtained from Beef and Beef-Based Products Deciphered Using Whole-Genome Sequencing. *Microorganisms* **2024**, *12*, 1166. [CrossRef] [PubMed]
26. Rahimi, E.; Yazdi, F.; Farzinezhadizadeh, H. Prevalence and Antimicrobial Resistance of *Listeria* Species Isolated from Different Types of Raw Meat in Iran. *J. Food Prot.* **2012**, *75*, 2223–2227. [CrossRef]

27. Chin, P.S.; Ang, G.Y.; Yu, C.Y.; Tan, E.L.; Tee, K.K.; Yin, W.F.; Chan, K.G.; Tan, G.Y.A. Prevalence, Antimicrobial Resistance, and Genetic Diversity of *Listeria* spp. Isolated from Raw Chicken Meat and Chicken-Related Products in Malaysia. *J. Food Prot.* **2018**, *81*, 284–289. [CrossRef]
28. Gradovska, S.; Šteingolde, Ž.; Kibilds, J.; Meistere, I.; Avsejenko, J.; Streikiša, M.; Alksne, L.; Terentjeva, M.; Berziņš, A. Genetic Diversity and Known Virulence Genes in *Listeria innocua* Strains Isolated from Cattle Abortions and Farm Environment. *Vet. Anim. Sci.* **2022**, *19*, 100276. [CrossRef]
29. Clayton, E.M.; Daly, K.M.; Guinane, C.M.; Hill, C.; Cotter, P.D.; Ross, P.R. Atypical *Listeria innocua* Strains Possess an Intact LIPI-3. *BMC Microbiol.* **2014**, *14*, 58. [CrossRef]
30. Kovacevic, J.; Sagert, J.; Wozniak, A.; Gilmour, M.W.; Allen, K.J. Antimicrobial Resistance and Co-Selection Phenomenon in *Listeria* spp. Recovered from Food and Food Production Environments. *Food Microbiol.* **2013**, *34*, 319–327. [CrossRef] [PubMed]
31. Moura, A.; Leclercq, A.; Vales, G.; Tessaud-Rita, N.; Bracq-Dieye, H.; Thouvenot, P.; Madec, Y.; Charlier, C.; Lecuit, M. Phenotypic and Genotypic Antimicrobial Resistance of *Listeria monocytogenes*: An Observational Study in France. *Lancet Reg. Health Eur.* **2023**, *37*, 10080. [CrossRef]
32. Caruso, M.; Fraccalvieri, R.; Pasquali, F.; Santagada, G.; Latorre, L.M.; Difato, L.M.; Miccolupo, A.; Normanno, G.; Parisi, A. Antimicrobial Susceptibility and Multilocus Sequence Typing of *Listeria monocytogenes* Isolated Over 11 Years from Food, Humans, and the Environment in Italy. *Foodborne Pathog. Dis.* **2020**, *17*, 284–294. [CrossRef]
33. da Rocha, L.S.; Gunathilaka, G.U.; Zhang, Y. Antimicrobial-Resistant *Listeria* Species from Retail Meat in Metro Detroit. *J. Food Prot.* **2012**, *75*, 2136–2141. [CrossRef] [PubMed]
34. Baquero, F.; Lanza, V.F.; Duval, M.; Coque, T.M. Ecogenetics of antibiotic resistance in *Listeria monocytogenes*. *Mol. Microbiol.* **2020**, *113*, 570–579. [CrossRef]
35. Duma, M.N.; Ciupescu, L.M.; Dan, S.D.; Crișan-Reget, O.L.; Tabaran, A. Virulence and Antimicrobial Resistance of *Listeria monocytogenes* Isolated from Ready-to-Eat Food Products in Romania. *Microorganisms* **2024**, *12*, 954. [CrossRef]
36. Escolar, C.; Gómez, D.; Rota García, M.C.; Conchello, P.; Herrera, A. Antimicrobial Resistance Profiles of *Listeria monocytogenes* and *Listeria innocua* Isolated from Ready-to-Eat Products of Animal Origin in Spain. *Foodborne Pathog. Dis.* **2017**, *14*, 357–363. [CrossRef]
37. Walsh, D.; Duffy, G.; Sheridan, J.J.; Blair, I.S.; McDowell, D.A. Antibiotic Resistance among *Listeria*, Including *Listeria monocytogenes*, in Retail Foods. *J. Appl. Microbiol.* **2001**, *90*, 517–522. [CrossRef] [PubMed]
38. Paillard, D.; Dubois, V.; Duran, R.; Nathier, F.; Guittet, C.; Caumette, P.; Quentin, C. Rapid Identification of Listeria Species by Using Restriction Fragment Length Polymorphism of PCR-Amplified 23S rRNA Gene Fragments. *Appl. Environ. Microbiol.* **2003**, *69*, 6386–6392. [CrossRef] [PubMed]
39. Li, F.; Ye, Q.; Chen, M.; Zhang, J.; Xue, L.; Wang, J.; Wu, S.; Zeng, H.; Gu, Q.; Zhang, Y.; et al. Multiplex PCR for the Identification of Pathogenic Listeria in Flammulina velutipes Plant Based on Novel Specific Targets Revealed by Pan-Genome Analysis. *Front. Microbiol.* **2021**, *11*, 634255. [CrossRef]
40. Kawacka, I.; Olejnik-Schmidt, A. Genoserotyping of *Listeria monocytogenes* Strains Originating from Meat Products and Meat Processing Environments. *Żywność Nauka Technol. Jak.* **2022**, *29*, 34–44. [CrossRef]
41. Clinical and Laboratory Standards Institute (CLSI). *M100 Performance Standards for Antimicrobial Susceptibility Testing*, 30th ed.; CLSI: Wayne, PA, USA, 2020.
42. Kawacka, I.; Pietrzak, B.; Schmidt, M.; Olejnik-Schmidt, A. *Listeria monocytogenes* Isolates from Meat Products and Processing Environment in Poland Are Sensitive to Commonly Used Antibiotics, with Rare Cases of Reduced Sensitivity to Ciprofloxacin. *Life* **2023**, *13*, 821. [CrossRef] [PubMed]
43. Wiśniewski, P.; Zakrzewski, A.J.; Zadernowska, A.; Chajęcka-Wierzchowska, W. Antimicrobial Resistance and Virulence Characterization of *Listeria monocytogenes* Strains Isolated from Food and Food Processing Environments. *Pathogens* **2022**, *11*, 1099. [CrossRef]
44. Ferri, G.; Lauteri, C.; Festino, A.R.; Vergara, A. ARGs Detection in *Listeria monocytogenes* Strains Isolated from the Atlantic Salmon (*Salmo salar*) Food Industry: A Retrospective Study. *Microorganisms* **2023**, *11*, 1509. [CrossRef]
45. Gana, J.; Gcebe, N.; Moerane, R.; Ngoshe, Y.; Tshuma, T.; Moabelo, K.; Adesiyun, A. Antimicrobial Resistance Profiles of *Listeria* Species Recovered from Retail Outlets in Gauteng Province, South Africa. *J. Food Prot.* **2024**, *87*, 100322. [CrossRef] [PubMed]

46. Babacan, O. Determination of the presence and antibiotic resistance of *Listeria* species and aerobic mesophilic bacteria count of cow milks. *Vet. Hekim. Dern. Derg.* **2021**, *92*, 16–23. [CrossRef]
47. Bertrand, S.; Huys, G.; Yde, M.; D'Haene, K.; Tardy, F.; Vrints, M.; Swings, J.; Collard, J.M. Detection and Characterization of *tet*(*M*) in Tetracycline-Resistant *Listeria* Strains from Human and Food-Processing Origins in Belgium and France. *J. Med. Microbiol.* **2005**, *54*, 1151–1156. [CrossRef] [PubMed]

genes

MDPI

Article

Preliminary Assessment of *Arnica montana* L. Extract: Antimicrobial Activity Against *Acinetobacter baumannii* and Biofilm-Related Gene Expression Profiling

Sylwia Andrzejczuk [1,*], Magdalena Sozoniuk [2,*] and Danuta Sugier [3]

1 Department of Pharmaceutical Microbiology, Medical University of Lublin, Chodźki 1, 20-093 Lublin, Poland
2 Institute of Plant Genetics, Breeding and Biotechnology, University of Life Sciences in Lublin, Akademicka 15 St., 20-950 Lublin, Poland
3 Department of Industrial and Medicinal Plants, University of Life Sciences in Lublin, Akademicka 15 St., 20-950 Lublin, Poland; danuta.sugier@up.edu.pl
* Correspondence: sylwia.andrzejczuk@umlub.edu.pl (S.A.); magdalena.sozoniuk@up.edu.pl (M.S.)

Abstract

Background/Objectives: *Arnica montana* L. is widely recognized for its diverse biological activities, including antimicrobial effects. This study aimed to evaluate the antimicrobial and antibiofilm activity of *A. montana* L. extracts against *Acinetobacter baumannii*, a pathogen of urgent public health concern due to its increasing antibiotic resistance and capacity for biofilm formation. **Methods**: The antimicrobial activity of ethanolic (EtE) and aqueous (AqE) extracts of *A. montana* flowers was evaluated via the broth microdilution method. The minimal inhibitory concentration (MIC) and minimal bactericidal concentration (MBC), and the MBC/MIC ratio were used. The effects of EtE on *A. baumannii* biofilm formation were assessed via a crystal violet assay. Additionally, transcriptional profiling of biofilm-associated genes following exposure to sub-MIC levels of the extract was conducted via RT-qPCR. **Results**: The anti-*Acinetobacter* activity of EtE was demonstrated (MIC = 234.4 and 468.75 μg/mL for *A. baumannii* ATCC BAA-3252 and ATCC 19606, respectively). The EtE exhibited bactericidal activity against both strains, whereas the AqE showed no activity. Additionally, EtE inhibited biofilm formation and significantly downregulated the expression of key biofilm-associated genes, including those of the *csu* operon and *ompA*. **Conclusions**: *Arnica montana* EtE demonstrated antimicrobial and antibiofilm activities against *A. baumannii* and inhibited biofilm development by suppressing the transcription of genes involved in pilus assembly and surface adherence, highlighting their essential role in biofilm formation.

Keywords: *Arnica montana* L.; antibacterial activity; sub-inhibitory MIC; biofilm; gene expression

1. Introduction

The present-day global problem of drug resistance among bacteria and fungi is escalating in severity. The use of plant-based products and natural extracts has emerged as a promising alternative to pharmacological interventions in the fight against multidrug-resistant (MDR) pathogens. In contemporary practice, their potential is most frequently utilized as antibiotic "enhancers" and/or biofilm inhibitors, with less frequent use as standalone drugs. Many secondary metabolites produced by plants provide chemical defenses against a wide range of infectious agents, including pathogenic microorganisms.

The rich diversity of these phytochemicals offers valuable resources for developing new antimicrobial agents to combat MDR-ESKAPE pathogens such as *Enterococcus* spp., *Staphylococcus aureus, Klebsiella pneumoniae, Acinetobacter baumannii, Pseudomonas aeruginosa*, and *Enterobacter* spp. [1]. A number of plant extracts and plant-derived compounds have been shown to possess antimicrobial properties. For instance, myristic acid (a fairly common form of saturated fatty acid) has been found to be effective against *S. aureus* [2]; quercetin-7-O-methyl ether and isoorientin, isolated from *Rhynchosia beddomei* Baker, have shown antibacterial properties against *P. aeruginosa* and *S. aureus* [3]; and Song et al. [4] reported that isopentenylated flavonoids (α-mangostin and isobavachalkone) were effective against antibiotic-resistant *E. faecium* and *S. aureus*. In their seminal study, Idris et al. [1] comprehensively summarized the extant literature on the subject, meticulously collecting data from numerous publications detailing the activity of various plant extracts against ESKAPE pathogens. The data presented by the authors revealed a paucity of phytochemicals with high potential for inhibiting the growth or killing of *A. baumannii* cells. However, an exception to this is observed with the ethanol extract from *Lavandula stoechas* flowers, for which MIC against planktonic bacteria of 35.9 μg/mL has been reported [5]. To date, several antimicrobial mechanisms of action of plant components have been described. Plant bioactive compounds have been demonstrated to function as inhibitors of efflux pumps (for instance, baikalin, catechins, and piperine) [6]. Natural compounds can act by modifying target proteins of selected pathogens, as exemplified by sulfur compounds from garlic (allicin/ajoene) and isothiocyanates (e.g., benzyl-ITC) [7].

As posited by numerous scientists, a further mechanism involves membrane disruption and osmotic stress (in the context of essential oils), with thymol, carvacrol, eugenol, cinnamaldehyde, and terpinen-4-ol being notable examples [8–11]. These substances have been demonstrated to compromise the integrity of bacterial and fungal cell membranes, thereby diminishing their capacity to establish biofilms and impeding the process of quorum sensing. In numerous models, these compounds are combined in vitro with antibiotics (e.g., amikacin), which results in a substantial reduction in the MIC against MDR bacteria of the genera *Klebsiella, Acinetobacter*, and *Staphylococcus* [10]. Compounds of natural origin have also been shown to possess antibacterial and/or antiadhesive properties, thereby impeding the adhesion of microorganisms to both biotic and abiotic surfaces [12]. Essential oils (e.g., cinnamon, clove, and lemongrass) and polyphenolic compounds (e.g., quercetin, kaempferol-3-rutinoside, and galangin) [12–14] have been shown to reduce various bacterial cell processes. For example, the processes of biofilm formation and maturation of *Vibrio parahaemolyticus,* which carry virulence-associated genes, are effectively inhibited at sub-MICs by lemongrass and cumin essential oils [13].

Arnica montana is a medicinal plant widely recognized for its antimicrobial, anti-inflammatory, antitumor, and analgesic properties. The multiple pharmacological applications of *A. montana* are attributed to its diverse bioactive compounds, including sesquiterpene lactones (e.g., helenalin, 11α,13-dihydrohelenalin), sesquiterpenes (e.g., caryophyllene, germacrene D), and monoterpenes (e.g., thymol, α-pinene), as well as eugenol, flavonoid glycosides, fatty acids, and oligofructozides, all of which collectively contribute to its broad therapeutic potential [15–17]. According to Thakur [16] and other researchers [18], arnica supports various functions in the body, including improving blood circulation, reducing muscle pain and swelling, and accelerating healing after surgery or injury. Arnica extracts have also demonstrated promising antibacterial activity against several pathogenic microorganisms [19–22], making them potential candidates for developing alternative prophylactic and novel antimicrobial therapies. However, the extant literature contains a paucity of studies that have examined the effects of *A. montana* extracts on *Acinetobacter* spp.

Acinetobacter baumannii is a Gram-negative, aerobic, opportunistic pathogen that poses a significant threat to patients with compromised immune systems (who experience prolonged hospitalization of more than 90 days). It is also a primary cause of various infections, including pneumonia (ventilator-associated pneumonia, community-acquired pneumonia), urinary tract infections, bloodstream infections (e.g., bacteremia, sepsis), nosocomial postneurosurgical meningitis, and wound and soft tissue infections [23,24]. As stated by many authors [23,25], this bacterium naturally inhabits water and soil environments, as well as food or medical devices. The pathogenesis is strongly attributed to a variety of virulence factors, including adhesins, lipooligosaccharides, outer membrane proteins, iron-chelating systems, protein secretion systems, and enzymes (phospholipases, proteases), among others [24,25]. One of the most significant factors contributing to the pathogenicity of *A. baumannii* is the ability of the bacterium to adhere to, form biofilms on, and colonize both biotic and abiotic surfaces, with a prevalence ranging from 32 to over 90% and simultaneously carrying multiple biofilm-related genes [24–26].

The primary threat posed by the presence of *A. baumannii* in the environment and human surroundings is the continuous development of antibiotic resistance, which is particularly alarming in the case of frequently used carbapenems (carbapenem-resistant *A. baumannii*, CRAB) and can lead to infections that are difficult to treat [24]. Consequently, *A. baumannii* is designated the fourth most critical pathogen by the World Health Organization in the critical group category. A significant constraint in efforts to combat these pathogens is the paucity of promising pharmaceutical agents currently under development. Furthermore, infections caused by critical pathogens can also be extremely difficult to prevent and are highly contagious [23,27]. Several comprehensive reviews have provided detailed insights into the mechanisms and stages of biofilm development in *A. baumannii* [25,26,28]. The inhibition of biofilm formation represents a promising therapeutic approach for combating *A. baumannii* infections, as biofilms significantly contribute to pathogen resistance to antibiotics and persistence in clinical environments. Targeting biofilm development can disrupt bacterial colonization and increase the efficacy of antimicrobial treatments.

This preliminary study evaluated the antibacterial and antibiofilm effects of *A. montana* flower extract (both aqueous and ethanolic) against two reference strains of *A. baumannii*, either planktonic or biofilm-forming cells, with a particular focus on transcriptional analysis of selected biofilm-associated genes to elucidate the molecular responses induced by the treatment and to provide insights into its antibiofilm activity.

2. Materials and Methods

2.1. Plant Material and Extraction Procedure

The plant material, consisting of *Arnica montana* L. flower heads in the full flowering phase, was collected from the experimental field of the University of Life Sciences in Lublin, eastern Poland (51°31′39.5″ N, 22°45′6.7″ E). After harvesting, the flower heads were air-dried at 40 °C, finely ground using a laboratory mill, and subsequently sieved (passing through a nylon sieve with a mesh size of 1 mm) to obtain homogeneous powdered plant material for further analyses. Five-gram portions of raw material were weighed and transferred into 250 mL Erlenmeyer flasks. Each sample was then supplemented with 50 mL of the appropriate solvent: either distilled water or 70% (v/v) ethanol. The flasks were tightly sealed and placed in a shaking water bath at 40 °C for 5 h. Upon completion of the extraction, the solutions were filtered through a plastic Büchner funnel fitted with cellulose filter paper. The resulting extracts were subjected to further analyses, while the plant material remaining after extraction was discarded.

2.2. Chemical Analyses

The content of sesquiterpene lactones was quantified via the chromatographic method described in the Polish Pharmacopoeia IX [29]. The results are expressed as a percentage of the total sesquiterpene lactone content, which was calculated as dihydrohelenalin tiglinate equivalents. The flavonoid content, expressed as quercetin equivalents, was determined via the spectrophotometric method specified in the Polish Pharmacopoeia VIII [30]. Each marking was performed in four replicates. Representative data are shown.

2.3. Microorganisms

Two reference strains, *Acinetobacter baumannii* ATCC 19606 and *A. baumannii* ATCC BAA-3252, obtained from the American Type Culture Collection (ATCC), were used. Bacteria were kept frozen at −70 ± 2 °C in brain heart infusion broth (BHI, Oxoid, Milan, Italy). They were cultured for 24 h at 35 ± 2 °C in Tryptic Soy Agar (TSA, Oxoid, UK) and then used to test the antimicrobial potential of the extracts.

2.4. Antimicrobial Activity Test

The objective of this study was to determine the antibacterial potential of ethanolic and aqueous extracts of *Arnica montana* L. against *Acinetobacter* spp. bacteria. The broth double microdilution method was utilized, as stipulated by the European Committee on Antimicrobial Susceptibility Testing (EUCAST), as the reference method for determining the antimicrobial potential of natural products and newly synthesized compounds, as well as compounds of an antibiotic and chemotherapeutic nature [31]. The quantitative activity of the tested products was measured with the following parameters: the minimum inhibitory concentration (MIC), defined as the lowest concentration of product tested that inhibits the in vitro visible growth of microorganisms expressed in μg/mL; the minimum bactericidal concentration (MBC), defined as the lowest concentration of product tested that results in the absence of visible growth of microorganisms, expressed in μg/mL. Furthermore, the MBC:MIC ratios were also calculated to evaluate the nature of the formulations' activity. The product was classified as bactericidal if the MBC:MIC values were ≤4 and as bacteriostatic if the MBC:MIC values were in the range of 4–32 [21,32].

The method involved the preparation of microbial suspensions at a standardized density (0.5 on the McFarland, McF) standard scale, corresponding to 1.5×10^8 CFU/mL and an optical density of $OD_{550} = 0.125$. The assay was performed within the concentration range of 7.32–30,000 μg/mL on polystyrene 96-well microplates (Medlab, Raszyn, Poland), with 200 μL of the initial test product concentration and 100 μL of sterile microbiological Mueller-Hinton broth medium (MHB, Biomaxima, Lublin, Poland) added to each well, as previously described [33]. Concurrently, the following controls were conducted: (a) a medium control consisting of 100 μL of sterile MHB medium devoid of microorganisms, (b) a microbial viability control comprising 100 μL of sterile medium with 2 μL of microbial suspension lacking extracts tested, (c) the *A. montana* extracts control involving serial twofold dilutions without added microorganisms (conducted in the same concentration ranges as the actual experiment; the OD_{600}_values read in these wells were used as a reference point for determining the MIC), and (d) amoxicillin (5 mg/mL, Glentham Life Sciences, Corsham, UK), as well as 70% ethanolic alcohol as a control of the solvent (to strictly adhere to the experimental protocol, the control conditions accurately reflected the procedure used for the bacterial strains under investigation, except for the addition of pure solvent in the same volumes as the extract. In this way, a series of double ethanol dilutions ranging from 0.03 to 70.0% was obtained, corresponding to dilutions of the tested extracts ranging from 7.32 to 30,000 μg/mL). The microtiter plates were then subjected to an incubation period of 18 ± 2 h at 35 ± 2 °C. A spectrophotometric assay was subsequently conducted on each well of the plate, with measurements recorded at 600 nm via an ELx800 plate reader (BioTek Instruments, Winooski, VT, USA). The MIC values were then determined in those wells where the extracts studied completely inhibited the growth of the reference microbial strains. The MBC values were determined by plating 5 μL of the contents of each well of the microtiter plates showing growth inhibition on the Mueller-Hinton agar (MHA, Biomaxima, Lublin, Poland) plates. The area on the plate where 99.9% inhibition of microbial growth (planktonic cells) was observed was identified as the lethal concentration. The experiment was performed in at least triplicate, and representative data are shown.

2.5. Anti-Biofilm Assay

The biofilm-forming ability of the tested *Acinetobacter* spp. strains and the antibiofilm potential of the ethanolic extract of *Arnica montana* L. were measured quantitatively via a modified 0.1% crystal violet (CV, Oxoid, UK) method based on the OD_{570} [34]. The biofilm assay was performed on Nunc™ MicroWell™ 96-well, Nunclon Delta-Treated, flat-bottom microplates (Thermo Fisher Scientific, Waltham, MA, USA). Bacterial suspensions were prepared in Tryptic Soy Agar (TSA, Oxoid, Basinstoke, UK) at a 0.5 McFarland density. A quantity of 200 μL of each arnica extract was added to the first wells in a row and then diluted as previously described to achieve the same dual dilutions. In the final stage, 2 μL of bacterial suspension (0.5 McF) was added to each well and then incubated for 24 h at 37 °C. The medium above the culture was then decanted. The microplates were meticulously washed multiple times with distilled water to remove nonadherent or loosely adherent cells. Thereafter, they were dried in an inverted position and stained with 200 μL of 0.1% CV. The plates were left to stand for 15 min, after which the dye was removed, and each well was rinsed several times with sterile distilled water to ensure the removal of non-adherent forms. Finally, the microtiter plates were dried, filled with 200 μL of 96% ethanol (POCH, Lublin, Poland), and the optical density (OD_{570}) was measured spectrophotometrically at a wavelength of 570 nm via an ELx 800 microplate reader (BioTek Instruments, Winooski, VT, USA) after 15 min [35]. Based on absorbance values, both *Acinetobacter* spp. strains were classified as biofilm-forming producers as described previously [36]. Analogue types of control were used: MHB control medium without the tested extract; control medium with extract added, but no bacterial inoculum added, and a series of dilutions (negative control). The OD_{570} values read from these wells were used as reference points to determine the minimal biofilm inhibitory concentration (MBIC) values. The viability of the tested strains, representing 100% growth without the addition of the tested extract, was checked independently. The experiment was performed in triplicate, and representative data are shown.

Based on the obtained data, the MBIC of the tested extract that inhibits biofilm formation by *A. baumannii* was determined. This value was determined in wells containing the lowest concentration of the extract (the highest dilution), in which a significant reduction or inhibition of biofilm formation compared to the untreated control was observed. The percentage of biofilm inhibition can be calculated using the formula $[(OD_{control} - OD_{sample})/OD_{control}] \times 100$ [37], where $OD_{control}$ was the absorbance of the positive control wells (untreated bacteria), and OD_{sample} was the absorbance of the wells treated with the extract. Biofilm inhibition was rated on a scale of 0–100%; values <0% indicated biofilm growth enhancement, while values between 0 and 50% showed weak anti-biofilm activity; values >50% represented good biofilm inhibition. All image analysis was performed using the ImageJ 1.54g software (US National Institutes of Health, Bethesda, MA, USA) to calculate the measured area occupied by biofilm (MAOB).

2.6. RNA Extraction and cDNA Synthesis

To assess transcriptional changes in genes associated with biofilm formation in *A. baumannii*, treatments included subinhibitory (sublethal) MIC concentrations: ½ MIC and ¼ MIC of *A. montana* extract, ethanol at equivalent concentrations (½ MIC and ¼ MIC), and an untreated control. The RNA extraction was performed in accordance with the manufacturer's instructions using TRIzol reagent (Invitrogen, Carlsbad, CA, USA). Immediately following isolation, the RiboLock RNase Inhibitor 40 U/µL (Thermo Scientific, Waltham, MA, USA) was added to the samples. The extracted RNA was then assessed spectrophotometrically with a NanoDrop2000 (Thermo Scientific, Waltham, MA, USA) and electrophoretically on a 1.5% agarose gel stained with ethidium bromide. The potential genomic DNA contamination was removed using RNase-free DNase I (EURx, Gdańsk, Poland) following the manufacturer's recommendations. The reverse transcription reactions were carried out using the NG dART RT Kit (EURx, Gdańsk, Poland) on 1.5 µg of RNA. The resulting cDNA was diluted and used as the template for subsequent RT-qPCR reactions.

2.7. RT-qPCR

Following a comprehensive review of the extant literature, six genes of interest (*bap*, *csuA*, *csuB*, *csuC*, *csuD*, and *ompA*) were selected for expression analysis via RT-qPCR. Both bacterial strains were tested for the presence of genes via classical PCR, visualized on a 1.5% agarose gel during electrophoresis, photographed using a Quantum ST4 system (Vilber Lourmat, France), documented, and archived via Quantum ST5 Xpress v 16.08g computer software (Vilber Lourmat, Marne-la-Vallée, France).

The primer sequences for the GOIs were designed using the PrimerBLAST tool [38]. The primers used for reference genes (*rpoB* and *rpoD*) and the *bap* gene were obtained from the literature [39,40]. The primer specifications are outlined in Table 1. The amplification specificity was then assessed by melting curve analysis. All primer pairs used in this study exhibited specific amplification, as confirmed by the presence of single peaks on the generated dissociation curves (Figure A1). The amplification efficiency was determined via analysis of standard curves prepared from serial dilutions of pooled cDNA. The RT-qPCR reactions were performed using SG/ROX on*Taq* qPCR Master Mix (2×) (EURx, Gdańsk, Poland) on 50 ng of cDNA with optimized primer concentrations (Table 1). The thermal cycling conditions were as recommended by the supplier (15 min at 95 °C for initial denaturation, followed by 40 cycles of 15 s at 94 °C, 30 s at 60 °C, and 30 s at 72 °C, followed by a dissociation curve step with continuous data collection from 60 °C to 95 °C). All reactions were carried out in three technical replicates along with no-template controls (NTCs) on a QuantStudio™ 6 Pro Real-Time PCR System (Applied Biosystems™, Thermo Fisher Scientific, Waltham, MA, USA). The relative expression levels of the target genes ($2^{-\Delta\Delta Ct}$ method) were calculated using the relative quantification module in Thermo Fisher Cloud (Thermo Fisher Scientific, Waltham, MA, USA).

Table 1. Primer sequences and characteristics.

Gene Symbol	Accession	Primer Sequence 5′ → 3′	Primer Concentration	Amplicon Size	R^2	Efficiency
bap	KT717635	F: GAGGGAACTTCTGCAAAACTTTC R: CAGACGTATGACTGCATTGGT	400 nM	108 bp	0.999	97.10%
csuA	AY241696.1	F: TTTTGGTGAAGCTACCACAGC R: ACCAGCACACTCGATCTGAAA	400 nM	91 bp	0.998	99.72%
csuB	AY241696.1	F: CACCTACAAACCGTCTGGCA R: CGGCCGTATGGGGTTCATTA	400 nM	78 bp	0.998	106.23%

Table 1. *Cont.*

Gene Symbol	Accession	Primer Sequence 5′ → 3′	Primer Concentration	Amplicon Size	R^2	Efficiency
csuC	AY241696.1	F: CCGACTTTCTGATGGCAACG R: AGCTCAGACTGGCCTTGTTG	100 nM	198 bp	0.996	90.97%
csuD	AY241696.1	F: TGAAGGTAACGCCGAGTCTG R: TCGAACAGAACTTCCCCACG	400 nM	145 bp	0.996	108.92%
ompA	OL347635.1	F: CGACAGCAAAATCAAGCCGT R: CAGAAAGCACCAACACCAGC	100 nM	132 bp	0.999	92.41%
rpoD	CP045110.1	F: GTTGCTGAAGAAGAAGCTGCTG R: ACTGTACCCATTTCACGCATGTA	100 nM	101 bp	0.997	85.85%
rpoB	CP045110.1	F: ACGCCTAAAGGTGAAACTCAGTTAA R: GTACCAGATGGAACACGTAAAGATG	100 nM	110 bp	0.995	96.21%

2.8. Statistical Analysis

Statistical analyses were performed to assess the significance of differences between the experimental groups. Homogeneity of variances was tested using Levene's test prior to parametric analyses. For parametric data with equal variances, one-way analysis of variance (ANOVA) followed by Tukey's post hoc test was used for multiple comparisons. For nonparametric data, the Mann–Whitney U test was applied to compare differences between groups. The results were considered statistically significant at a significance level of $p < 0.05$. The experimental data are shown as the means ± standard deviations (SDs). The statistical tests were performed via Statistica 13.1 software (StatSoft, Inc., Tulsa, OK, USA), and the graphs were created using GraphPad Prism 8 version 8.4.2 (GraphPad Software, LLC, San Diego, CA, USA).

3. Results

3.1. Chemical Analysis of Extracts

The ethanolic extract of *A. montana* flowers contained approximately 2.5-fold higher concentrations of sesquiterpene lactones (STLs) compared to the aqueous extract (1.645% versus 0.670%, respectively). Furthermore, the flavonoid content was also greater in the ethanolic extract, reaching 0.725%, while in the aqueous extract, it amounted to 0.680%. Figure A1 in Appendix A presents an example HPLC chromatogram of STLs in aquatic (A) and ethanolic (B) extracts.

3.2. Antibacterial Activity Assay Against A. baumannii Planktonic Cells

The findings outlined in Table 2 were derived from experimental trials employing the broth double dilution method. The MIC values for the planktonic cells of the reference strains of *A. baumannii* in the presence of the ethanolic extract ranged from 234.4 to 468.75 µg/mL, while the MBC values were the same for both strains at 937.5 µg/mL (Table 2). The alcoholic extract from arnica flowers demonstrated bactericidal properties against both *Acinetobacter* spp. reference strains, with MBC/MIC ratios of 2 and 4, for *A. baumannii* ATCC 19606 and ATCC BAA-3252, respectively. Conversely, the aquatic extract did not present any potential activity against the strains studied. The MIC values exceeded the concentration range used.

Table 2. Results of the quantitative antibacterial assay against *Acinetobacter baumannii* planktonic cells.

Concentrations [μg/mL]	***A. baumannii* ATCC 19606**			***A. baumannii* ATCC BAA-3252**		
Type of Solution	**AqE**	**EtE**	**EtOH**	**AqE**	**EtE**	**EtOH**
MIC	>30,000	468.75	937.5	>30,000	234.4	468.75
MBC	n.d.	937.5	937.5	n.d.	937.5	1875
MBC/MIC ratio	n.d.	2	1	n.d.	4	4
type of activity (+) bactericidal (−) bacteriostatic	n.d.	(+)	(+)	n.d.	(+)	(+)

Explanatory notes: AqE—aqueous extract; EtE—ethanolic extract; EtOH—ethanol (a series of double ethanol dilutions ranging from 0.03 to 70.0%, corresponding to dilutions of the tested extracts); n.d.—not defined; the MIC values presented are the average of at least three replicates; no standard deviation was observed, and all results equaled their average.

The MIC values showed that arnica ethanol extract was active against planktonic cells of the reference strains of the species *A. baumannii*. The multiplication of both strains of *A. baumannii* increased in the same way when the extract was present (Figure 1). They were gradually inhibited, and the extent of this inhibition was dependent on the concentration of the ethanolic extract. At low concentrations ranging from 14.6 to 58.6 μg/mL, bacterial growth was comparable to that of the growth control (the strain not exposed to the extract, 100% growth). The OD_{600} values for *A. baumannii* ATCC 19606 ranged from 1.032 ± 0.040 to 0.962 ± 0.063, and for *A. baumannii* ATCC BAA-3252 ranged from 1.019 ± 0.016 to 0.943 ± 0.038 (Table A1). There was a important decrease in bacterial proliferation just before a very high concentration of the extract (between 234.4 and 468.8 μg/mL). An inhibition of planktonic cells' growth of both strains was reached at those concentrations of extract, the levels between 34.2–89.9 ± 0.007% for *A. baumannii* ATCC 19606, and between 69.6–93.1 ± 0.020% for ATCC 3252 (difference between strains was statistically significant with $p < 0.05$; Figure 1). A strong reduction in bacterial growth was observed at MIC values exceeding the ethanolic extract concentration of 468.8 μg/mL. The growth of planktonic cells was inhibited in both *Acinetobacter* spp. strains, with an estimated mean level of 88.8–92.4 ± 0.0140 for ATCC 19606 strain and 89.7–92.4 ± 0.0109% for ATCC BAA-3252 strain ($p = 0.999$). Statistically significant differences were observed at sublethal concentrations ($p < 0.0001$). Moderate inhibition of the growth of planktonic cells of these bacteria was caused by the presence of >468.8 μg/mL of the extract, with growth inhibition ranging from 87.4 ± 0.0046% to 96.6 ± 0.0136% for *A. baumannii* ATCC 19606 and from 95.4 ± 0.0005% to 96.9 ± 0.0042% for *A. baumannii* ATCC BAA-3252 (statistically significant difference was noted at every extract concentrations except 30,000 μg/mL; $p < 0.0001$).

It is important to note that the MIC values for pure ethanol were 2-fold higher than the MIC values (937.5 and 1875.0 μg/mL for *A. baumannii* ATCC 19606 and ATCC 3252, respectively). Furthermore, the growth of bacterial planktonic cells was also caused by the presence of ethanol (at concentrations corresponding to an MIC of >468.8 μg/mL of the extract). Growth inhibition ranged from 89.7 ± 0.013% to 96.9 ± 0.004% for *A. baumannii* ATCC 19606, and from 81.9 ± 0.045% to 95.3 ± 0.003% for *A. baumannii* ATCC 3252. A statistically significant difference was observed between ethanol-treated and non-treated cells of both strains ($p < 0.0001$).

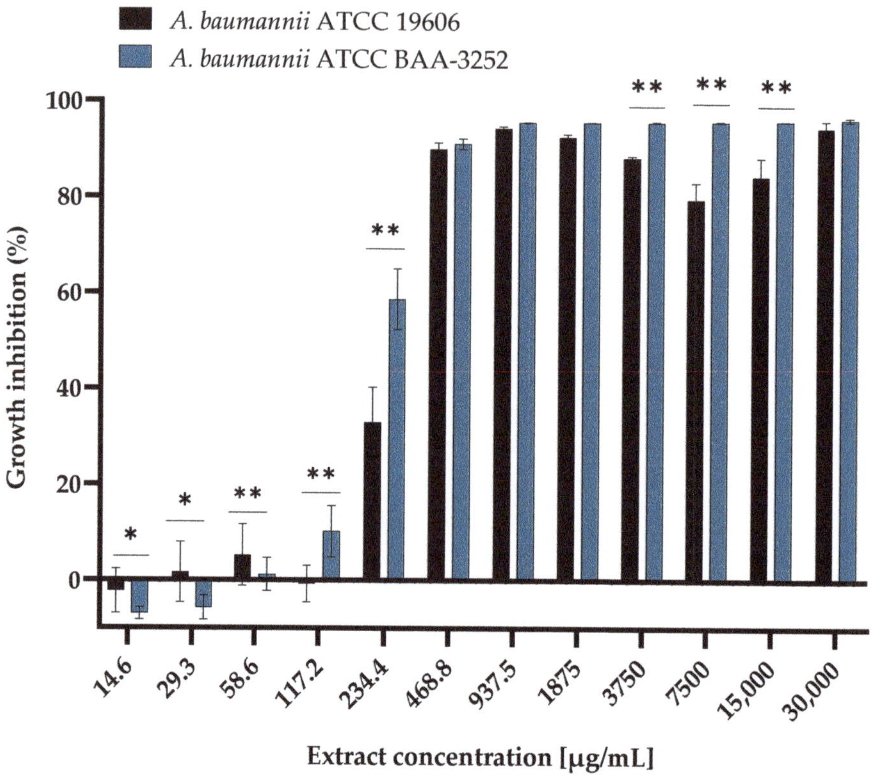

Figure 1. Reduction in planktonic cells growth of *Acinetobacter baumannii* reference strains ATCC 19606 and ATCC BAA-3252 treated with different concentrations of the *Arnica montana* L. ethanolic extract after 24 h of incubation at a temperature of 37 °C in aerobic conditions. Explanatory notes: mean $OD_{600} \pm$ SD values for both strains control without extract (100% growth of the non-treated bacterial strains) were 1.0206 ± 0.019 and 1.0295 ± 0.028 for *A. baumannii* ATCC 19606 and *A. baumannii* ATCC BAA-3252. Statistical data were analyzed using the Mann–Whitney U nonparametric test (with continuity correction). The results marked with an asterisk (*) are statistically significant at $p < 0.05$; (**) are significant at $p < 0.01$. Data were based on ten replicates for each bacterial strain.

3.3. *Effects of Arnica Ethanolic Extract on A. baumannii Biofilm Cells*

The arnica extract (initial concentration of 30,000 μg/mL) was tested on biofilm cells at twelve different concentrations (corresponding to dilutions at the antimicrobial assay), including subinhibitory concentrations of ½ and ¼ MIC, as well as 2- and 4-fold the respective planktonic MIC (2× and 4× MIC), which significantly inhibited biofilm formation across both reference *A. baumannii* ATCC 19606 and ATCC BAA-3252 strains (Figure 2). The presence of arnica ethanol extract in the culture medium had a similar effect on the biofilm formation of *A. baumannii* strains ATCC 19606 and ATCC BAA-3252.

	¼ MIC	½ MIC	MIC	2× MIC	4× MIC	non-treated cells
Conc. [μg/mL]	**117.2**	**234.4**	**468.75**	**937.5**	**1875**	**n.d.**
$OD_{570} \pm SD$	0.636± 0.072	0.341 ± 0.056	0.228 ± 0.0336	0.156 ± 0.012	0.358 ± 0.054	1.142 ± 0.223
A. baumannii ATCC 19606	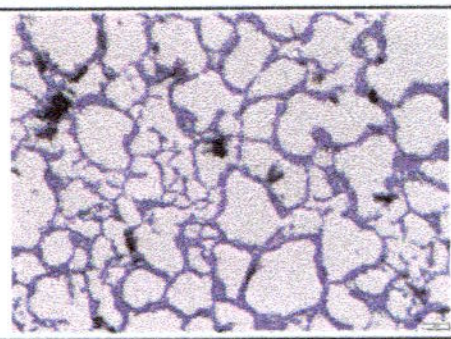	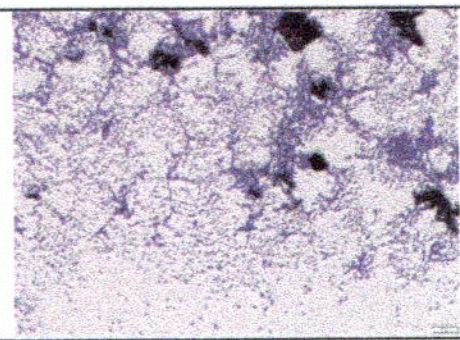	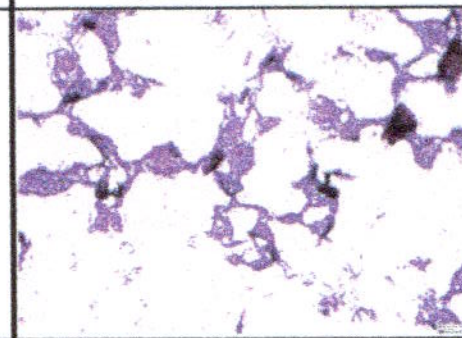	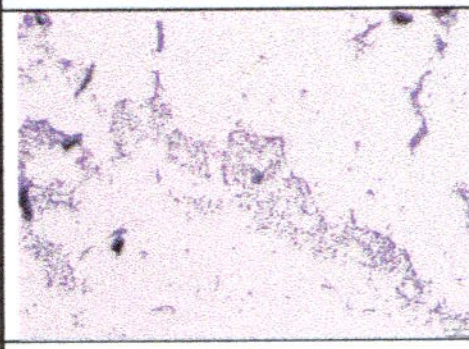	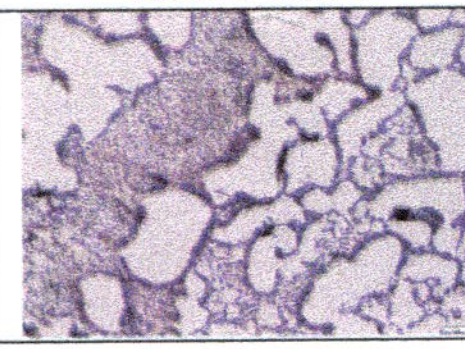	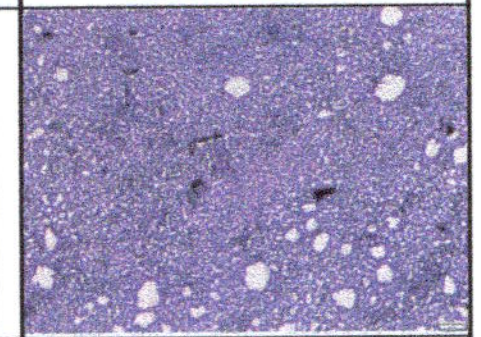
MAOB (%)	**25.86**	**22.61**	**19.55**	**7.03**	**26.79**	**65.54**
Conc. [μg/mL]	**58.6**	**117.2**	**234.4**	**468.8**	**937.5**	**n.d.**
$OD_{570} \pm SD$	1.549 ± 0.113	1.206 ± 0.178	0.778 ± 0.062	0.246 ± 0.074	0.279 ± 0.104	1.318 ± 0.220
A. baumannii ATCC BAA-3252	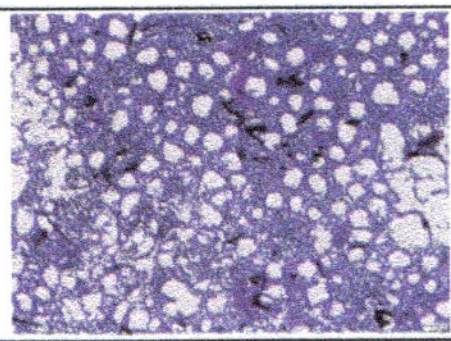	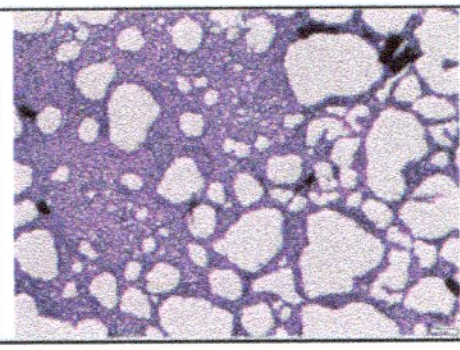	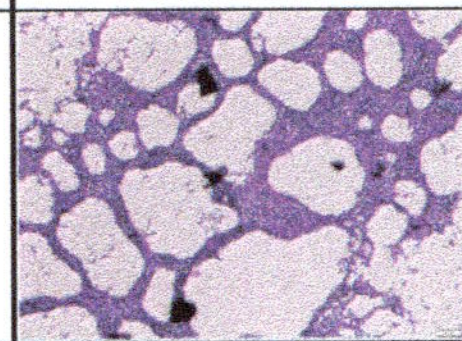	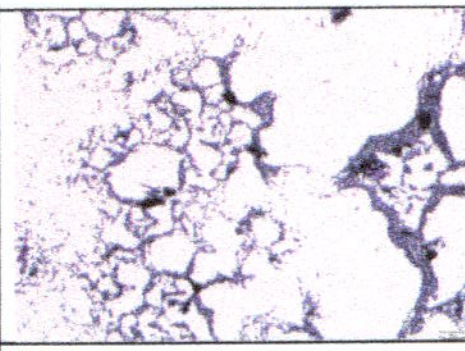	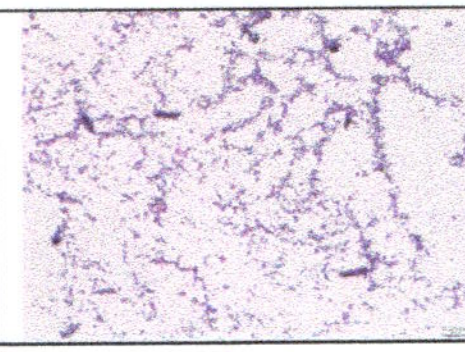	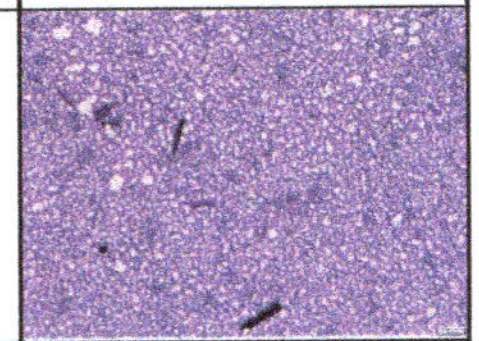
MAOB (%)	56.24	43.56	33.78	17.22	12.63	55.79

Figure 2. Light inverted microscopy observation of the subinhibitory MIC antibiofilm effects of the *Arnica montana* ethanolic extract on *Acinetobacter baumannii* reference strains (ATCC 19606 and ATCC BAA-3252 under stationary conditions (2000× magnification, 20 μm scale bar, Olympus DP22 automated inverted light microscope, CellSens Dimensions 2.3 software, Olympus Corporation, Tokio, Japan). Explanatory notes: $OD_{570} \pm SD$—median values of OD_{570} of four replicates with standard deviation; MAOB—measured area occupied by biofilm (%) calculated via ImageJ 1.54g software (US National Institutes of Health, Bethesda, Maryland, USA); strain control without extract means the 100% growth of the non-treated bacterial strain.

The ethanolic Arnica extract had various levels of biofilm inhibitory activity against both *Acinetobacter* spp. strains. The extracts showed more than 50% biofilm inhibition at the concentrations of 468–3750.0 µg/mL and 234.4–1875 µg/mL for ATCC 19606 and ATCC BAA-3252, respectively. On the other hand, there was no biofilm formation inhibition > 50% by the plant extract at the concentrations of below 58.6 µg/mL and above 7500 µg/mL (Table 3). Some differences between strains were statistically significant (Figure 3).

Table 3. The effect of arnica alcohol extract on the growth of *Acinetobacter baumannii* biofilm-forming cells determined based on minimal biofilm inhibitory concentration (MBIC) values and biofilm inhibition.

Strain	**ATCC 19606**				**ATCC BAA-3252**			
Treatment	**EtE**		**EtOH**		**EtE**		**EtOH**	
MBIC	1875		3750		937.5		7500	
MBIC/MIC ratio	8		4		2		16	
Conc. [µg/mL]	biofilm inhibition (%)	±SD	biofilm inhibition (%)	±SD	biofilm inhibition (%)	±SD	biofilm inhibition (%)	±SD
14.6	−87.9 **	0.256	−89.8	0.076	−191.8 **	0.121	−155.9	0.121
29.3	−40.8	0.064	−59.2	0.091	−73.1 **	0.161	−16.2	0.095
58.6	−45.4	0.060	−87.5	0.060	−50.9	0.195	−83.3	0.206
117.2	14.2 ***	0.081	−52.0	0.047	34.6 **	0.071	−14.5	0.056
234.4	39.5 **	0.049	−69.1	0.155	73.9	0.043	−34.3	0.038
468.7	74.9 *	0.063	2.2	0.160	77.8	0.026	−32.5	0.102
937.5	85.0 **	0.032	−29.4	0.113	90.3	0.008	−0.7	0.120
1875	84.1 *	0.053	−55.5	0.204	67.4	0.045	−51.0	0.085
3750	57.7 ****	0.005	71.8	0.039	−26.2 ****	0.061	20.4	0.035
7500	−40.6 ***	0.016	76.2	0.029	−77.7	0.097	81.3	0.016
15,000	−87.8 ****	0.158	83.9	0.025	−180.6	0.084	82.5	0.017
30,000	−89.7 **	0.050	51.5	0.054	−118.8	0.199	63.7	0.014

Explanatory notes: AqE—aqueous extract; EtE—ethanolic extract; EtOH—ethanol (concentration *v/v* in the medium corresponding to a series of dilutions); SD—standard deviation calculated from four replicates; biofilm inhibition rated on a scale of 0–100%; values < 0% indicated bacterial biofilm growth enhancement; values of 0–50% indicated weak anti-biofilm activity; values > 50% indicated good biofilm inhibition. Low inhibition values for biofilm cell growth are generally caused by the presence of insoluble solid particles and residues from the extract. High concentrations of these particles may increase the available surface area for colonization. Statistical data were analyzed using the Mann–Whitney U test. The asterisks mean statistically significant differences among one bacterial strain between extract-treated biofilm cells and ethanol-treated biofilm cells as follows: (*) are statistically significant at $p < 0.05$; (**) are significant at $p < 0.01$; (***) are significant at $p < 0.001$; (****) are significant at $p < 0.0001$. Data were based on four replicates.

The effect of ethanol on the biofilm formation by bacteria was also evaluated for both strains, with the ethanol diluted in a *v/v* ratio analogous to that of the extract. The MBIC values obtained were two to four times higher (3750–7500 µg/mL) than those obtained for the untreated strains. The MBIC/MIC ratio indicates that both the alcohol extract and ethanol inhibited biofilm formation, particularly concerning *A. baumannii* ATCC BAA-3252 cells. Statistically significant differences were observed between cells treated with the extract or ethanol at most concentrations (Table 3) for the ATCC 19606 strain, particularly at a subinhibitory concentration of ¼ MIC (117.2 µg/mL; $p = 0.0001$). Similar observations were noted for the ATCC BAA-3252 strain at a concentration of ½ MIC (117.2 µg/mL; $p = 0.0017$).

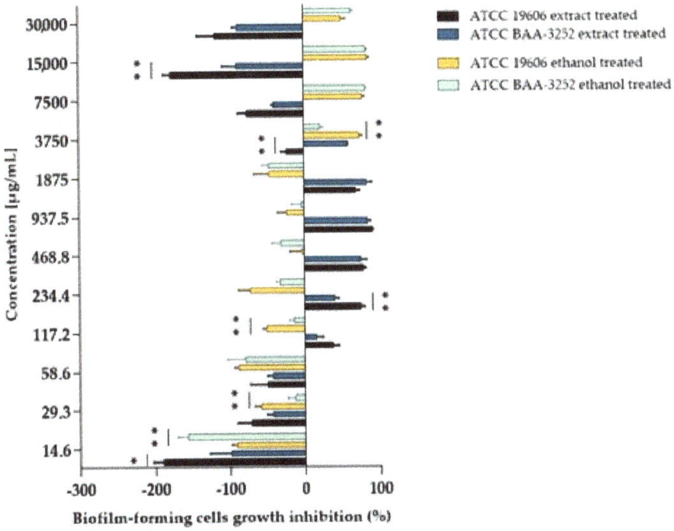

Figure 3. The inhibition of biofilm-forming cell growth depends on the *Acinetobacter baumannii* strain and the extract and/or alcohol used. Explanatory notes: biofilm inhibition rated on a scale of 0–100%; values < 0% indicated bacterial biofilm growth enhancement; values of 0–50% indicated weak anti-biofilm activity; values > 50% indicated good biofilm inhibition; Statistical data were analyzed using the Mann–Whitney U test between bacterial strains treated with plant extract or between bacterial strains treated with ethanol. The asterisks mean statistically significant differences as follows: (*) are statistically significant at $p < 0.05$; (**) are significant at $p < 0.01$. Data were based on four replicates.

3.4. Biofilm-Related Genes

The presence of biofilm-related genes was determined via PCR with specific primer pairs followed by agar gel electrophoresis. Both reference strains of *A. baumannii* presented the same set of genes, except for *csuA*, which was present only in the *A. baumannii* ATCC 19606 strain.

3.5. Effects of Sub-MICs on Biofilm Formation Gene Expression

A. baumannii ATCC 19606 exhibited significant transcriptional alterations in all tested biofilm-related genes, except for *bap*, whose expression remained unchanged regardless of the treatment (Figure 4). The expression of the *csuA-C* genes was markedly downregulated following treatment with either arnica extract or ethanol (Figure A2). For example, compared with the control, exposure to sublethal ½ MIC of the *A. montana* flower extract decreased *csuA-C* transcript levels by approximately 6- to 8-fold. Likewise, ¼ MIC of the extract caused a 3- to 6-fold decrease in their transcription. The changes induced by ethanol treatment were analogous and showed no statistically significant difference compared to those caused by the extract. The decrease observed in *csuD* expression was less pronounced and reached statistical significance compared to the control only at ½ MIC of both the extract and ethanol treatments. In the case of *ompA*, its expression decreased significantly in response to the ½ MIC of the extract and ethanol (by 8- and 9-fold), while the reduction at the ¼ MIC of the extract did not reach statistical significance.

While the mRNA levels of *bap*, *csuD*, and *ompA* exhibited a modest reduction in *A. baumannii* ATCC BAA-3252 following the treatments, the changes were not statistically significant (Figure 5). Exposure to ½ MIC and ¼ MIC of *A. montana* flower extract resulted in strong transcriptional inhibition of *csuB* (~13-fold), with ½ MIC ethanol treatment producing a similar effect. In contrast, ¼ MIC ethanol did not significantly affect *csuB* expression. A similar trend was observed for *csuC*, with mRNA levels decreasing approximately 5- to 7-fold in response to ½ MIC and ¼ MIC of extract, as well as ½ MIC ethanol. In contrast, ¼ MIC ethanol caused only a minor, nonsignificant decrease.

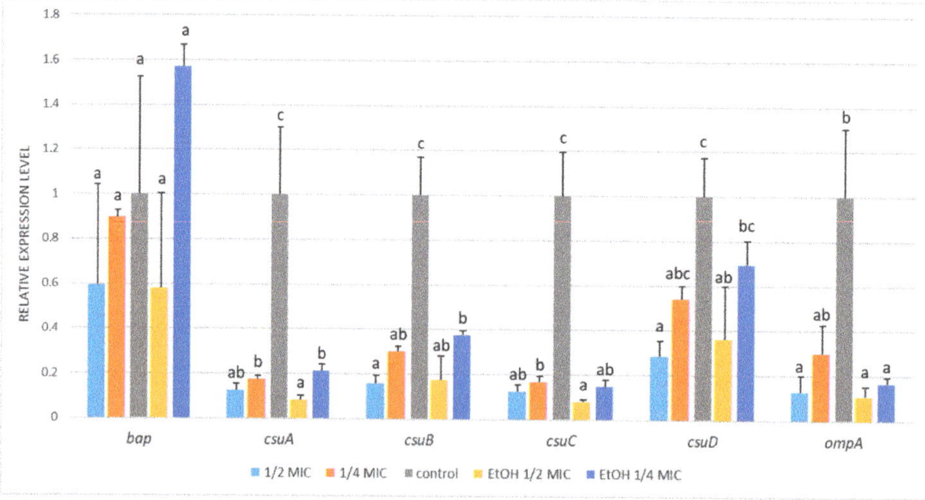

Figure 4. Relative expression of target genes in *Acinetobacter baumannii* ATCC 19606 subjected to ½ MIC (½ MIC of alcoholic *A. montana* flower extract); ¼ MIC (¼ MIC of alcoholic *A. montana* flower extract), EtOH ½ MIC (½ MIC of ethanol), EtOH ¼ MIC (¼ MIC of ethanol). The data represent the means ± SDs ($n = 3$). Means concerning the same gene, followed by different letters, are significantly different according to Tukey's test at $p < 0.05$.

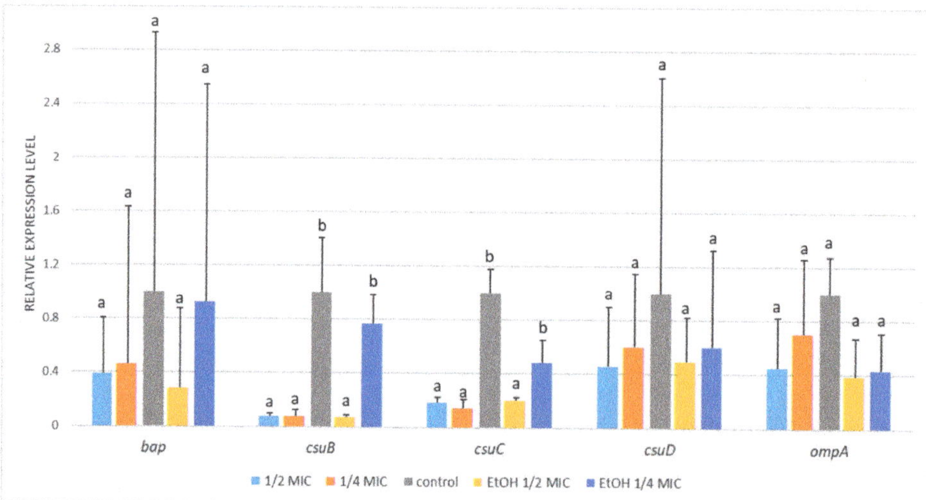

Figure 5. Relative expression of target genes in *Acinetobacter baumannii* ATCC BAA-3252 subjected to ½ MIC (½ MIC of alcoholic *A. montana* flower extract), ¼ MIC (¼ MIC of alcoholic *A. montana* flower extract), EtOH ½ MIC (½ MIC of ethanol), EtOH ¼ MIC (¼ MIC of ethanol). The data represent the means ± SDs ($n = 3$). Means concerning the same gene, followed by different letters, are significantly different according to Tukey's test at $p < 0.05$.

4. Discussion

Arnica montana L. extracts have been the subject of extensive research over the course of many years. Their antibacterial properties have been demonstrated to be particularly effective against Gram-positive bacteria, such as *Staphylococcus aureus*, including methicillin-resistant ATCC 25923 MRSA strains, as well as Gram-negative bacteria, such as *Escherichia coli* ATCC 25922, and yeasts, such as *Candida albicans* ATCC 60193 [19]. The authors demonstrated that all of the tested extracts successfully inhibited microorganisms, with *S. aureus* exhibiting the greatest sensitivity (MIC ranging from 0.10 to 0.31 mg/mL), *E. coli* exhibiting intermediate sensitivity (MIC ranging from 1.23 to 2.58 mg/mL), and *C. albicans* exhibiting the greatest resistance (MIC ranging from 1.41 to 5.16 mg/mL). These results varied depending on extraction conditions such as temperature (36–64 °C) and pressure (5.9–34.1 MPa). Kryvtsova and Koščová [22] studied ethyl and methyl extracts produced from the inflorescences of *A. montana* L. against staphylococci, including the biofilm-forming strain *S. aureus* CCM 4223. The authors concluded that the arnica extract demonstrated antimicrobial effects, including activity against clinical MRSA and biofilm-producing *S. aureus* strains. At the same time, the activities of the ethyl and methyl extracts were comparable. It was also demonstrated that introducing a 0.1% extract solution reduced biofilm formation by 92%. As the extract concentration decreased, the antibiofilm activity gradually declined. Yalgi and Bhat [20] reported good antibacterial activity of *A. montana* tincture against *S. mutans* and *E. faecalis*, with MIC values of 62.5 mg/mL and 8 mg/mL, respectively. Meanwhile, Amato et al. [21] tested the antimicrobial activity of an *A. montana* ethanolic extract against various bacteria and *Candida* spp. yeasts. They have demonstrated a growth of all bacterial strains at different extract concentrations, except for *B. subtilis* and *P. aeruginosa*. Under these experimental conditions, *A. montana* did not demonstrate any activity that would suggest its use as an antimicrobial agent. In addition, it has been demonstrated that arnica extracts possess the capacity to impede the development of bacterial biofilms [22]. This property holds significant promise for applications in the field of topical antibiotic and antifungal therapy, as well as in the management of various forms of oral and oropharyngeal inflammation. However, the available literature does not provide strong evidence that Arnica is effective against *Acinetobacter* spp. bacteria. Treatment of *Acinetobacter* infections currently focuses on novel antibiotic therapies or other methods rather than traditional remedies, such as arnica. In this study, the ethanolic extract of *A. montana* flowers exhibited strong bactericidal properties against *Acinetobacter* spp. reference strains, with MIC values ranging from 234.4 to 468.75 μg/mL, and the MBC values at the same level of 937.5 μg/mL. The MIC values showed that arnica ethanol extract was active against planktonic cells of the reference strains of the species *A. baumannii*. We have observed a decreased bacterial growth when the concentration of the ethanolic extract exceeded 468.8 μg/mL. For *Acinetobacter* ATCC 19606, planktonic cell growth was inhibited by 81.3–94.3%, while for ATCC 3252, inhibition ranged from 88.4–95.0%. The presence of ethanol inhibited the growth of planktonic cells of these bacteria (at concentrations corresponding to MIC of >468.8 μg/mL of the extract), with growth inhibition ranging from 91.5 to 96.3% for *A. baumannii* ATCC 19606 and from 95.6% to 96.3% for *A. baumannii* ATCC 3252. The extract was 2-fold as effective at inhibiting bacterial growth as pure ethanol at a lower concentration. This suggests that the arnica extract could work by combining or strengthening the effects of the solvent and the substances it contains. At high concentrations, the solvent itself was more toxic to bacteria than the extract containing plant bioactive compounds (e.g., STLs or flavonoids). Moreover, subinhibitory concentrations of the extract inhibited biofilm formation by *A. baumannii*, likely affecting processes such as surface colonization, adhesion, and biofilm development. As demonstrated in the extant literature, subinhibitory concentrations of plant extracts have the capacity to inhibit biofilm formation in many bacterial strains, including Gram-negative rods. Several studies have reported that certain plant extracts, such as *Piper betle* [41], have the capacity to substantially reduce the growth of biofilms and even eradicate mature biofilms at concentrations of ¼ MIC or ½ MIC. This activity is attributed to the extract's ability to modify bacterial surface proteins, disrupt the formation of the biofilm polysaccharide matrix, and potentially interfere with *quorum-sensing* mechanisms. Furthermore, several other mechanisms of action have been postulated. The presence of plant extracts has been demonstrated to lead to the formation of an unfavorable layer on biotic and abiotic surfaces, thereby preventing bacteria from adhering to and colonizing new surfaces. Medicinal plant extracts rich in essential oils (e.g., *Thymus zygis*, *Rosmarinus officinalis*, *Origanum majorana*, and *Lippia graveolens*) have the capacity to penetrate and destabilize the intricate polysaccharide matrix that serves as a structural basis for bacterial biofilms, thereby inducing their detachment [42,43]. Moreover, these extracts disrupt communication signals via *quorum sensing* employed by bacteria to orchestrate the formation of robust and resilient bacterial biofilms, thereby impeding the progression of mature biofilms [10,42,43].

Biofilm formation plays a critical role in the emergence and persistence of multidrug-resistant *Acinetobacter* infections, as it protects bacteria from environmental stressors and antimicrobial agents, thereby facilitating their prolonged presence in clinical settings [25,26,44]. Key *A. baumannii* genes involved in biofilm formation include *bap*, which encodes a biofilm-associated protein; the *csu* operon, which is responsible for building pili for attachment; and genes such as *ompA*, *pgaA*/*pgaB*, *epsA*, *bfmR-bfmS*, and *ptk*, which contribute to biofilm structure, maturation, and interaction with surfaces [45–50]. These genes are often associated with antibiotic resistance and the persistence of *A. baumannii* in hospital environments, making them important targets for understanding and combating infection.

Bap (biofilm-associated protein) is a large protein expressed on the bacterial cell surface. It is assumed to be crucial for the intercellular adhesion process, accumulation and aggregation of bacterial cells, biofilm development, and the formation of complex, three-dimensional structures [44]. Many *A. baumannii* isolates carrying the *bap* gene form biofilms on both biotic and abiotic surfaces under stationary conditions [51]. The essential role of the *bap* gene in biofilm formation was confirmed in *Staphylococcus xylosus*, where all biofilm-producing strains contained *bap*, whereas the only biofilm-negative strain, TMW 2.1602, carried a truncated *bap* sequence [45].

The *csu* operon, comprising six genes (*csuA/B*, *csuA*, *csuB*, *csuC*, *csuD*, and *csuE*), is conserved in most clinically significant *Acinetobacter* spp. species and encodes a chaperone-usher pili assembly system important for surface adherence and biofilm formation [52]. An early study by Tomaras et al. [53] demonstrated that mutation of *csuC* or *csuE* in *A. baumannii* ATCC 19606 led to the loss of pili observed in the parental strain and biofilm deficiency. Another study reported that mutants of *A. baumannii* ATCC 17978 lacking *csuD* also did not produce Csu pili and were completely incapable of forming visible biofilm structures [52]. Findings from a different study revealed that major epidemic *A. baumannii* ST191 clones presented significantly greater biofilm masses than minor epidemic and sporadic clones did. This enhanced biofilm formation correlated with increased expression levels of the *csuC*, *csuD*, and *csuE* genes in major epidemic clones. Although *bap* expression was elevated in both major and minor epidemic clones relative to sporadic clones, it did not differ significantly between the two epidemic groups. These results suggest that the differential expression of specific *csu* operon genes plays a key role in biofilm formation variability among epidemic clones [54]. The carbapenem-resistant *A. baumannii* mutant 37662RM2, which lacks capsule synthesis but exhibits markedly enhanced biofilm formation, showed significant overexpression of the *csu* operon and its regulatory genes *bfmR-bfmS* compared to the wild-type 37662 strain. In contrast, the expression of *bap* was only marginally altered. These data imply that the increased biofilm phenotype in 37662RM2 is driven primarily by the upregulation of the *csu* operon and its regulators, highlighting their critical role in biofilm development independent of capsule production [55]. Ahmad et al. [47] highlighted the multifaceted role of Csu pili in *A. baumannii* pathogenicity. The Csu pili were shown to facilitate bacterial adherence to lung epithelial cells and contribute to the formation of structurally complex biofilms. Kishii et al. [56] investigated the relationship between biofilm formation and the presence and transcriptional activity of biofilm-associated genes in clinical *A. baumannii* strains collected from hospitals in Japan. They found that 73.7% of the isolates carried the complete *csu* operon, whereas the remainder lacked these genes. In contrast, *ompA* and *abaI* (*quorum-sensing* autoinducer) were present in all the isolates. Strains lacking the *csu* operon exhibited weak biofilm formation, whereas *csu*-positive strains expressing the operon genes showed strong biofilm production. Interestingly, some *csu*-positive strains did not express the operon, yet formed intermediate biofilms. This suggests that pili may not be strictly required for biofilm development and that other adhesion factors may compensate for the absence of the functional Csu system. OmpA (outer membrane protein A) is widely recognized for its role in bacterial adhesion and biofilm formation, contributing to the overall stability of the biofilm structure. Moreover, the role of *ompA* in the multidrug resistance phenotype of *A. baumannii* has been demonstrated, as disruption of the *ompA* gene resulted in decreased MICs for chloramphenicol, aztreonam, and nalidixic acid [50]. In another study [57], a Δ*ompA* mutant of *A. nosocomialis* ATCC 17903 exhibited significantly reduced biofilm formation and decreased adherence to human epithelial cells. Deletion of the *ompA* gene significantly reduced bacterial dissemination between organs and impaired the development of secondary pneumonia in a murine infection model [58]. A separate study demonstrated a link between elevated *ompA* expression and the adaptive response of *A. baumannii* persister cells to meropenem exposure, indicating that upregulation may enhance bacterial survival during antibiotic treatment and contribute to persistence mechanisms [59]. Furthermore, strains with elevated protein expression were more invasive [60]. Na et al. [61] screened small molecules targeting *ompA* promoter activity in *A. baumannii*. The study identified three chemical compounds that inhibited expression in the *A. baumannii*. Encinales et al. [58] proposed that blocking OmpA could serve as a novel therapeutic strategy against Gram-negative bacilli infections, particularly those caused by *A. baumannii*. These findings highlighted that overexpression of the OmpA protein significantly correlates with higher risks of pneumonia, bacteremia, and increased mortality, underscoring the pivotal role of the OmpA protein in disease severity and clinical outcomes.

Taken together, the genes listed above contribute to *A. baumannii*'s ability to form biofilms, which protect the bacteria from host immune responses and antibiotics, leading to chronic and difficult-to-treat infections. In a study by Kasperski et al. [62], 72% of 100 tested *A. baumannii* isolates obtained from hospitalized patients were strong biofilm producers and consistently harbored the *bap*, *csuE*, and *ompA* genes. Another study involving 107 *A. baumannii* isolates from both clinical and environmental sources reported that 31.2% of clinical strains and 58.7% of environmental strains were strong biofilm producers, with *ompA* detected in all isolates and *bap* present in 89 strains [50]. Saadati et al. [48] reported a high prevalence of biofilm-associated genes in 100 clinical *A. baumannii* isolates, with all strains positive for *ompA* and 89 harboring the *bap* gene. Additionally, all the isolates were MDR, and 79% were extensively drug resistant (XDR). Yang et al. [44] found that 144 of 154 *A. baumannii* strains produced biofilms, with 79.2% positive for *bap*, 68.8% for *csuE*, and 91.6% for *ompA*. The study demonstrated that strains harboring *bap*, *ompA*, and *csuE* genes formed stronger biofilms compared to those lacking these genes. Interestingly, all ten *A. baumannii* nonbiofilm-forming strains contained the *ompA* gene, whereas *bap* and *csuE* were detected in six and seven of these strains, respectively. Amin et al. [63] analysed 64 *A. baumannii* isolates obtained from burn infection cases, revealing important insights into the presence and expression of biofilm-associated genes. While genes such as *ompA* and *abaI* were detected in both biofilm-producing and nonproducing isolates, the *bap* gene was exclusively found among biofilm producers. Furthermore, expression levels of all biofilm-related genes were significantly higher in strong biofilm producers compared to moderate or weak producers. These findings indicate that both the presence and expression levels of biofilm-associated genes crucially impact the bacteria's ability to form biofilms.

In this study, in addition to the assessment of the antibacterial and antibiofilm activities of the *A. montana* ethanolic extract, its influence on the expression of biofilm-related genes in *A. baumannii* was studied. Transcriptional regulation of biofilm-associated genes (*bap, csuABCD, ompA*) was evaluated following exposure to subinhibitory ½ MIC and ¼ MIC of *A. montana* extract, as well as in solvent controls at corresponding concentrations, and compared with those of untreated bacteria. In most cases, the treatment resulted in pronounced downregulation of the target genes, except *bap*, whose expression remained unaffected.

In contrast to our findings, the methanolic extract of *Nothoscordum bivalve*, which inhibited biofilm formation in *A. baumannii*, slightly upregulated *bap* expression at 50% MBC = 3.74 mg/mL and 75% MBC = 5.61 mg/mL. Yet, bacteria treated only with solvent (methanol at 50 and 75% of the MBC) showed no changes in *bap* expression, which is congruent with our results [64]. A related study on *A. baumannii* examined the impact of nanoparticles on biofilm formation and the expression of the *bap* and *csu* genes. Compared with the control, treatment with silver nanoparticles at the MIC (6.25 µg/mL) led to reduced expression of *bap, csuC*, and *csuE*. In contrast, gold nanoparticles conjugated with vancomycin at MIC (0.625 µg/mL) upregulated *csuC* and *csuE* but downregulated *bap* relative to controls. Despite these differing transcriptional responses, both nanoparticle types demonstrated inhibitory effects on biofilm formation, suggesting distinct underlying mechanisms of action [65].

Various plant extracts and compounds vary in their inhibition of bacterial biofilm formation, underscoring the complexity and specificity of their antibiofilm actions. Choudhary et al. [66] demonstrated that eugenol and geraniol exhibit antibiofilm properties in *A. baumannii*, concurrently leading to a downregulation of *csuE* and inhibition of biofilm formation by interfering with pilus assembly mechanisms. A recent investigation of the *Origanum vulgare* ethanolic extract demonstrated its impact on virulence-associated genes in dental plaque bacteria. In *Streptococcus mitis* SA5, *bapA1* expression was significantly decreased at the highest extract concentration (0.781 mg/mL), with no significant effect at lower doses [67]. In *S. epidermidis*, propolis interfered with biofilm development by reducing the transcription of biofilm-associated genes [68]. Similarly, *Cleome droserifolia* extract was found to repress transcription of biofilm-related genes in 43% of the tested *S. aureus* isolates, providing a molecular basis for its antibiofilm activity [69]. In the biofilm-forming *Proteus mirabilis* isolates, an extract of the traditional medicinal plant *Alhagi maurorum* was shown to decrease biofilm production while displaying concentration-dependent antiadhesive and antiquorum-sensing activities. These findings were supported by transcriptomic evidence indicating the downregulation of genes associated with biofilm regulation [70]. The exposure of *S. aureus* to 0.5 × MIC *Thesium chinense* ethyl acetate extract led to transcriptional reprogramming, with 531 genes upregulated and 340 genes downregulated. Notably, the suppressed genes encompassed those involved in biofilm formation [71]. Ranfaing et al. [72] investigated the transcriptomic effects of cranberry proanthocyanidins (190 µg/mL), propolis (102.4 µg/mL), and their combination on a clinical uropathogenic

Escherichia coli strain via microarray analysis. Propolis potentiated cranberry proanthocyanidins' effects, leading to downregulation of more adhesion- and biofilm-related genes. In another study, Liu et al. [73] demonstrated that the methanol-phase extract of *Rumex madaio* exerts species- and strain-specific transcriptional changes. Specifically, *Vibrio alginolyticus* ATCC 17749 upregulated *quorum-sensing*-related genes, whereas *Vibrio parahaemolyticus* B4-10 predominantly downregulated their transcription. In contrast, *quorum sensing* was not among the significantly affected metabolic pathways in *V. parahaemolyticus* ATCC 17802, highlighting strain-dependent transcriptional variability. These findings underscore the complexity of plant extract–bacteria interactions, where conserved virulence traits such as *QS* may be targeted in various ways depending on the species and strains, ultimately influencing the ability of these pathogens to initiate and sustain biofilm formation.

This is congruent with our findings, which also indicate strain-dependent responses to plant-derived antimicrobials. Our study utilized two *A. baumannii* strains: ATCC 19606, a well-established reference strain commonly employed in biofilm and virulence research due to its relative antibiotic susceptibility, and ATCC BAA-3252, a clinically derived multidrug-resistant isolate that reflects the therapeutic challenges posed by contemporary clinical infections. While sub-MIC concentrations of *A. montana* extract significantly downregulated *csuD* and *ompA* in *A. baumannii* ATCC 19606, no comparable transcriptional changes were detected in ATCC BAA-3252. This variability underscores the heterogeneity of virulence regulation among different isolates and highlights the importance of evaluating plant-based antibiofilm agents across multiple strains.

These results are in line with the observations of Khosravy et al. [49], who analyzed extensively drug-resistant (XDR) *A. baumannii* strains isolated from intensive care patients. They found that clinical isolates showed upregulation of biofilm genes compared to the reference strain ATCC 19606, including approximately 12-fold higher *bap* expression. Kakavan et al. [46] also reported a significant association between *bap* overexpression and antibiotic resistance in a study of 100 clinical *A. baumannii* isolates obtained from hospitalized patients. Their findings demonstrated a direct correlation between biofilm production capacity and *bap* expression, with strong biofilm-producing isolates exhibiting the highest *bap* expression levels. These observations suggest that the *bap* gene may represent a promising target for the development of novel therapeutic strategies against *A. baumannii*. Although the *bap* gene plays a critical role in biofilm formation, our findings indicate that the antibiofilm activity of *A. montana* extract may not be directly mediated through transcriptional regulation of this gene, as its expression remained unchanged during treatment. This implies the extract inhibits biofilm development by targeting other virulence factors linked to biofilm formation. According to the obtained results, *A. baumannii* ATCC 19606 exhibited downregulation of the *csuABCD* operon as well as the *ompA* gene, indicating a broader repression of genes related to biofilm formation and outer membrane structure. In contrast, strain ATCC BAA-3252 displayed downregulation limited to *csuB* and *csuC*, suggesting a more selective impact on biofilm-associated genes. This differential gene expression implies that ATCC 19606 may undergo a more comprehensive alteration in biofilm assembly and cell surface properties compared to ATCC BAA-3252. Furthermore, *A. baumannii* ATCC 19606 and ATCC BAA-3252 exhibited similarities and differences in their antimicrobial susceptibility profiles. Both strains shared the same MBC of 937.5 μg/mL, indicating a comparable ethanolic extract concentration required to kill the planktonic cells of bacteria. However, MIC values differed: ATCC 19606's MIC (468.75 μg/mL) was twice that of ATCC BAA-3252 (234.4 μg/mL). This suggests that while both strains require the same bactericidal level of the arnica ethanolic extract, ATCC BAA-3252 is inhibited at a lower concentration, indicating slightly higher susceptibility. This difference in MIC values highlights strain-specific variation in antimicrobial efficacy, despite similar bactericidal thresholds.

A comprehensive understanding of these differences is crucial for the development and optimization of targeted therapeutic strategies against *A. baumannii* and requires further detailed investigations.

In the available literature, studies explicitly comparing the effects of plant extracts and corresponding solvent controls on bacterial gene expression are scarce. Our study uniquely addresses this gap by including ethanol solvent controls matched to the extract concentrations. Antibacterial activity assay demonstrated that for both *A. baumannii* strains tested, the MIC for the solvent control was two times higher compared to the MIC of the Arnica extract. These results clearly highlight that the inhibitory effects observed are predominantly mediated by bioactive compounds present in the extract rather than the ethanol alone. However, our findings also confirm that ethanol, particularly at elevated concentrations, can contribute significantly to growth inhibition.

In strain ATCC 19606, the solvent control elicited gene expression changes similar to those induced by the extract, suggesting solvent-driven transcriptional influence. Conversely, in strain ATCC BAA-3252, ethanol at ¼ MIC failed to suppress expression of *csuB* and *csuC*, whereas the Arnica ethanolic extract at equivalent concentration did. This discrepancy emphasizes that in strain ATCC BAA-3252, gene expression modulation is principally attributed to phytochemicals in the extract rather than solvent effects.

These observations highlight a limitation of using ethanol as a solvent, as its impact may vary with strain sensitivity and can mask or mimic gene regulatory effects. Possible explanations for the different responses include strain-specific ethanol tolerance or adaptation, as well as the presence of synergistic or antagonistic phytochemicals in the extract influencing transcriptional regulation. The limitations of this study include possible confounding effects of the solvent on gene regulation and the need to isolate active compounds to clarify their precise roles. Further research using solvent-free formulations or purified compounds is needed to elucidate the mechanisms underlying these observations. Nevertheless, our findings emphasize that the inhibitory effect on *csu* operon expression in ATCC BAA-3252 stems from active phytochemicals within the Arnica extract, since ethanol alone failed to elicit comparable gene suppression. This highlights the contribution of extract components beyond the solvent in modulating bacterial gene expression

As stated above, most published research focuses on the effects of extracts themselves, often overlooking or inadequately controlling for solvent influences. Our study uniquely contributes by including an ethanol solvent control at concentrations equivalent to those in the extract treatments. This approach enables a clearer differentiation between gene expression changes caused by bioactive compounds and those potentially induced by the solvent. The findings underscore the importance of incorporating such controls to accurately assess extract-specific transcriptional effects and avoid misinterpretations arising from solvent artifacts. Incorporation of solvent controls thus represents a critical methodological consideration for future investigations aiming to elucidate extract-mediated gene expression modulation.

5. Conclusions

The ethanolic extract of *A. montana* L. flowers contains bioactive compounds that exhibit bactericidal effects against *A. baumannii* and inhibit its biofilm formation. This bactericidal activity, combined with the downregulation of key biofilm-associated genes, specifically the *csu* operon and *ompA*, underscores the extract's dual mode of action in targeting both bacterial viability and biofilm development. By disrupting pilus assembly and bacterial adhesion mechanisms, the extract impairs biofilm formation and increases bacterial susceptibility to antimicrobial agents. These findings highlight the potential of ethanolic *A. montana* L. extracts as a complementary therapeutic strategy against infections caused by antibiotic-resistant *A. baumannii* strains. However, further studies are required to comprehensively elucidate the antimicrobial mechanisms involved.

Author Contributions: Conceptualization, S.A. and M.S.; methodology, S.A., M.S. and D.S.; software, M.S.; validation, S.A. and M.S.; formal analysis, S.A. and M.S.; investigation, S.A., M.S. and D.S.; resources, S.A. and M.S.; data curation, S.A. and M.S.; writing—original draft preparation, S.A. and M.S.; writing—review and editing, S.A., M.S. and D.S.; visualization, S.A. and M.S.; supervision, S.A. and M.S.; project administration, S.A. and M.S.; funding acquisition, S.A. and M.S. All authors have read and agreed to the published version of the manuscript.

Funding: This research was conducted as part of the 3rd (Magdalena Sozoniuk) and 4th (Sylwia Andrzejczuk) editions 'Staż za miedzą' internship program, organized under the framework of the Union of Lublin Universities (ZUL) between the Medical University of Lublin and the University of Life Sciences in Lublin, Poland.

Institutional Review Board Statement: Not applicable.

Informed Consent Statement: Not applicable.

Data Availability Statement: The raw data supporting the conclusions of this article will be made available by the authors on request.

Conflicts of Interest: The authors declare no conflicts of interest.

Abbreviations

The following abbreviations are used in this manuscript:

MDR	Multi-drug resistant pathogen
MIC	Minimum inhibitory concentration
MBC	Minimum bactericidal concentration
XDR	Extensively drug-resistant pathogen
QS	*Quorum-sensing*
DEGs	Differentially expressed genes
CRAB	carbapenem-resistant *Acinetobacter baumannii*

Appendix A

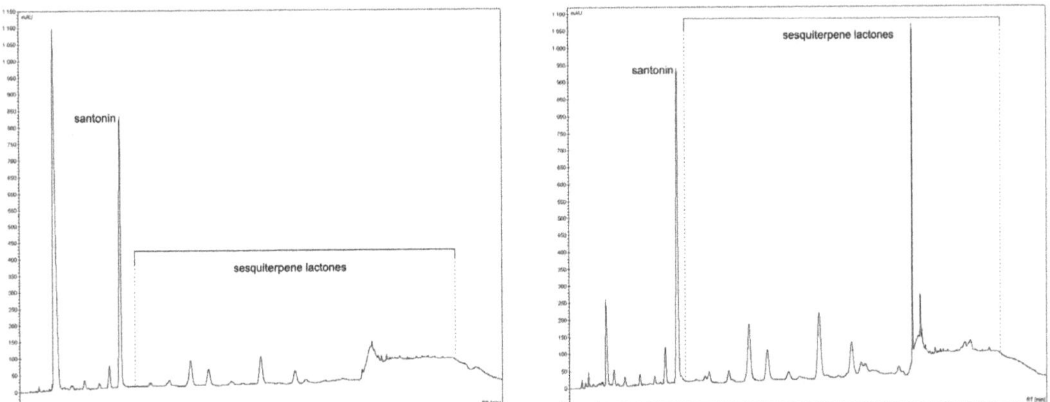

(a) (b)

Figure A1. The HPLC at 225 nm chromatogram of sesquiterpene lactones in (**a**) aquatic and (**b**) ethanolic extracts of *Arnica montana* L.

Table A1. Reduction in planktonic cells growth of *Acinetobacter baumannii* reference strains treated with different concentrations of the *Arnica montana* L. ethanolic extract for 24 h.

Type of Solution / Strain \ Parameters		Ethanolic Extract-Treated Cells			Non-Treated Cells		Ethanol-Treated Cells		
	Conc. [µg/mL]	Mean OD_{600}	±SD	Growth Inhibition (%)	Mean OD_{600}	±SD	Mean OD_{600}	±SD	Growth Inhibition (%)
ATCC 19606	14.6	1.032	0.0402	−0.3	1.010	0.002	1.0838	0.0117	−7.4
	29.3	0.9971	0.0530	1.3	1.213	0.002	1.0504	0.0352	−5.6
	58.6	0.9618	0.0638	5.4	1.141	0.001	1.0028	0.0311	1.3
	117.2	1.0089	0.0411	−1.3	1.157	0.003	0.8475	0.1010	12.1
	234.4	0.6624	0.0855	34.3	1.017	0.007	0.3696	0.0881	58.9
	468.8	0.1043	0.0119	89.4	1.008	0.001	0.0859	0.0122	91.8
	937.5	0.0604	0.0057	94.5	1.028	0.001	0.0448	0.0009	95.5
	1875	0.0784	0.0067	92.7	1.018	0.001	0.0443	0.0007	95.6
	3750	0.122	0.0049	88.1	1.032	0.005	0.0444	0.0017	95.6
	7500	0.191	0.0404	80.0	1.023	0.009	0.0429	0.0005	95.7
	15,000	0.154	0.0365	83.7	1.088	0.001	0.0421	0.0006	95.8
	30,000	0.058	0.0134	93.7	1.021	0.004	0.0382	0.0041	96.1
ATCC BAA-3252	14.6	1.019	0.0157	−7.1	1.018	0.0010	1.0601	0.0137	21.5
	29.3	1.0204	0.0333	−6.6	1.032	0.0081	1.0533	0.0216	2.6
	58.6	0.9433	0.0375	1.7	1.039	0.0006	1.0106	0.0202	21.1
	117.2	0.7346	0.1130	9.3	1.046	0.0012	0.8305	0.0416	22.3
	234.4	0.2566	0.0817	58.4	1.023	0.0010	0.1726	0.0416	79.5
	468.8	0.0738	0.0063	91.1	1.003	0.0006	0.0618	0.0046	88.1
	937.5	0.0544	0.0012	95.5	1.001	0.0010	0.0458	0.0057	90.3
	1875	0.0749	0.0019	95.6	1.007	0.0003	0.0443	0.0006	89.4
	3750	0.1213	0.0060	95.6	1.009	0.0042	0.0436	0.0005	88.5
	7500	0.1596	0.0416	95.7	1.029	0.0544	0.0431	0.0009	91.9
	15,000	0.1312	0.0189	95.8	1.024	0.0322	0.0424	0.0010	95.3
	30,000	0.0501	0.0085	96.1	1.004	0.0162	0.0373	0.0043	94.7

Explanatory notes: SD—standard deviation calculated from ten replicates; median growth inhibition (%) was reported from ten replicates for both strains in each concentration; median control of broth medium OD_{570} = 0.048 ± 0.007438638; median extract control without bacteria OD_{570} = 0.0655 ± 0.039811963.

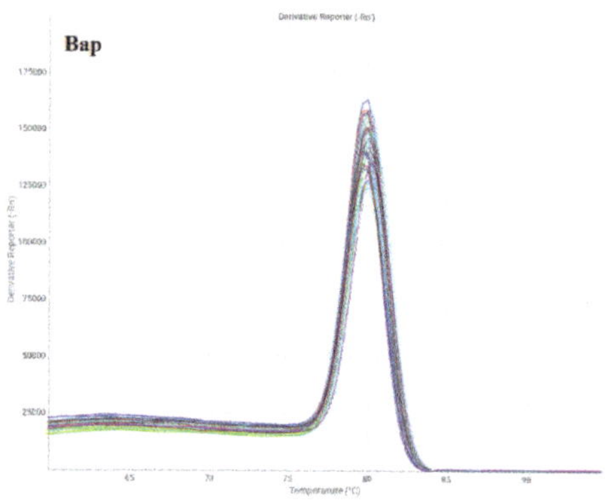

bap

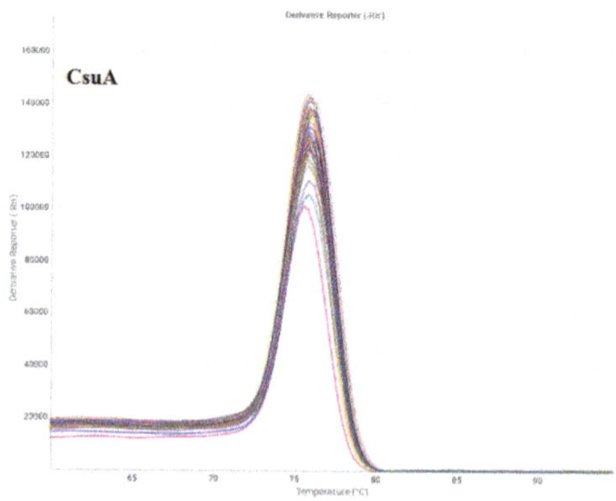

csuA

Figure A2. *Cont.*

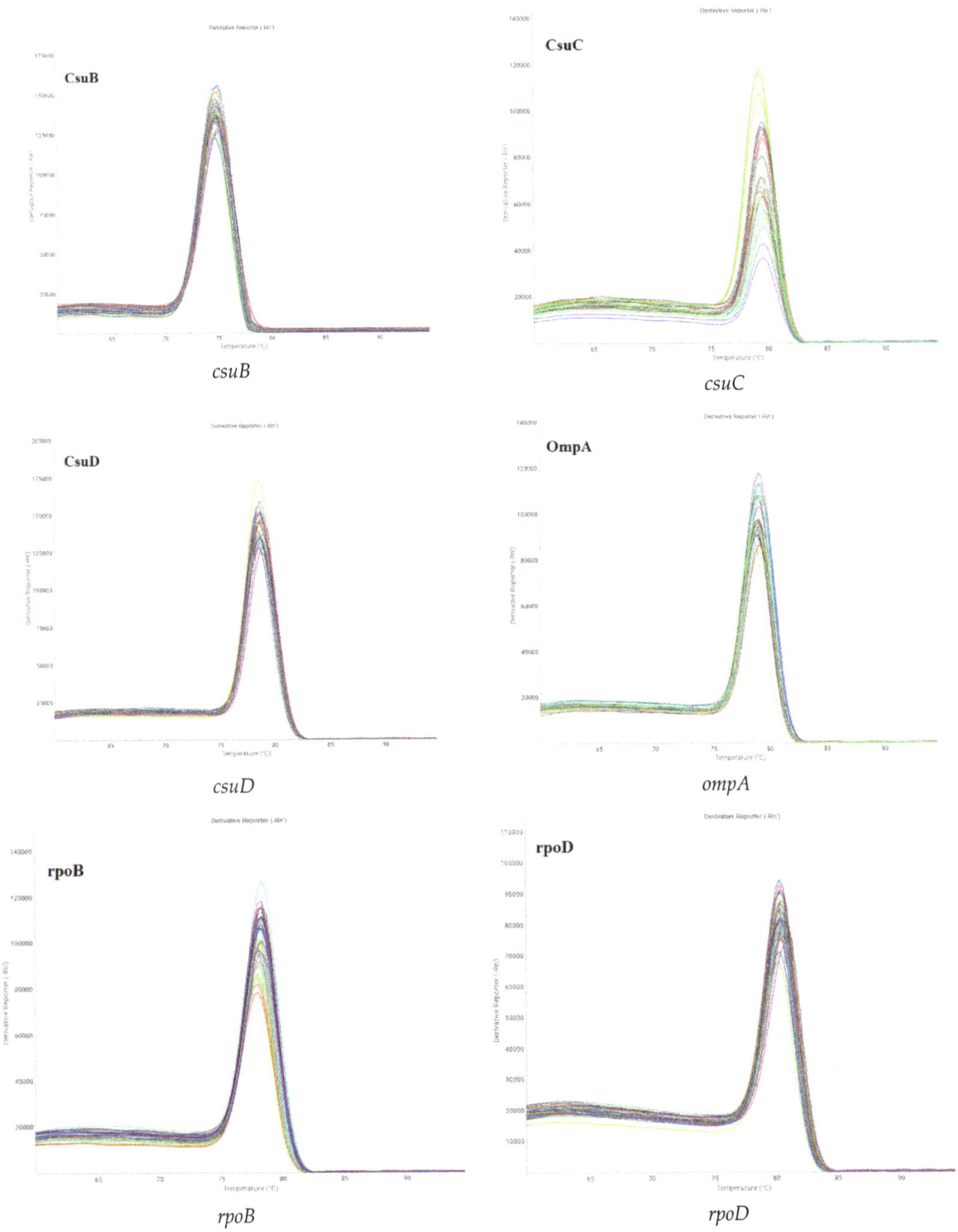

Figure A2. Melting-curve analysis of all selected biofilm-related genes in *Acinetobacter baumannii* ATCC 19606.

References

1. Idris, F.N.; Nadzir, M.M. Multi-Drug Resistant ESKAPE Pathogens and the Uses of Plants as Their Antimicrobial Agents. *Arch. Microbiol.* **2023**, *205*, 115. [CrossRef]

2. Narasimhan, B.; Mourya, V.; Dhake, A. Design, Synthesis, Antibacterial, and QSAR Studies of Myristic Acid Derivatives. *Bioorg. Med. Chem. Lett.* **2006**, *16*, 3023–3029. [CrossRef]
3. Rammohan, A.; Bhaskar, B.V.; Venkateswarlu, N.; Rao, V.L.; Gunasekar, D.; Zyryanov, G.V. Isolation of Flavonoids from the Flowers of *Rhynchosia beddomei* Baker as Prominent Antimicrobial Agents and Molecular Docking. *Microb. Pathog.* **2019**, *136*, 103667. [CrossRef]
4. Song, M.; Liu, Y.; Li, T.; Liu, X.; Hao, Z.; Ding, S.; Panichayupakaranant, P.; Zhu, K.; Shen, J. Plant Natural Flavonoids Against Multidrug Resistant Pathogens. *Adv. Sci.* **2021**, *8*, 2100749. [CrossRef]
5. Canlı, K.; Yetgin, A.; Benek, A.; Bozyel, M.E.; Murat Altuner, E. In Vitro Antimicrobial Activity Screening of Ethanol Extract of *Lavandula stoechas* and Investigation of Its Biochemical Composition. *Adv. Pharmacol. Pharm. Sci.* **2019**, *2019*, 3201458. [CrossRef]
6. Chan, B.C.; Ip, M.; Lau, C.B.; Lui, S.L.; Jolivalt, C.; Ganem-Elbaz, C.; Litaudon, M.; Reiner, N.E.; Gong, H.; See, R.H.; et al. Synergistic Effects of Baicalein with Ciprofloxacin against NorA Over-Expressed Methicillin-Resistant Staphylococcus Aureus (MRSA) and Inhibition of MRSA Pyruvate Kinase. *J. Ethnopharmacol.* **2011**, *137*, 767–773. [CrossRef]
7. Khameneh, B.; Iranshahy, M.; Soheili, V.; Fazly Bazzaz, B.S. Review on Plant Antimicrobials: A Mechanistic Viewpoint. *Antimicrob. Resist. Infect. Control* **2019**, *8*, 118. [CrossRef]
8. El-Demerdash, A.S.; Alfaraj, R.; Farid, F.A.; Yassin, M.H.; Saleh, A.M.; Dawwam, G.E. Essential Oils as Capsule Disruptors: Enhancing Antibiotic Efficacy against Multidrug-Resistant Klebsiella Pneumoniae. *Front. Microbiol.* **2024**, *15*, 1467460. [CrossRef] [PubMed]
9. Gan, C.; Langa, E.; Wang, G.; Van Bambeke, F.; Ballestero, D.; Pino-Otín, M.R. Mechanisms of Action and Resistance Prevention of Synergistic Thymol and Carvacrol Combinations with Antibiotics in Staphylococcus Aureus and *Acinetobacter baumannii*. *Nat. Prod. Bioprospect.* **2025**, *15*, 36. [CrossRef]
10. Touati, A.; Mairi, A.; Ibrahim, N.A.; Idres, T. Essential Oils for Biofilm Control: Mechanisms, Synergies, and Translational Challenges in the Era of Antimicrobial Resistance. *Antibiotics* **2025**, *14*, 503. [CrossRef] [PubMed]
11. Sugier, P.; Sugier, D.; Miazga-Karska, M.; Nurzyńska, A.; Król, B.; Sęczyk, Ł.; Kowalski, R. Effect of Different *Arnica montana* L. Plant Parts on the Essential Oil Composition, Antimicrobial Activity, and Synergistic Interactions with Antibiotics. *Molecules* **2025**, *30*, 3812. [CrossRef]
12. Qureshi, K.A.; Parvez, A.; Ismatullah, H.; Almahasheer, H.; Al Rugaie, O. Exploring the Antimicrobial and Antibiofilm Potency of Four Essential Oils against Selected Human Pathogens Using in Vitro and in Silico Approaches. *PLoS ONE* **2025**, *20*, e0315663. [CrossRef]
13. Alasgah, A.A.; Ahmed, G.E.; El Bayomi, R.M.; Mahmoud, A.F.A.; El-tahlawy, A.S.; Abdallah, K.M.E.; Elabbasy, M.T.; Ibrahem, U.M.; Morshdy, A.E.M.A. Antibiofilm Potential of Cumin and Lemongrass Essential Oils against Multidrug-Resistant *Vibrio parahaemolyticus* in Retailed Fish Samples. *Food Control* **2025**, *172*, 111162. [CrossRef]
14. De Rossi, L.; Rocchetti, G.; Lucini, L.; Rebecchi, A. Antimicrobial Potential of Polyphenols: Mechanisms of Action and Microbial Responses—A Narrative Review. *Antioxidants* **2025**, *14*, 200. [CrossRef]
15. Judentienė, A.; Būdienė, J. Analysis of the Chemical Composition of Flower Essential Oils from *Arnica montana* of Lithuanian Origin. *Chemija* **2009**, *20*, 190–194.
16. Thakur, T. Ethnobotanical, Phytochemical, Pharmacological and Homoeopathic Review of *Arnica montana* Linn. *World J. Pharm. Med. Res.* **2017**, *3*, 152–157.
17. Sugier, D.; Sugier, P.; Jakubowicz-Gil, J.; Gawlik-Dziki, U.; Zając, A.; Król, B.; Chmiel, S.; Kończak, M.; Pięt, M.; Paduch, R. Nitrogen Fertilization and Solvents as Factors Modifying the Antioxidant and Anticancer Potential of *Arnica montana* L. Flower Head Extracts. *Plants* **2022**, *12*, 142. [CrossRef]
18. Toma, C.-C.; Marrelli, M.; Puticiu, M.; Conforti, F.; Statti, G. Effects of Arnica Phytotherapeutic and Homeopathic Formulations on Traumatic Injuries and Inflammatory Conditions: A Systematic Review. *Plants* **2024**, *13*, 3112. [CrossRef] [PubMed]
19. Žitek, T.; Postružnik, V.; Knez, Ž.; Golle, A.; Dariš, B.; Knez Marevci, M. *Arnica montana* L. Supercritical Extraction Optimization for Antibiotic and Anticancer Activity. *Front. Bioeng. Biotechnol.* **2022**, *10*, 897185. [CrossRef] [PubMed]
20. Yalgi, V.S.; Bhat, K.G. Antibacterial Activity of Homoeopathic Tinctures on Bacterial Strains of *Streptococcus mutans* and *Enterococcus faecalis*: An In Vitro Study. *J. Clin. Diagn. Res.* **2019**, *13*, ZC01–ZC04. [CrossRef]
21. Amato, A.L.; Carvalho, A.C.; Coutinho, S.D. In Vitro Antimicrobial Activity of *Arnica montana*. *Int. J. High Dilution Res.* **2006**, *5*, 16.
22. Kryvtsova, M.; Koščová, J. Antibiofilm-Forming and Antimicrobial Activity of Extracts of *Arnica montana* L., *Achillea millefolium* L. on Staphylococcus Genus Bacteria. *Biotechnol. Acta* **2020**, *13*, 30–37. [CrossRef]
23. Howard, A.; O'Donoghue, M.; Feeney, A.; Sleator, R.D. *Acinetobacter baumannii*: An Emerging Opportunistic Pathogen. *Virulence* **2012**, *3*, 243–250. [CrossRef]
24. Singh, S.; Singh, S.; Trivedi, M.; Dwivedi, M. An Insight into MDR *Acinetobacter baumannii* Infection and Its Pathogenesis: Potential Therapeutic Targets and Challenges. *Microb. Pathog.* **2024**, *192*, 106674. [CrossRef]
25. Gedefie, A.; Demsis, W.; Ashagrie, M.; Kassa, Y.; Tesfaye, M.; Tilahun, M.; Bisetegn, H.; Sahle, Z. *Acinetobacter baumannii* Biofilm Formation and Its Role in Disease Pathogenesis: A Review. *Infect. Drug Resist.* **2021**, *14*, 3711–3719. [CrossRef]

26. Mendes, S.G.; Combo, S.I.; Allain, T.; Domingues, S.; Buret, A.G.; Da Silva, G.J. Co-Regulation of Biofilm Formation and Antimicrobial Resistance in *Acinetobacter baumannii*: From Mechanisms to Therapeutic Strategies. *Eur. J. Clin. Microbiol. Infect. Dis.* **2023**, *42*, 1405–1423. [CrossRef] [PubMed]
27. World Health Organization, *WHO Bacterial Priority Pathogens List, 2024: Bacterial Pathogens of Public Health Importance, to Guide Research, Development, and Strategies to Prevent and Control Antimicrobial Resistance*; World Health Organization: Geneva, Switzerland, 2024; ISBN 924009346X.
28. Upmanyu, K.; Haq, Q.M.R.; Singh, R. Factors Mediating *Acinetobacter baumannii* Biofilm Formation: Opportunities for Developing Therapeutics. *Curr. Res. Microb. Sci.* **2022**, *3*, 100131. [CrossRef]
29. The European Directorate for the Quality of Medicines & HealthCare. *European Pharmacopoeia IX*, 9th ed.; EDQM: Strasbourg, France, 2016.
30. The European Directorate for the Quality of Medicines & HealthCare. *European Pharmacopoeia VIII*, 8th ed.; EDQM: Strasbourg, France, 2014.
31. European Committee on Antimicrobial Susceptibility Testing. European Committee on Antimicrobial Susceptibility Testing Breakpoint Tables for Interpretation of MICs and Zone Diameters. *EUCAST Clinical Breakpoint Tables* 2025, 15.0. Available online: https://www.eucast.org/bacteria/clinical-breakpoints-and-interpretation/clinical-breakpoint-tables/ (accessed on 25 November 2025).
32. O'donnell, F.; Smyth, T.J.P.; Ramachandran, V.N.; Smyth, W.F. A Study of the Antimicrobial Activity of Selected Synthetic and Naturally Occurring Quinolines. *Int. J. Antimicrob. Agents* **2010**, *35*, 30–38. [CrossRef]
33. Janowski, M.; Janowska, S.; Andrzejczuk, S.; Kosikowska, U.; Jasiński, R.; Mirosław, B.; Feldo, M.; Wujec, M.; Demchuk, O.M. New 4-(Morpholin-4-Yl)-3-Nitrobenzhydrazide Based Scaffold: Synthesis, Structural Insights, and Biological Evaluation. *Molecules* **2025**, *30*, 3343. [CrossRef] [PubMed]
34. Kosikowska, U.; Malm, A.; Pitucha, M.; Rajtar, B.; Polz-Dacewicz, M. Inhibitory Effect of N-Ethyl-3-Amino-5-Oxo-4-Phenyl-2, 5-Dihydro-1 H-Pyrazole-1-Carbothioamide on *Haemophilus* spp. Planktonic or Biofilm-Forming Cells. *Med. Chem. Res.* **2014**, *23*, 1057–1066. [CrossRef] [PubMed]
35. Kosikowska, U.; Korona-Głowniak, I.; Niedzielski, A.; Malm, A. Nasopharyngeal and Adenoid Colonization by *Haemophilus influenzae* and *Haemophilus parainfluenzae* in Children Undergoing Adenoidectomy and the Ability of Bacterial Isolates to Biofilm Production. *Medicine* **2015**, *94*, e799. [CrossRef]
36. Stepanović, S.; Vuković, D.; Hola, V.; Bonaventura, G.D.; Djukić, S.; Ćirković, I.; Ruzicka, F. Quantification of Biofilm in Microtiter Plates: Overview of Testing Conditions and Practical Recommendations for Assessment of Biofilm Production by Staphylococci. *Apmis* **2007**, *115*, 891–899. [CrossRef]
37. Adeyemo, R.O.; Famuyide, I.M.; Dzoyem, J.P.; Lyndy Joy, M. Anti-biofilm, Antibacterial, and Anti-quorum Sensing Activities of Selected South African Plants Traditionally Used to Treat Diarrhoea. *Evid.-Based Complement. Altern. Med.* **2022**, *2022*, 1307801. [CrossRef]
38. Ye, J.; Coulouris, G.; Zaretskaya, I.; Cutcutache, I.; Rozen, S.; Madden, T.L. Primer-BLAST: A Tool to Design Target-Specific Primers for Polymerase Chain Reaction. *BMC Bioinform.* **2012**, *13*, 134. [CrossRef] [PubMed]
39. de Oliveira, P.A.A.; Baboghlian, J.; Ramos, C.O.A.; Mançano, A.S.F.; Porcari, A.d.M.; Girardello, R.; Ferraz, L.F.C. Selection and Validation of Reference Genes Suitable for Gene Expression Analysis by Reverse Transcription Quantitative Real-Time PCR in *Acinetobacter baumannii*. *Sci. Rep.* **2024**, *14*, 3830. [CrossRef]
40. Al-Shamiri, M.M.; Zhang, S.; Mi, P.; Liu, Y.; Xun, M.; Yang, E.; Ai, L.; Han, L.; Chen, Y. Phenotypic and Genotypic Characteristics of *Acinetobacter baumannii* Enrolled in the Relationship among Antibiotic Resistance, Biofilm Formation and Motility. *Microb. Pathog.* **2021**, *155*, 104922. [CrossRef] [PubMed]
41. Saeloh, D.; Visutthi, M. Efficacy of Thai Plant Extracts for Antibacterial and Anti-Biofilm Activities against *Pathogenic bacteria*. *Antibiotics* **2021**, *10*, 1470. [CrossRef]
42. Lagha, R.; Ben Abdallah, F.; Al-Sarhan, B.O.; Al-Sodany, Y. Antibacterial and Biofilm Inhibitory Activity of Medicinal Plant Essential Oils against *Escherichia coli* Isolated from UTI Patients. *Molecules* **2019**, *24*, 1161. [CrossRef] [PubMed]
43. Tapia-Rodriguez, M.R.; Cantu-Soto, E.U.; Vazquez-Armenta, F.J.; Bernal-Mercado, A.T.; Ayala-Zavala, J.F. Inhibition of *Acinetobacter baumannii* Biofilm Formation by Terpenes from Oregano (*Lippia graveolens*) Essential Oil. *Antibiotics* **2023**, *12*, 1539. [CrossRef]
44. Yang, C.-H.; Su, P.-W.; Moi, S.-H.; Chuang, L.-Y. Biofilm Formation in *Acinetobacter baumannii*: Genotype-Phenotype Correlation. *Molecules* **2019**, *24*, 1849. [CrossRef]
45. Schiffer, C.; Hilgarth, M.; Ehrmann, M.; Vogel, R.F. Bap and Cell Surface Hydrophobicity Are Important Factors in *Staphylococcus xylosus* Biofilm Formation. *Front. Microbiol.* **2019**, *10*, 1387. [CrossRef]
46. Kakavan, M.; Gholami, M.; Ahanjan, M.; Ebrahimzadeh, M.A.; Salehian, M.; Roozbahani, F.; Goli, H.R. Expression of Bap Gene in Multidrug-Resistant and Biofilm-Producing *Acinetobacter baumannii* Clinical Isolates. *BMC Microbiol.* **2025**, *25*, 108. [CrossRef]

47. Ahmad, I.; Nadeem, A.; Mushtaq, F.; Zlatkov, N.; Shahzad, M.; Zavialov, A.V.; Wai, S.N.; Uhlin, B.E. Csu Pili Dependent Biofilm Formation and Virulence of *Acinetobacter baumannii*. *NPJ Biofilms Microbiomes* **2023**, *9*, 101. [CrossRef]
48. Saadati, M.; Rahbarnia, L.; Farajnia, S.; Naghili, B.; Mohammadzadeh, R. The Prevalence of Biofilm Encoding Genes in Multidrug-Resistant *Acinetobacter baumannii* Isolates. *Gene Rep.* **2021**, *23*, 101094. [CrossRef]
49. Khosravy, M.; Hosseini, F.; Razavi, M.R.; Khavari, R.A. Expression of Biofilm-Related Genes in Extensively Drug-Resistant *Acinetobacter baumannii*. *Jundishapur J. Microbiol.* **2023**, *16*, 4. [CrossRef]
50. Bardbari, A.M.; Arabestani, M.R.; Karami, M.; Keramat, F.; Alikhani, M.Y.; Bagheri, K.P. Correlation between Ability of Biofilm Formation with Their Responsible Genes and MDR Patterns in Clinical and Environmental *Acinetobacter baumannii* Isolates. *Microb. Pathog.* **2017**, *108*, 122–128. [CrossRef]
51. Fattahian, Y.; Rasooli, I.; Gargari, S.L.M.; Rahbar, M.R.; Astaneh, S.D.A.; Amani, J. Protection against *Acinetobacter baumannii* Infection via Its Functional Deprivation of Biofilm Associated Protein (Bap). *Microb. Pathog.* **2011**, *51*, 402–406. [CrossRef] [PubMed]
52. Moon, K.H.; Weber, B.S.; Feldman, M.F. Subinhibitory Concentrations of Trimethoprim and Sulfamethoxazole Prevent Biofilm Formation by *Acinetobacter baumannii* through Inhibition of Csu Pilus Expression. *Antimicrob. Agents Chemother.* **2017**, *61*, 10–1128. [CrossRef] [PubMed]
53. Tomaras, A.P.; Dorsey, C.W.; Edelmann, R.E.; Actis, L.A. Attachment to and Biofilm Formation on Abiotic Surfaces by *Acinetobacter baumannii*: Involvement of a Novel Chaperone-Usher Pili Assembly System. *Microbiology* **2003**, *149*, 3473–3484. [CrossRef]
54. Selasi, G.N.; Nicholas, A.; Jeon, H.; Na, S.H.; Kwon, H.I.; Kim, Y.J.; Heo, S.T.; Oh, M.H.; Lee, J.C. Differences in Biofilm Mass, Expression of Biofilm-Associated Genes, and Resistance to Desiccation between Epidemic and Sporadic Clones of Carbapenem-Resistant *Acinetobacter baumannii* Sequence Type 191. *PLoS ONE* **2016**, *11*, e0162576. [CrossRef]
55. Gao, Q.; Meng, X.; Gu, H.; Chen, X.; Yang, H.; Qiao, Y.; Guo, X. Two Phenotype-Differentiated *Acinetobacter baumannii* Mutants That Survived in a Meropenem Selection Display Large Differences in Their Transcription Profiles. *Front. Microbiol.* **2019**, *10*, 2308. [CrossRef]
56. Kishii, K.; Hamada, M.; Aoki, K.; Ito, K.; Onodera, J.; Ishii, Y.; Tateda, K. Differences in Biofilm Formation and Transcription of Biofilm-Associated Genes among *Acinetobacter baumannii* Clinical Strains Belonging to the International Clone II Lineage. *J. Infect. Chemother.* **2020**, *26*, 693–698. [CrossRef] [PubMed]
57. Kim, S.W.; Oh, M.H.; Jun, S.H.; Jeon, H.; Kim, S.I.; Kim, K.; Lee, Y.C.; Lee, J.C. Outer Membrane Protein A Plays a Role in Pathogenesis of *Acinetobacter nosocomialis*. *Virulence* **2016**, *7*, 413–426. [CrossRef] [PubMed]
58. Sánchez-Encinales, V.; Álvarez-Marín, R.; Pachón-Ibáñez, M.E.; Fernández-Cuenca, F.; Pascual, A.; Garnacho-Montero, J.; Martínez-Martínez, L.; Vila, J.; Tomás, M.M.; Cisneros, J.M. Overproduction of Outer Membrane Protein A by *Acinetobacter baumannii* as a Risk Factor for Nosocomial Pneumonia, Bacteremia, and Mortality Rate Increase. *J. Infect. Dis.* **2017**, *215*, 966–974. [PubMed]
59. Schmitt, B.L.; Leal, B.F.; Leyser, M.; de Barros, M.P.; Trentin, D.S.; Ferreira, C.A.S.; de Oliveira, S.D. Increased OmpW and OmpA Expression and Higher Virulence of *Acinetobacter baumannii* Persister Cells. *BMC Microbiol.* **2023**, *23*, 157. [CrossRef]
60. Liu, H.; Cao, C.Y.; Qiu, F.L.; Huang, H.N.; Xie, H.; Dong, R.; Shi, Y.Z.; Hu, X.N. Iron-Rich Conditions Induce OmpA and Virulence Changes of *Acinetobacter baumannii*. *Front. Microbiol.* **2021**, *12*, 725194. [CrossRef]
61. Na, S.H.; Jeon, H.; Oh, M.H.; Kim, Y.J.; Lee, J.C. Screening of Small Molecules Attenuating Biofilm Formation of *Acinetobacter baumannii* by Inhibition of OmpA Promoter Activity. *J. Microbiol.* **2021**, *59*, 871–878. [CrossRef]
62. Kasperski, T.; Romaniszyn, D.; Jachowicz-Matczak, E.; Pomorska-Wesołowska, M.; Wójkowska-Mach, J.; Chmielarczyk, A. Extensive Drug Resistance of Strong Biofilm-Producing *Acinetobacter baumannii* Strains Isolated from Infections and Colonization Hospitalized Patients in Southern Poland. *Pathogens* **2023**, *12*, 975. [CrossRef]
63. Amin, M.; Navidifar, T.; Shooshtari, F.S.; Rashno, M.; Savari, M.; Jahangirmehr, F.; Arshadi, M. Association between Biofilm Formation, Structure, and the Expression Levels of Genes Related to Biofilm Formation and Biofilm-Specific Resistance of *Acinetobacter baumannii* Strains Isolated from Burn Infection in Ahvaz, Iran. *Infect. Drug Resist.* **2019**, *12*, 3867–3881. [CrossRef]
64. Alejandro, H.-M.D.; Humberto, M.-O.M.; Fidel, G.-L.; Antonio, G.-L.M.; Eduardo, S.-G. Inhibition of *Acinetobacter baumannii* Biofilm Formation by Methanolic Extract of Nothoscordum Bivalve. *Adv. Microbiol.* **2018**, *8*, 422. [CrossRef]
65. Rezania, N.; Rahmati, P.; Noorbakhsh, F.; Farhadyar, N.; Lotfali, E. Investigation the Effects of Silver Nanoparticles and Gold Nanoparticles on Expression of Bap and Csu Genes in Biofilm Formation of *Acinetobacter baumannii*. *Iran. J. Microbiol.* **2022**, *14*, 510. [CrossRef]
66. Choudhary, M.; Shrivastava, R.; Vashistt, J. Eugenol and Geraniol Impede Csu-Pilus Assembly and Evades Multidrug-Resistant *Acinetobacter baumannii* Biofilms: In-Vitro and in-Silico Evidence. *Biochem. Biophys. Res. Commun.* **2022**, *636*, 10–17. [CrossRef]
67. Idir, F.; Van Ginneken, S.; Coppola, G.A.; Grenier, D.; Steenackers, H.P.; Bendali, F. *Origanum vulgare* Ethanolic Extracts as a Promising Source of Compounds with Antimicrobial, Anti-Biofilm, and Anti-Virulence Activity against Dental Plaque Bacteria. *Front. Microbiol.* **2022**, *13*, 999839. [CrossRef]

68. Ong, T.H.; Chitra, E.; Ramamurthy, S.; Ling, C.C.S.; Ambu, S.P.; Davamani, F. Cationic Chitosan-Propolis Nanoparticles Alter the Zeta Potential of *S. Epidermidis*, Inhibit Biofilm Formation by Modulating Gene Expression and Exhibit Synergism with Antibiotics. *PLoS ONE* **2019**, *14*, e0213079. [CrossRef] [PubMed]
69. Alqahtani, J.; Negm, W.A.; Elekhnawy, E.; Hussein, I.A.; Hassan, H.S.; Alanzi, A.R.; Moglad, E.; Ahmed, R.; Ibrahim, S.; El-Sherbeni, S.A. Potential Surviving Effect of *Cleome droserifolia* Extract against Systemic Staphylococcus Aureus Infection: Investigation of the Chemical Content of the Plant. *Antibiotics* **2024**, *13*, 450. [CrossRef]
70. Mirzaei, A.; Nasr Esfahani, B.; Ghanadian, M.; Moghim, S. *Alhagi maurorum* Extract Modulates Quorum Sensing Genes and Biofilm Formation in Proteus Mirabilis. *Sci. Rep.* **2022**, *12*, 13992. [CrossRef] [PubMed]
71. Wei, J.; Zhang, C.; Ma, W.; Ma, J.; Liu, Z.; Ren, F.; Li, N. Antibacterial Activity of *Thesium chinense* Turcz Extract against Bacteria Associated with Upper Respiratory Tract Infections. *Infect. Drug Resist.* **2023**, *16*, 5091–5105. [CrossRef] [PubMed]
72. Ranfaing, J.; Dunyach-Remy, C.; Louis, L.; Lavigne, J.-P.; Sotto, A. Propolis Potentiates the Effect of Cranberry (*Vaccinium macrocarpon*) against the Virulence of Uropathogenic Escherichia Coli. *Sci. Rep.* **2018**, *8*, 10706. [CrossRef]
73. Liu, Y.; Yang, L.; Liu, P.; Jin, Y.; Qin, S.; Chen, L. Identification of Antibacterial Components in the Methanol-Phase Extract from Edible Herbaceous Plant Rumex Madaio Makino and Their Antibacterial Action Modes. *Molecules* **2022**, *27*, 660. [CrossRef]

genes

MDPI

Review

Impact of Heavy Metal and Resistance Genes on Antimicrobial Resistance: Ecological and Public Health Implications

Carlos G. Sánchez-Corona [1], Luis Uriel Gonzalez-Avila [1], Cecilia Hernández-Cortez [2], Jorge Rojas-Vargas [3], Graciela Castro-Escarpulli [1,*] and Hugo G. Castelán-Sánchez [4,*]

1 Laboratorio de Investigación Clínica y Ambiental, Departamento de Microbiología, Escuela Nacional de Ciencias Biológicas, Instituto Politécnico Nacional, Prolongación de Carpio y Plan de Ayala s/n, Col. Casco de Santo Tomas, Miguel Hidalgo, Ciudad de Mexico 11340, Mexico; gabrielwaldorf.cc@gmail.com (C.G.S.-C.); u_gza@hotmail.com (L.U.G.-A.)

2 Laboratorio de Bioquímica Microbiana, Departamento de Microbiología, Escuela Nacional de Ciencias Biológicas, Instituto Politécnico Nacional, Prolongación de Carpio y Plan de Ayala s/n, Col. Casco de Santo Tomas, Miguel Hidalgo, Ciudad de Mexico 11340, Mexico; cecihercor@hotmail.com

3 Departments of Biology and Microbiology & Immunology, Western University, London, ON N6A 3K7, Canada; jarvar86@gmail.com

4 Department of Pathology and Laboratory Medicine, Western University, London, ON N6A 3K7, Canada

* Correspondence: chelacastro@hotmail.com (G.C.-E.); hcastela@uwo.ca (H.G.C.-S.)

Abstract: Heavy metals (HMs) are widespread pollutants that can exert selection pressure on microbial populations due to their toxicity and persistence, leading to the evolution of heavy metal resistance genes (HMRGs). These genes are part of the resistome, and their spread often occurs via mobile genetic elements that allow co-selection with antibiotic and biocide resistance genes. Such processes have an impact on microbial biodiversity, biogeochemical cycling and public health in agriculture, industry and urban areas. The selection pressure exerted by HM promotes the spread of multidrug-resistant strains and thus increases ecological and health risks. This review discusses the interaction between HMRGs and genetic determinants such as virulence genes that influence biofilm formation, cellular homeostasis and oxidative stress. It also discusses the dual role of HMRGs in promoting ecological functions such as bioremediation while potentially limiting them by reducing microbial diversity. Understanding such interactions contributes significantly to targeting different systems to overcome the challenges associated with antimicrobial resistance (AMR).

Keywords: heavy metal resistance; heavy metal resistome; co-selection of resistance

1. Introduction

There is an important relationship that exists between heavy metals and microorganisms. It influences the evolutionary dynamics of ecosystems and is also one of the most important issues related to environmental aspects and public health. The ecological relevance arises from the ubiquity of the heavy metal, whereby it selects microbial communities and promotes their resistance mechanisms for the occurrence of mutations and horizontal gene transfer [1].

Chemical speciation, i.e., the chemical forms or states of HMs in the environment, controls mobility, persistence and toxicity. These processes are regulated by physicochemical factors such as pH, redox potential and organic material and influence the uptake into an organism and the toxic effects they can cause [2]. Selection pressure from HMs has driven the evolution of a range of microbial resistance mechanisms, from active excretion

https://doi.org/10.3390/genes16060625

of metals using efflux pumps to conversion to less harmful chemical forms using specific enzymatic processes. Most of these mechanisms are mediated by genes specialized for metal resistance, mostly operons, e.g., *mer*, *ars*, *chr*, *cad* and *pbr*, which encode transport proteins, detoxification proteins and regulatory proteins [3,4]. Furthermore, HMRGs not only ensure the survival of microbes in polluted environments but are also associated with the resistance of microbes to other antimicrobial agents such as antibiotics and biocides through co-selection mechanisms. The transfer of HMRGs between microbial species can not only increase the prevalence of resistance but also increase the pathogenicity and adaptability of microorganisms to hostile environments [5,6]. In addition, HMs and HMRGs are involved in the disruption of important ecological processes, such as the reduction of microbial biodiversity in soils, affecting the decomposition of organic matter and the recycling of essential nutrients [7], and altering the ecological balance in aquatic environments, promoting phenomena such as eutrophication and affecting the health of aquatic biota and, ultimately, humans through bioaccumulation in food chains [7,8].

Against this background, the present work presents an interdisciplinary analysis of the role of HMRGs in microbial resistance and homeostasis in relation to public health and global ecosystem dynamics. Our aim is to contribute to strategies for the ecologically sustainable management of pollution and AMR through a comprehensive analysis of the genetic and biochemical mechanisms of resistance.

1.1. Heavy Metals: Speciation and Bioavailability

Heavy metals (HMs) are an ambiguous group of elements that include some transition metals, metalloids, lanthanides and actinides with different physicochemical properties but which are considered ubiquitous agents in the environmental context, either as part of the Earth's crust or as a product of anthropogenic activities that exert selective pressure on microorganisms and influence their genetic variation mechanisms [1].

HMs have different chemical forms or states in the environment that depend on physicochemical factors such as temperature, pH or redox potential and on the interaction with other compounds such as organic material. These chemical forms determine their bioavailability, toxicity, persistence and mobility, which are also regulated by mechanical, physicochemical and biological migration mechanisms, which, in turn, depend on factors such as fluid dynamics, gravity, metabolism and trophic networks [9].

The chemical speciation of HMs generalizes their presence and increases their availability for absorption, use in essential physiological functions or to cause toxic effects on microorganisms. At low concentrations, metals such as zinc (Zn) and copper (Cu) act as enzymatic cofactors and are involved in the regulation of antioxidant functions; others such as cadmium (Cd) and mercury (Hg), even at low concentrations, can interfere with protein synthesis and enzymatic function, with negative effects on viability and growth capacity [2].

In the atmosphere, HMs occur as gasses (e.g., elemental mercury and methylarsenes), particles and aerosols, mainly from industrial and vehicle emissions. Their persistence and dispersion depend on reactions such as oxidation and photooxidation, as well as climatic and topographical factors, and generally ends with their deposition (dry or wet) in water bodies, agricultural soils, forest soils, etc. [10,11].

The speciation and mobility of HMs in soil is determined by soil composition, with the most active components being clay minerals, organic matter, microorganisms and their derived complexes. HMs can occur as solid precipitates, adsorption complexes or free ions in solutions resulting from processes influenced by redox potential (Eh) and pH, e.g., adsorption–desorption, oxidation–reduction, methylation–demethylation or chelation [12]. In vegetation, dissolved HMs such as chromium (Cr), lead (Pb), arsenic (As), nickel (Ni),

Cd and Hg are transported by diffusion or convection into the rhizosphere, where they are taken up by the roots and then bioaccumulated from their entry into the cellular level through phosphate transport channels or by diffusion through aquaporins [13,14]. The action of rainwater or irrigation can mobilize HMs, either as soil-bound particles or in their dissolved form, which can lead to their accumulation in groundwater [10].

In lentic and lotic waters, HMs can be found on the surface or in bottom sediments in dissolved or suspended form, bound to colloids or to organic and inorganic complexes. The speciation of HMs and the physicochemical conditions of the environment determine their bioavailability, so that they can enter aquatic biota through sorption or absorption processes. This leads to the bioaccumulation of HMs in these organisms and increases their concentration at higher trophic levels [7,15].

1.2. Heavy Metal Resistance

HMs have multiple effects on microorganisms, which can be both toxic and adaptive, depending on the concentration, type of metal and exposure time. Some of these mechanisms include substitution by metal–ligand bonds affecting the biological function of target molecules; the generation of reactive oxygen species (ROS) by covalent and ionic reduction–oxidation (redox) reactions of metal ions with cellular thiols or by Fenton-type reactions with transition metals (iron (Fe), Cu and Ni); disruption of ion regulation in specific membrane transporters and effects on DNA composition [16,17].

The cellular stress exerted by HMs on microorganisms has promoted the evolution of genes that regulate the expression of adaptive proteins that enable resistance to these toxic compounds. Common mechanisms include efflux and sequestration systems that remove metals from the cell or encapsulate them in safe compartments, as well as transformation into more tolerable forms [18,19].

1.3. Heavy Metal Resistome

The resistome was originally described on the basis of clinical studies on human pathogens and is defined as the totality of genes present in genomes or mobilomes that confer direct (phenotypic resistance genes) and indirect (silent and proto-resistance genes) resistance to a specific group of antimicrobial agents in a given microorganism at a given time point [20]. This collection of genes is the result of chromosomal mutations and gene transfers that have taken place over millions of years and have been influenced by various biotic and abiotic factors [21]. HM resistance and the corresponding resistome are considered one of the oldest and most widespread microbial adaptation strategies, as shown by their wide phylogenetic distribution in the Archaea, Bacteria and Eukarya domains [22,23]. They are also thought to have evolved around 2.4 billion years ago after the Great Oxidation Event, when heavy metals became bioavailable [24,25].

The heavy metal resistome consists of genes encoding proteins responsible for the detoxification, transformation or exclusion of a specific metal, located mainly in the cell membrane and cytoplasm (Figure 1). Other reactions mediated by non-specialized proteins also take place in other cellular environments [26].

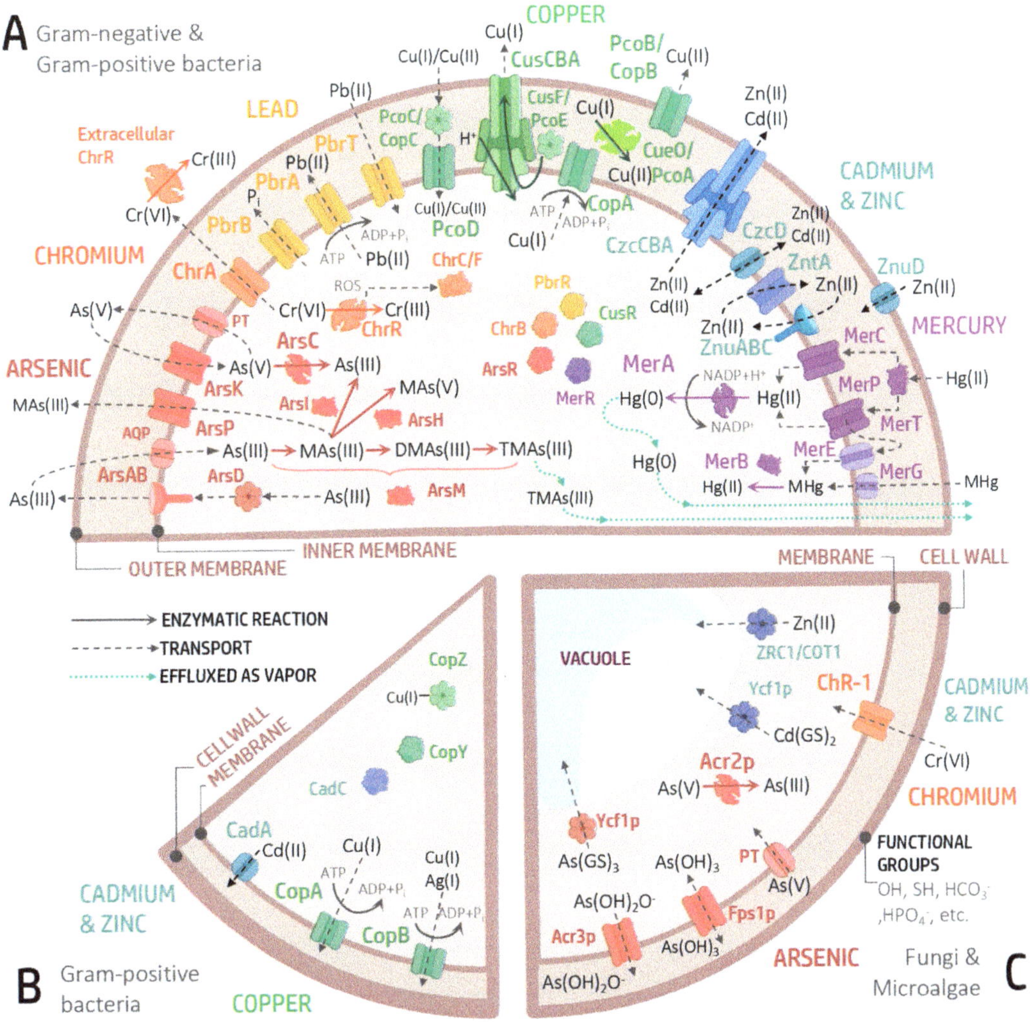

Figure 1. Heavy metals resistance systems for prokaryotes and eukaryotes. (**A**) The resistance systems to As, Cr, Pb and Hg correspond to both Gram-negative and Gram-positive organisms and reside in the *ars*, *chr*, *pbr* and *mer* operons, respectively, while the system for Cu includes the *cue*, *cus*, *pco* and *cop* operons. The resistance systems exclusively for Gram-negative bacteria to Zn include the *znu* and *zur* operons, in addition to the ZnT proteins and the *czc* system, which is shared with Co and Cd. (**B**) For Gram-positive bacteria, the systems for Cu and Cd exclusively include homologs of *cop* and the *cad* operon, respectively. (**C**) Eukaryotic systems include fungal membrane proteins resistant to As, Cr, Pb, Cd and Zn and vacuolar transporters for Cd, Zn and As, as well as reactive compounds in the cytoplasm and functional groups in the cell wall of fungi and microalgae that react with HMs. Figure created from BioRender.

2. Non-Specific Systems

2.1. Extracellular Medium

In this environment, metals are susceptible to conversion and/or adsorption by reaction with excreted metabolites or products of extracellular enzymes. For example, Pb^{2+} can form metal–ligand complexes with the siderophores pyochelin and pyoverdine, which are chelators that are excreted as intermediates in Fe uptake [27]. Cr^{6+} can be reduced to Cr^{3+} by H_2S produced by sulfide adenosyl transferase and adenylyl sulfate reductase, as well as by Fe^{2+} generated by Fe^{3+}-reducing microorganisms [28]. On the other hand, Hg^{2+} can be adsorbed by sulfide or organosulfur compounds [4]. These interactions are even more prominent when metals react with exopolysaccharides (EPS), a group of carbohydrate polymers of wide compositional diversity and with a predominant anionic charge, which are present free in the extracellular medium or deposited on the cell wall surface [29,30]. Furthermore, EPS are related to biofilm development in bacteria, fungi, cyanobacteria and microalgae, playing a crucial role in cell adhesion, cohesion, aggregation, nutrient bioavailability and protection from the external environment, which is relevant for the cellular resistance system [31–35].

2.2. Cell Wall

Compositional differences in the cell wall of bacteria, fungi, algae and other microorganisms can be translated into various proportions of functional groups, such as amide (-CO-NR_2), amine (-NH_2), carboxyl (-COOH), carbonyl (C=O) and hydroxyl (-OH). These functional groups can establish covalent wall–metal bonds or exchange ions with cations or anions of wall compounds, such as EPS [16]. This results in a higher metal accumulation capacity in Gram-positive bacteria compared to Gram-negative bacteria due to their higher proportion of peptidoglycan and the presence of anionic compounds such as teichoic acid and teichuronic acid [36].

2.3. Cell Membrane

Metal entry into cells is adventitious and occurs primarily through competition with essential compounds that are structurally or reactively similar, using transmembrane transport pathways. For example, Cr enters through sulfate permeases [37], while As is transported by phosphate transporters and via aquaglyceroporins [38]. Pb and Cd enter cells through divalent metal transporters, such as Mn and Zn, or through Fe transport-specific ligands, such as siderophores [39,40]. Cr compounds, in contrast, predominate as insoluble complexes that are inaccessible to the cell. However, transient products of extracellular Cr^{6+} reduction, such as hydroxyl complexes or NAD^+-Cr^{3+}, can enter the cell through mechanisms that have not yet been clarified [41,42].

In fungi, metal entry is also carried out by transporters or by passive diffusion due to the high permeability that the membrane can present in some species [43]. In algae, metal transport is not completely understood, but it is believed to also be based on sulfate, phosphate and divalent ion transporters [44].

2.4. Cytoplasm

In this region, interactions focus on the intracellular access of metals through active transport in the membrane and their subsequent compartmental accumulation in organelles, with the vacuole being the main center of cellular homeostasis in eukaryotes [45,46]. The effectiveness of this organelle is due, in part, to vacuolar ATPases (V-ATPase), which translocate protons and act as Ca^{2+}/H^+ exchangers in the vacuolar membrane. This maintains a proton gradient that allows HM sequestration [47–49]. The vacuole also hosts a variety of proteases and peptidases, including serine carboxypeptidases involved in the synthesis of phytochelatins [50,51]. These are cysteine-rich polypeptides that can form complexes with HMs due to the partial negative charge of the sulfhydryl groups of these residues. This metal sequestration mechanism, based on a high sulfur content, is also present in glutathione (GSH), a tripeptide composed of glutamate, cysteine and glycine, related to the control of oxidative stress, as well as in metallothioneins, cysteine-rich proteins with specific binding to metals (Cd, Pb, Zn and Cu), responsible for the transport of metal ions from the cytosol to the organelles [52,53]. Some cyanobacteria have demonstrated the ability to synthesize metallothioneins and siderophores, Fe-chelating compounds [54].

3. Specific Systems

3.1. Arsenic

The *ars* operon (*arsRDABC* or *arsRBC*) comprises the genes responsible for the detoxification of As and is found in both plasmids and bacterial chromosomes (Figure 1A). As can enter the cells as arsenate (AsO_4^{3-}) or arsenite (AsO_3^{3-}) via phosphate or glycerol transporters. The efflux of arsenite (As^{3+}) occurs via the ArsAB complex or directly by ArsB, a membrane-bound efflux pump that can function alone, although its activity is enhanced by ArsA, an intracellular ATPase. In other cases, ArsB can be replaced or supplemented by Acr3, a functional homolog found in bacteria and fungi. ArsK is another transporter that removes metals such as antimony (Sb^{3+}) and bismuth (Bi^{3+}) [48,55]. In the cytosol, ArsC, an arsenate reductase, converts arsenate (As^{5+}) into arsenite (As^{3+}), a less toxic form. This process involves a redox cascade mediated by thiolate nucleophiles. ArsC is coupled to glutaredoxin in species such as *Staphylococcus aureus* and *Bacillus subtilis* or to thioredoxin in *Escherichia coli* (plasmid R773). In eukaryotes (Figure 1C) such as *Saccharomyces cerevisiae*, the Acr2p reductase fulfills similar functions for As^{5+} [56]. Arsenite is removed from the cytoplasm by being exported by the permease Acr3p or transported into the vacuole by the ABC transporter Ycf1p as glutathione conjugates [$As(GS)_3$], which also functions for complexes with Cd [$Cd(GS)_2$]. The aquaglyceroporin Fps1p can control the influx and efflux of As^{3+} [55].

The methyltransferase ArsM can convert As^{3+} into methylarsenite (MAs^{3+}), which can be effluxed through ArsP, ArsK or channels such as GlpF. In the second reaction, ArsM produces dimethylarsenite ($DMAs^{3+}$) and, finally, trimethylarsenite ($TMAs^{3+}$), a non-toxic and volatile compound. Under aerobic conditions, these products are rapidly oxidized to non-toxic pentavalent forms [26]. ArsI, an extradiol dioxygenase capable of cleaving the C–As bond, demethylates MAs^{3+} to As^{3+} and other trivalent aromatic arsenic compounds, such as Rox(III) (roxarsone), Nit(III) (nitarsone) and p-ASA(III) (p-arsanilic acid) (Figure 1A) [57].

In addition, ArsD (Figure 1A), a polypeptide chaperone, transfers arsenite to ArsA, while ArsR acts as a negative regulatory repressor and represses basal and maximal expression of the operon [58,59].

Other related operons are *aio*, which is involved in arsenite oxidation, and *arr*, which is involved in anaerobic arsenate respiration. The *aio* operon (*aioSRABcytC* or *aioRSABC*) encodes the arsenite oxidase AioAB, the histidine kinase AioS, the transcriptional regulator AioR and the As^{3+}-binding protein AioX [60]. The *arr* operon (*arrSRABD*) contains the ArrAB protein, which belongs to the dimethyl sulfoxide reductase family and acts as the final electron acceptor by transferring electrons from the respiratory chain to arsenate [61].

3.2. Chromium

The *chr* operon, which comprises the genes involved in Cr resistance (Figure 1A), is found in both plasmids and chromosomes of diverse organisms, including Eukarya, Bacteria and Archaea. This operon encodes proteins such as ChrA, a chromate (Cr^{6+}) efflux pump belonging to the CHR superfamily, and ChrB, a chromate-sensitive transcriptional regulator. In addition, the proteins ChrC and ChrF, which correspond to Fe and manganese superoxide dismutases (Fe-SOD and Mn-SOD), participate in the reduction of ROS generated during Cr detoxification [15,62]. ChrA is responsible for extruding chromate out of the cell, while, in some organisms, such as fungi of the phylum Ascomycota, its homolog

CHR-1 transports chromate into the cytosol, facilitating its intracellular accumulation [41] (Figure 1C). An additional key mechanism for Cr detoxification involves the reduction of Cr^{6+} to Cr^{3+}, catalyzed by diverse chromate reductases that vary in their mechanism of action and cofactors (Figure 1A). The main enzymes involved are:

- ChrR/YieF/NemA: Flavoproteins that reduce chromate and its intermediates simultaneously. ChrR transfers one or two electrons from NADH, forming Cr^{5+} or Cr^{4+}, which are then converted to Cr^{3+}. During this process, ROS can be generated, but ChrR also exhibits activity that protects the cell against these reactive species [63]. YieF reduces Cr^{6+} directly to Cr^{3+}, generating less ROS, while NemA uses NADPH as an electron donor to reduce nitroaromatic compounds and Cr^{6+} [64].
- NfsA/NfsB: Flavoproteins that transfer electrons to reduce nitrogen compounds and quinones and are also involved in the reduction of Cr^{6+} [65].
- NfoR: This enzyme reduces Cr^{6+} indirectly through the reduction of FMN. This non-catalytic mechanism can be enhanced in the presence of Cu^{2+} [66].

Other enzymes, such as glutathione reductase (GRd), dihydrolipoamide dehydrogenase (LpDH) and C-type cytochromes in sulfate-reducing bacteria, also participate in the reduction of Cr^{6+} by electron transfer, generating H_2O_2, which subsequently reacts to form hydroxyl radicals, thus completing the detoxification process [67].

3.3. *Lead*

Cupriavidus metallidurans CH34 is a model organism from which several genes related to Pb detoxification have been identified, including three *pbr* loci, which encode Pb^{2+} response regulators: PbrR, PbrR2 (previously PbrR691) and PbrR3 (previously PbrR710). In addition, the *pbrUTRABCD* operon, which expression depends on PbrR and the divergent pbr promoter, acts as a repressor of the transcription of opposing structural genes [68].

Regarding the proteins encoded in *pbrUTRABCD* (Figure 1A), they include PbrT, a transmembrane permease responsible for Pb^{2+} uptake; PbrA, a P-type ATPase that transports Zn^{2+}, Cd^{2+} and Pb^{2+} from the cytoplasm to the periplasmic space; PbrB, an undecaprenyl pyrophosphate phosphatase, which function is to release phosphate and, when combined with Pb^{2+}, forms a precipitate in the periplasmic space; PbrC, an aspartic peptidase that could act as a signal peptidase for the PbrB precursor; PbrD, a high-affinity metallochaperone for Pb^{2+}, rich in cysteine, serine and proline and finally, PbrU, which function is still unknown and has been found exclusively in *C. metallidurans* CH34 [35,58,60]. Pb^{2+} efflux can be mediated by homologous proteins associated with the transport of other divalent ions such as Cd^{2+} and Zn^{2+}, including CadA, ZntA and PbrA, P-type ATPases that transport Pb ions against the concentration gradient using ATP hydrolysis as an energy source [69,70]. The regulation of these transporters includes genes such as *pbrR*, *zntR*, *cadR*, *cueR* and *merR*, belonging to the Mer family, initially identified in Hg^{2+} resistance systems [70].

3.4. *Copper*

Several Cu resistance systems have been described (Figure 1A), some of which are encoded in chromosomes. One of these is the *cue* system, composed of the genes *cueR* and *cueO*. CueR is a regulator of cueO and copA expression and is homologous to MerR. CueO is a periplasmic multicopper oxidase that converts Cu^{1+} to Cu^{2+} and reduces O_2 to water in an electron transfer process. Another system is *cus* (*cusCFBA*), which consists of the proteins CusA, CusB, CusC and CusF. CusCBA is an efflux complex for silver (Ag) and Cu that uses the proton gradient across the inner membrane as an energy source to expel excess of these metals into the extracellular space. CusA, a trimeric protein of the Resistance-Nodulation-Division (RND) family, is in the inner membrane and facilitates the efflux of Cu^{1+} and Ag^{1+}. CusB, a soluble periplasmic protein of the Membrane Fusion Protein (MFP) family, stabilizes the CusCBA complex and, due to its structural flexibility and substrate-binding capacity, could act as a switch of the complex function. CusF is a periplasmic metallochaperone that transfers Cu to CusCBA via CusB, thereby improving detoxification efficiency [71,72].

In contrast, other Cu efflux systems are found in plasmids, such as *pcoABCDRSE* and *copABCDRS* (Figure 1A), two homologous systems identified in *E. coli* isolated from pigs fed Cu-supplemented diets and in *Pseudomonas syringae* from tomato plants exposed to Cu-based fungicides, respectively [73]. These systems consist mainly of four structural genes (*pcoABCD/copABCD*) and the regulators PcoRS/CopRS. PcoA/CopA are soluble periplasmic proteins homologous to multicopper oxidases such as CueO. PcoB/CopB are localized in the outer membrane and facilitate Cu import. CopC/PcoC, located in the periplasm functions as chaperones that bind Cu and transfer it to PcoD/CopD, present in the inner membrane and responsible for Cu transport. PcoRS/CopRS form a two-component regulatory system, where PcoS/CopS is a membrane-associated sensor and PcoR/CopR is a transcriptional activator. The *pco* system also includes PcoE, a periplasmic protein that transports Cu to PcoB, which is highly inducible by Cu, although not essential for the functioning of the system [12,72].

In Gram-positive bacteria such as *Enterococcus hirae* and *Lactococcus lactis* (Figure 1B), the *cop* system includes *copA*, *copB*, *copY* and *copZ*. CopA captures Cu for the metalation of a cuproenzyme at the cytoplasmic membrane, while CopB effluxes excess Cu^{+1} and Ag^{+1}. CopY encodes a Cu-inducible transcriptional repressor, and CopZ is a Cu chaperone [74,75].

3.5. Cadmium

The bacterial resistance system to Cd is mainly based on the *cad* and *czc* operons. The *cad* operon (Figure 1B), identified in Gram-positive bacteria such as *S. aureus*, includes CadA, an efflux pump belonging to the P-type ATPase superfamily, and CadC, a regulator of the ArsR/SmtB family, together with CadR, a transcriptional regulator of the MerR family. CadA transports Cd^{2+} into the periplasm and is regulated by CadR, which has a high affinity for Cd^{2+}, as well as a relative affinity for Zn^{2+} and Pb^{2+}. Other proteins of the operon include the transporter CadD, the regulator CadX and the carbonic anhydrase CadW, which can sequester Cd^{2+} [76,77].

The *czc* operon (Figure 1A), identified in Gram-negative bacteria such as *Pseudomonas aeruginosa*, consists of up to eight genes. Some of these encode the cation-proton transmembrane transporter CzcCBA, the inducible two-component system CzcRS and the transporter CzcD, which belongs to the Cation Diffusion Facilitator (CDF) protein family. The CzcCBA complex includes CzcA, an RND protein in the inner membrane; CzcC, which is in the outer membrane and could modify the specificity of the complex to include sensitivity to Cd and cobalt (Co) and CzcB, a membrane fusion protein responsible for expelling excess ions from the cytoplasm and periplasm to the outside of the cell [78,79]. The CzcRS system is composed of the sensor histidine kinase CzcS and the response regulator CzcR, which acts through a phosphorylation cascade. Other genes in the operon, such as *czcI* and *czcN*, are being studied to determine their functions and products [80].

3.6. Zinc

Intracellular regulation of the Zn levels is mediated by several molecular systems, mainly transporters (Figure 1A). Zn entry and exit in bacterial cells is regulated by high- and low-affinity uptake systems, depending on intracellular Zn levels. The ZnuABC complex, consisting of ZnuA, ZnuB and ZnuC, belongs to the ATP-binding cassette transporter family and acts as a high-affinity Zn uptake regulatory system. In this complex, ZnuA is a Zn-binding protein located in the periplasmic space, while ZnuB, located in the inner membrane, interacts with ZnuA, facilitating Zn transport across the cytoplasmic membrane. ZnuC, an ABC-type ATPase, drives active Zn transport into the cytoplasm by ATP hydrolysis [81]. In bacteria such as *Neisseria meningitidis*, this system has been associated with ZnuD, an extracellular Zn transport protein located in the outer membrane [55].

In addition to transporters, Zn homeostasis is also maintained by uptake regulators such as Zur, a transcriptional regulator responsive to reversible Zn binding, which belongs to the FUR (Ferric Uptake Regulator) superfamily of metal-sensitive transcriptional regulators. Zur proteins possess high-affinity and extremely sensitive binding sites, capable of detecting intracellular Zn concentrations in the femtomolar (fM) range. This mechanism allows a gradual response in gene expression to fluctuations in Zn levels, reflected in the successive formation of oligomers and multimers as the concentration increases. Zur has been reported in Gram-negative bacteria such as *E. coli* and Gram-positive bacteria such as *B. subtilis* [82]. In the latter, the import of Zn mediated by PfeT or ZosA, a P1B-type ATPase (like *copA* and *cadA*), which expression is induced by H_2O_2 and Fe, has also been documented, with proposed functions in protection against oxidative stress and Fe efflux [78].

ZnT proteins, which belong to the CDF family, constitute a group of transporters widely distributed in prokaryotes and eukaryotes. In bacteria, several of these have been identified, such as YiiP, a Zn^{2+}/H^+ antiporter; ZitB, which promotes Zn^{2+} efflux across the cytoplasmic membrane; ZntA, a P-ATPase that transports Zn^{2+}; ZntB, a Zn^{2+}/H^+ symporter and CzcD, a Cd^{2+}, Co^{2+} and Zn^{2+}/H^+-K^+ antiporter, which is involved in maintaining intracellular divalent cation and potassium homeostasis [83] (Figure 1A). In yeasts such as *S. cerevisiae*, ZRC1 and COT1 (vacuolar ZnTs) have been identified, which act as Zn^{2+}/H^+ antiporters and regulate Zn^{2+} homeostasis, transporting and storing it in the vacuole [84] (Figure 1C).

3.7. Mercury

The Hg resistance system is based on the *mer* operon, which has diverse structural configurations and contains several coding genes (Figure 1A), The most common and well known are *merA*, *merT*, *merP* and *merR*. This mechanism begins with the specific uptake of Hg. For inorganic Hg, MerP, a receptor protein located in the periplasm, captures and transfers the Hg^{2+} ion to MerF or MerT. MerF facilitates the entry of Hg^{2+} into the cytosol as a transmembrane carrier protein. Organic Hg, on the other hand, enters through the proteins MerG and MerE, located in the periplasmic and inner membrane, respectively. MerT, located in the inner membrane, is a transport protein for organic and inorganic Hg with three transmembrane domains. Other identified transporters include MerC, which transports organic and inorganic Hg, and MerH, which transports mercuric ions ($Hg_2{}^{2+}$), both crossing the inner membrane [3,4].

Once in the cytosol, MerA, a mercury(II) reductase that uses NADPH as an electron source, transforms Hg^{2+} into Hg^0, which is released from the cell as vapor. MerB, an organomercurial lyase, demethylates organic Hg to generate Hg^{2+}, which is processed by MerA [85].

The regulation of the *mer* operon is carried out by MerR, an activator and repressor protein dependent on Hg^{2+} (narrow spectrum) and, occasionally, organomercurial compounds (broad spectrum). Furthermore, MerD is a MerR antagonist protein that negatively regulates the expression of operon genes by binding to the MerO operator region, controlling the positive and negative expression of *mer* genes [86].

4. Heavy Metals Homeostasis, Resistance and Virulence

Of the metals reviewed here, Zn, Ni, Cr, Cu and Co can be considered essential micronutrients for cellular functions and components of relevant proteins such as RNA and DNA polymerases, urease, cytochrome and cytochrome c-oxidase. Although they are manifestly toxic at high concentrations, they are considered toxic elements with metabolic relevance [87]. Faced with this potential toxicity, bacteria seek to maintain the homeostasis of these metal ions—that is, a delicate balance in their intracellular levels that are sufficient to guarantee their functions while avoiding intoxication [88]. These mechanisms not only allow survival in hostile environments but are also linked to the regulation of resistance and virulence genes [89].

Co-Selection of Resistance

The presence of multiple antimicrobial agents in the environment, in addition to HMs—for example, antibiotics, polycyclic aromatic hydrocarbons (PAHs), microplastics, nanoparticles and residues from disinfectants and personal care products—facilitates the selection and spread of common resistance genes, either due to similarities between molecular resistance mechanisms or due to the presence of genes expressing different resistance phenotypes in the same mobile genetic element (MGE), such as plasmids, transposons and integrons, facilitating their mobilization through horizontal gene transfer (HGT), which involves the exchange of genes between genomes of the same or different microbial species [87,90,91]. This process, called co-selection, includes different models such as co-resistance, since the selective pressure of at least one of these antimicrobial agents is sufficient to cause the transfer of the entire MGE, as the resistance genes are physically linked; cross-resistance, where the molecular pathways of action of the different antimicrobial agents converge towards a common target, stimulating resistance to both agents in a single event and co-regulation, as these genes can respond to the same regulator or set of regulators, allowing an organism to coordinate its response to different environmental stressors together [92–94].

Other transfer mechanisms include transformation, where the presence of stressors, such as HMs, can stimulate the direct incorporation of extracellular DNA, including HMRGs, and transduction, where lysogenic phages can transfer HMRGs from the viral genome to infected bacteria, enhancing viral host resistance against HMs toxicity [67].

The presence of HMRGs in MGEs is broad and widespread, and other resistance determinants such as virulence genes or antimicrobial resistance genes (ARGs) for antibiotics or biocides are also commonly found (Table 1). In *E. coli*, the Tn21 transposon contains the *mer* operon, while, in *Acidithiobacillus caldus*, it contains the *ars* operon [95,96]; plasmids such as pMOL30, which confer resistance to Cd, Zn and Co, have been reported in *C. metallidurans* CH34 [97] and HMRGs such as *chrA* have been found in class 1 integrons (intI1–dfrA12–orf–aadA2–qacEΔ1–sul1–chrA–padR) in *Aeromonas aquatica* MX16A [98]. On the other hand, recent research has pointed to HMs resistance through transposable elements in eukaryotes such as the fungus *Paecilomyces variotii*, where the putative transposon Hφ (HEPHAESTUS) was identified, which participates in tolerance to metals such as Cu, Zn and Pb [99].

Table 1. Examples of MGEs with HMRGs and ARGs.

Bacteria	MGEs	HMRGs	ARGs	References
Staphylococcus aureus	*pPA3*	CadR (Cd) As	*ermR* (erythromycin) *tetR* (tetracycline) *pCH11* (bacteriocine)	[100]
Escherichia fergusonii	*p280_40A*	Te, Cu, Hg and Ag operons	*APH(4)-Ia* (hygromycin B) *AAC(3)-Iva* (apramycin) *strA* (streptomycin) *straB* (streptomycin) *sul1* (sulfonamide) *sul2* (sulfonamide) $bla_{CTX-M-2}$ (β-lactamase) *tetA* (tetracycline) *tetB* (tetracycline)	[101]
Listeria spp.	*pLM80*	cadA2C2 (Cd)	*bcrABC* (benzalkonium chloride) *tmr* (crystal violet and malachite green)	[102]
Shigella flexneri	Tn21	*mer* operon (Hg)	*aadA1* (streptomycin and spectinomycin)	[95]
Salmonella enterica	*IncFIB*	*mer* operon (Hg)	*qacEdelta1* (antiseptics)	[103]
Pseudomonas aeruginosa BJ86	*pBJ86*	*mer* operon (Hg)	bla_{DIM-1} (β-lactamase) *qnrVC6* (fluoroquinolone) *mexCD−oprJ* (fluoroquinolones, tetracyclines and chloramphenicol)	[104]

5. Ecological Homeostasis

At the ecosystem level, antimicrobial resistance (AMR) is a fundamental phenomenon for genetic evolution and microbial adaptation. Many of the genes carried by MGEs exhibit parasitic, beneficial and cooperative expression traits, which are relevant in microbial ecological dynamics [105].

ARGs, including HMRGs, can be considered global-reaching factors that coexist with inducing antimicrobial agents. On the one hand, the co-selective pressure of multiple antimicrobial agents can produce hysteresis, allowing resistance mechanisms and their dissemination to persist even after their elimination [106]. On the other hand, ARGs have a high capacity for mobilization and encoding across phylogenetically distant species, thus surpassing the reach capacity of antimicrobial-resistant bacteria (ARBs) [94]. This phenomenon is directly linked to human activities and can affect the microbiomes of natural or anthropogenic ecosystems (Figure 2) [34].

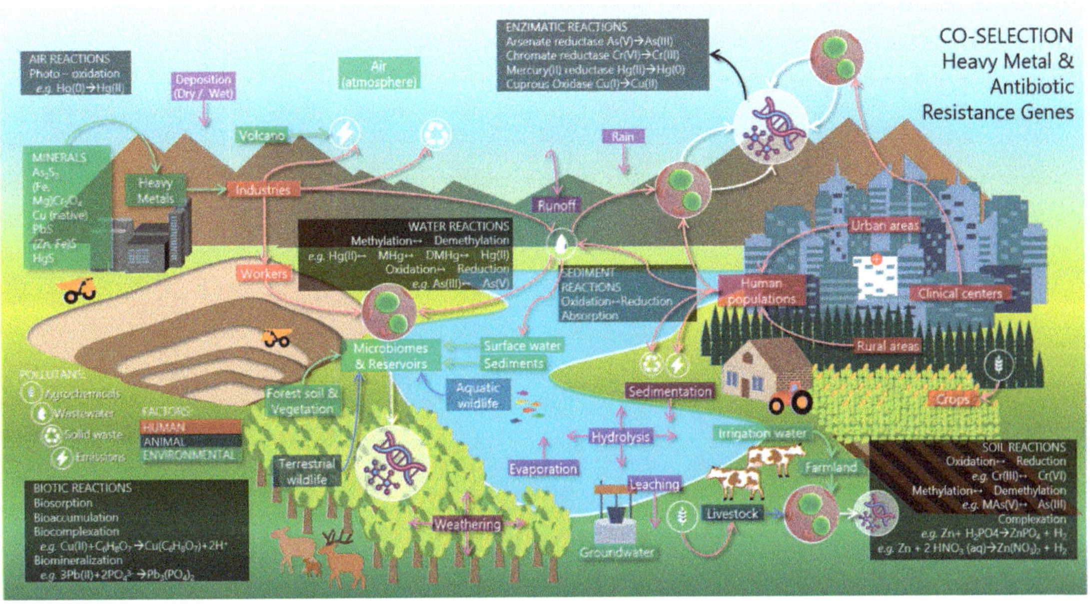

Figure 2. Development and dissemination of HMRGs in natural and anthropogenic environments, influenced by the environmental chemical speciation of the HMs and by the presence of other sources of pollution (wastewater, solid waste, atmospheric emissions from combustion or energy consumption and agrochemicals). Three main factors are recognized: a human factor, which includes human populations and anthropogenic environments such as agricultural, industrial areas and clinical centers; an animal factor, which includes wild and farmed organisms and an environmental factor, made up of natural resources and microbial reservoirs, such as human and animal microbiota, and the microbiomes of soil and water bodies. It is in these reservoirs where HMRGs mainly reside and where AMR is promoted. Figure created from WPS Slides.

Microbiomes present in different environmental settings play an essential role in biogeochemical cycles, ecosystem health and interaction with higher organisms. They are complex and dynamic communities containing diverse microorganisms, including bacteria, archaea, fungi, microalgae, protozoa and viruses [5].

5.1. Natural Environments

5.1.1. Soils

Soil bacteria such as *Burkholderia* spp., *Pseudomonas* spp., *Streptomyces* spp. and some species of *Rhizobium* spp. stand out for their ability to tolerate high concentrations of HMs and possess HMRGs, such as efflux pumps (*czc* for Zn) and detoxification systems (*cop* and *cus* for Cu). These mechanisms not only allow the survival of these bacteria in toxic environments but also contribute to their role in the decomposition of organic matter, promotion of plant growth and the carbon cycle, allowing symbiotic interactions to be established between fungi, (e.g., *Trichoderma* spp. and *Mycorrhizae* spp.) and plants to improve nutrient uptake [107]. Furthermore, the study of soil microbiota using multi-omics approaches has shown that HMRGs are also associated with a decrease in microbial biodiversity, as they favor the proliferation of resistant strains while sensitive ones are displaced. This can lead to a loss of ecological functions critical to soil health, such as the degradation of organic compounds and the recycling of essential nutrients, which can further be aggravated by the presence of other contaminants, such as antibiotics and pesticides [108–110].

5.1.2. Plants

The interaction between soil bacteria, plants and HMRGs has significant effects on ecosystems. Plant growth-promoting bacteria (PGPRs), such as *Pseudomonas* spp., *Bacillus* spp. and *Rhizobium* spp., are crucial for nutrient acquisition and the production and secretion of auxins, gibberellins and cytokinins (plant growth-promoting substances); phytohormone production; nutrient solubilization (PO^{4-}, Fe^{2+} and Fe^{3+}); metabolic stimulation of roots; increased disease resistance and reduced toxicity of HMs, thus reducing plant stress and facilitating plant growth in contaminated soils [111–113].

5.1.3. Wildlife

The microbiota of wild animals generally does not contain ARBs, given the isolation from their hosts; however, they can also interact with human populations or impacted sites, motivating the uptake of ARBs and ARGs (including HMRGs) and subsequently their short- and long-distance spread, through local interactions or migratory movements, respectively [114]. Honeybees and their microbiota are an example of this dissemination, acting as ARB vectors and as a link between natural and anthropic environments, as they are affected by the use of pesticides and antibiotics for prolonged periods [115].

5.1.4. Water Bodies and Aquatic Wildlife

In the current global context, many water bodies act as concentration centers for wastewater of diverse compositions, including HMs and other antimicrobials, as well as ARGs and HMRGs [108]. The contact of aquatic organisms with contaminated water can affect the health of many other organisms through the food chain, including people [8,87].

The proliferation of ARBs can also alter the ecological balance of aquatic bodies, enhancing processes such as eutrophication, where the excessive growth of algae and bacteria reduces dissolved oxygen, affecting aquatic life, degrading water quality [91,109] and altering the native microbiome, which, in freshwater bodies, is dominated by *Cyanobacteria* spp., *Betaproteobacteria* spp. and *Bacteroidetes* spp., while saltwater environments are rich in *Prochlorococcus* spp. and *Pelagibacter* spp., microorganisms essential for the global carbon cycle [116].

In groundwater, microbial diversity is less abundant and diverse than in other environments; therefore, the introduction of antimicrobial agents or ARGs can alter this diversity and generate dangerous metabolites [17,117]. It is also a potential source of contamination of drinking water and contributes to the development of antibiotic resistance in humans, increasing virulence, pathogenicity and the incidence of outbreaks of bacterial infections [118].

5.2. *Anthropic Environments*

5.2.1. Agricultural Environments

In agriculture and livestock farming, the use of compounds containing HMs (Cu, Cr and Zn) is common, as components of medicines, growth promoters and additives in livestock feed, although some, such as As, present in compounds such as Rox(III), Nit(III) and *p*-ASA(III), are being abandoned because they are not considered safe [119].

In any case, the use of these compounds and practices such as the application of livestock manure as fertilizer facilitate the generation and dissemination of HMRGs in environmental microbiomes [120,121] and can modify the enzymatic activity of the microbiota of agricultural soils, where the genera *Bacillus* spp. and *Pseudomonas* spp. Predominate, and *Nitrosomonas* spp., which are involved in processes such as nitrogen fixation and the degradation of xenobiotic compounds [122].

Other practices with the same effect in plant agriculture are irrigation with wastewater and the application of fertilizers and pesticides in agricultural soils, which introduce metals such as Cu, Cd and Pb, which can generate a potential risk to the health of consumers by promoting their bioaccumulation in livestock tissues, milk, eggs and other by-products [98].

5.2.2. Industrial Environments

Industrial activities, such as mining, metallurgy and inappropriate waste management, are significant sources of HMs pollution in soil, water and air. These activities introduce high concentrations of metals, which exert selective pressure on local microbial communities and contribute to the mobility of resistance genes in microbial communities. These dynamics pose significant challenges in resistance management in the environment and public health due to the persistence and mobility of these genes in the environment [123,124]. The microbiomes of industrial environments where HMs predominate include highly resistant species, including bacteria from the phyla Proteobacteria, Acidobacteria, Firmicutes and Bacteroidetes and fungi from the phyla Ascomycota and Ciliophora, capable of not only resisting toxic metals but also contributing to biogeochemical cycles, such as the degradation of contaminants [125,126].

5.2.3. Hospital Environments

Hospitals represent a significant reservoir of microorganisms, including multidrug-resistant pathogens such as *S. aureus*, *Acinetobacter baumannii*, *Klebsiella pneumoniae* and *P. aeruginosa*, many of them carrying antibiotic and HMRGs [127]. The intensive use of antibiotics and disinfectants, hospital waste management and sanitation strategies exert selective pressures on microorganisms present on surfaces, medical devices and hospital wastewater, allowing the survival of resistant strains and promoting the co-selection of resistance genes, since some genetic mechanisms can confer resistance to both HMs and antibiotics and biocides in bacteria such as *A. baumannii* and *P. aeruginosa*, facilitating dispersion in the environment through the wastewater flow [5,6].

Several studies have linked exposure to HMs with the selection of ARBs. However, epidemiological studies examining this relationship are scarce. For example, in 2020, Bazzi et al. [128] reported that multidrug-resistant *A. baumannii* (MDRAb) was the main pathogen isolated from wounded US soldiers during the Iraq War. In these conflict areas, high levels of HMs (Hg, Zn, Cu, Ni, Pb and Cr) were detected in soil and debris, which may have favored resistance to HMs and antibiotics, contributing to severe infections in field hospitals. Likewise, in 2021, Eggers et al. [129] found that, in a human population from urban areas with high HM pollution (Wisconsin, USA), elevated urinary Pb levels correlated with a higher prevalence of ARBs, including methicillin-resistant *S. aureus* (MRSA), vancomycin-resistant enterococci (VRE), fluoroquinolone-resistant Gram-negative bacilli (RGNB) and *Clostridium difficile*.

6. Conclusions

HMRGs are a key component of microbial resistance and homeostasis. They play a fundamental role in the adaptation of microorganisms to environments with toxic concentrations of HMs and in the stabilization of essential metabolic functions. However, their importance goes beyond the microbial sphere, as they have a direct impact on the development and spread of AMR, especially for antibiotics.

The relationship between HMRGs and ARGs is complex and multilayered. It is linked by processes such as co-selection, which not only facilitates HGT between bacteria, including pathogens, but also enhances the ability of microorganisms to adapt and survive in different ecosystems. In clinical settings, this phenomenon contributes to the emergence of multidrug-resistant bacteria such as *P. aeruginosa* and *S. aureus*, which poses a major public health challenge by limiting available therapeutic options and increasing the burden of infectious diseases.

From an ecological perspective, HMRGs also affect natural dynamics by promoting changes in microbial biodiversity and altering critical functions in biogeochemical cycles, which can also have negative consequences for agricultural productivity and water quality. In addition, the spread of HMRGs in natural and anthropogenic environments may facilitate the transfer of resistance to previously unexposed microbial reservoirs, increasing the risk of genetic pollution on a global scale.

In the long term, the persistence and spread of HMRGs could exacerbate the problems associated with AMR and worsen their impact on human, animal and environmental health. This phenomenon underscores the importance of pursuing integrated approaches such as the One Health perspective that take into account the link between natural ecosystems, production systems and public health. In addition, concerted efforts are needed to monitor the presence and mobility of HMRGs, develop strategies to mitigate their impact and promote sustainable practices that minimize the release of heavy metals and antibiotics into the environment.

HMRGs are critical microbial adaptation factors that simultaneously pose an emerging threat to public health and environmental sustainability. Addressing this problem requires coordinated action that combines scientific research, environmental management and public policy. The aim is to reduce the associated risks and ensure the resilience of ecosystems in the face of current and future global challenges.

Author Contributions: C.G.S.-C.: Writing—review and editing, Writing—original draft, Methodology, Investigation and Conceptualization. L.U.G.-A.: Methodology and Investigation. C.H.-C.: Methodology and Investigation. J.R.-V.: Methodology, Investigation and Writing—review and editing. G.C.-E.: Writing—review and editing, Supervision, Methodology, Funding acquisition and Conceptualization. H.G.C.-S.: Writing—review and editing, Supervision, Methodology and Conceptualization. All authors have read and agreed to the published version of the manuscript.

Funding: This study was approved by the "Secretaría de Investigación y Posgrado del Instituto Politécnico Nacional" (SIPs: 20232102 and 20240986). The doctoral studies of C.G.S.-C. were supported by the National Council of Humanities, Sciences and Technologies (CONAHCyT, Mexico) through scholarship number 935376.

Acknowledgments: C.G.S.-C. received grant-aided support from the "Consejo Nacional de Humanidades Ciencias y Tecnologías (CONAHCyT, México)", while L.U.G.-A., C.H.-C. and G.C.-E. received support from the "Sistema Nacional de Investigadores (SNI)" from CONAHCyT. G.C.-E. also received support from the Programa de Estímulos al Desempeño de los Investigadores, Comisión de Operación y Fomento de Actividades Académicas (Instituto Politécnico Nacional.

Conflicts of Interest: The authors declare no conflicts of interest.

References

1. Margaryan, A.; Panosyan, H.; Birkeland, N.-K. Heavy Metal Resistance in Prokaryotes: Mechanism and Application. In *Microorganisms for Sustainability*; Springer: Singapore, 2021; pp. 273–313.
2. He, Z.; Shen, J.; Li, Q.; Yang, Y.; Zhang, D.; Pan, X. Bacterial Metal(Loid) Resistance Genes (MRGs) and Their Variation and Application in Environment: A Review. *Sci. Total Environ.* **2023**, *871*, 162148. [CrossRef] [PubMed]
3. Naguib, M.M.; El-Gendy, A.O.; Khairalla, A.S. Microbial Diversity of Mer Operon Genes and Their Potential Rules in Mercury Bioremediation and Resistance. *Open Biotechnol. J.* **2018**, *12*, 56–77. [CrossRef]
4. Priyadarshanee, M.; Chatterjee, S.; Rath, S.; Dash, H.R.; Das, S. Cellular and Genetic Mechanism of Bacterial Mercury Resistance and Their Role in Biogeochemistry and Bioremediation. *J. Hazard. Mater.* **2022**, *423*, 126985. [CrossRef] [PubMed]

5. Berendonk, T.U.; Manaia, C.M.; Merlin, C.; Fatta-Kassinos, D.; Cytryn, E.; Walsh, F.; Bürgmann, H.; Sørum, H.; Norström, M.; Pons, M.-N.; et al. Tackling Antibiotic Resistance: The Environmental Framework. *Nat. Rev. Microbiol.* **2015**, *13*, 310–317. [CrossRef]
6. Zhang, S.; Huang, J.; Zhao, Z.; Cao, Y.; Li, B. Hospital Wastewater as a Reservoir for Antibiotic Resistance Genes: A Meta-Analysis. *Front. Public Health* **2020**, *8*, 574968. [CrossRef]
7. Squadrone, S. Water Environments: Metal-Tolerant and Antibiotic-Resistant Bacteria. *Environ. Monit. Assess.* **2020**, *192*, 238. [CrossRef]
8. Zhao, Y.; Xu, R.; Cox, S.F.; Qiao, M.; Guo, H. Metals Are Overlooked in the Evolution of Antibiotic Resistance. *Soil Ecol. Lett.* **2024**, *6*, 240244. [CrossRef]
9. Deng, H.; Tu, Y.; Wang, H.; Wang, Z.; Li, Y.; Chai, L.; Zhang, W.; Lin, Z. Environmental Behavior, Human Health Effect, and Pollution Control of Heavy Metal(Loid)s toward Full Life Cycle Processes. *Eco -Environ. Health* **2022**, *1*, 229–243. [CrossRef]
10. Bishop, K.; Shanley, J.B.; Riscassi, A.; de Wit, H.A.; Eklöf, K.; Meng, B.; Mitchell, C.; Osterwalder, S.; Schuster, P.F.; Webster, J.; et al. Recent Advances in Understanding and Measurement of Mercury in the Environment: Terrestrial Hg Cycling. *Sci. Total Environ.* **2020**, *721*, 137647. [CrossRef]
11. Luo, X.; Bing, H.; Luo, Z.; Wang, Y.; Jin, L. Impacts of Atmospheric Particulate Matter Pollution on Environmental Biogeochemistry of Trace Metals in Soil-Plant System: A Review. *Environ. Pollut.* **2019**, *255*, 113138. [CrossRef]
12. Li, P.; Nayeri, N.; Górecki, K.; Becares, E.R.; Wang, K.; Mahato, D.R.; Andersson, M.; Abeyrathna, S.S.; Lindkvist-Petersson, K.; Meloni, G.; et al. PcoB Is a Defense Outer Membrane Protein That Facilitates Cellular Uptake of Copper. *Protein Sci.* **2022**, *31*, e4364. [CrossRef]
13. Al-Makishah, N.H.; Taleb, M.A.; Barakat, M.A. Arsenic Bioaccumulation in Arsenic-Contaminated Soil: A Review. *Chem. Pap.* **2020**, *74*, 2743–2757. [CrossRef]
14. Kim, R.-Y.; Yoon, J.-K.; Kim, T.-S.; Yang, J.E.; Owens, G.; Kim, K.-R. Bioavailability of Heavy Metals in Soils: Definitions and Practical Implementation—A Critical Review. *Environ. Geochem. Health* **2015**, *37*, 1041–1061. [CrossRef] [PubMed]
15. Rahman, Z.; Singh, V.P. The Relative Impact of Toxic Heavy Metals (THMs) (Arsenic (As), Cadmium (Cd), Chromium (Cr)(VI), Mercury (Hg), and Lead (Pb)) on the Total Environment: An Overview. *Environ. Monit. Assess.* **2019**, *191*, 419. [CrossRef]
16. Leong, Y.K.; Chang, J.-S. Bioremediation of Heavy Metals Using Microalgae: Recent Advances and Mechanisms. *Bioresour. Technol.* **2020**, *303*, 122886. [CrossRef]
17. Hemme, C.L.; Deng, Y.; Gentry, T.J.; Fields, M.W.; Wu, L.; Barua, S.; Barry, K.; Tringe, S.G.; Watson, D.B.; He, Z.; et al. Metagenomic Insights into Evolution of a Heavy Metal-Contaminated Groundwater Microbial Community. *ISME J.* **2010**, *4*, 660–672. [CrossRef] [PubMed]
18. Baker-Austin, C.; Wright, M.S.; Stepanauskas, R.; McArthur, J.V. Co-Selection of Antibiotic and Metal Resistance. *Trends Microbiol.* **2006**, *14*, 176–182. [CrossRef]
19. Pal, A.; Bhattacharjee, S.; Saha, J.; Sarkar, M.; Mandal, P. Bacterial Survival Strategies and Responses under Heavy Metal Stress: A Comprehensive Overview. *Crit. Rev. Microbiol.* **2022**, *48*, 327–355. [CrossRef]
20. Wright, G.D. The Antibiotic Resistome. *Expert Opin. Drug Discov.* **2010**, *5*, 779–788. [CrossRef]
21. Perry, J.A.; Westman, E.L.; Wright, G.D. The Antibiotic Resistome: What's New? *Curr. Opin. Microbiol.* **2014**, *21*, 45–50. [CrossRef]
22. Banu, H.; Prasad, K.P. Role of Plasmids in Microbiology. *J. Aquac. Res. Dev.* **2017**, *8*, 466. [CrossRef]
23. Biswas, R.; Halder, U.; Kabiraj, A.; Mondal, A.; Bandopadhyay, R. Overview on the Role of Heavy Metals Tolerance on Developing Antibiotic Resistance in Both Gram-Negative and Gram-Positive Bacteria. *Arch. Microbiol.* **2021**, *203*, 2761–2770. [CrossRef] [PubMed]
24. Chen, S.-C.; Sun, G.-X.; Yan, Y.; Konstantinidis, K.T.; Zhang, S.-Y.; Deng, Y.; Li, X.-M.; Cui, H.-L.; Musat, F.; Popp, D.; et al. The Great Oxidation Event Expanded the Genetic Repertoire of Arsenic Metabolism and Cycling. *Proc. Natl. Acad. Sci. USA* **2020**, *117*, 10414–10421. [CrossRef] [PubMed]
25. Chi Fru, E.; Rodríguez, N.P.; Partin, C.A.; Lalonde, S.V.; Andersson, P.; Weiss, D.J.; El Albani, A.; Rodushkin, I.; Konhauser, K.O. Cu Isotopes in Marine Black Shales Record the Great Oxidation Event. *Proc. Natl. Acad. Sci. USA* **2016**, *113*, 4941–4946. [CrossRef]
26. Li, L.-G.; Xia, Y.; Zhang, T. Co-Occurrence of Antibiotic and Metal Resistance Genes Revealed in Complete Genome Collection. *ISME J.* **2017**, *11*, 651–662. [CrossRef]
27. Tiquia-Arashiro, S.M. Lead Absorption Mechanisms in Bacteria as Strategies for Lead Bioremediation. *Appl. Microbiol. Biotechnol.* **2018**, *102*, 5437–5444. [CrossRef]
28. Long, D.; Hashmi, M.Z.; Su, X.; Pongpiachan, S. Cr(VI) Reduction by an Extracellular Polymeric Substance (EPS) Produced from a Strain of *Pseudochrobactrum saccharolyticum*. *3 Biotech* **2019**, *9*, 111. [CrossRef]
29. Islam, S.T.; Lam, J.S. Synthesis of Bacterial Polysaccharides via the Wzx/Wzy-Dependent Pathway. *Can. J. Microbiol.* **2014**, *60*, 697–716. [CrossRef]
30. Osemwegie, O.O.; Adetunji, C.O.; Ayeni, E.A.; Adejobi, O.I.; Arise, R.O.; Nwonuma, C.O.; Oghenekaro, A.O. Exopolysaccharides from Bacteria and Fungi: Current Status and Perspectives in Africa. *Heliyon* **2020**, *6*, e04205. [CrossRef]

31. Bhunia, A.; Lahiri, D.; Nag, M.; Upadhye, V.; Pandit, S. Bacterial Biofilm Mediated Bioremediation of Hexavalent Chromium: A Review. *Biocatal. Agric. Biotechnol.* **2022**, *43*, 102397. [CrossRef]
32. Kumawat, T.K.; Kumawat, V.; Sharma, S.; Kandwani, N.; Biyani, M. Applications of EPS in Environmental Bioremediations. In *Microbial Exopolysaccharides as Novel and Significant Biomaterials*; Nadda, A.K., Sajna, K.V., Sharma, S., Eds.; Springer Series on Polymer and Composite Materials; Springer International Publishing: Cham, Switzerland, 2021; pp. 285–302, ISBN 978-3-030-75288-0.
33. Morais, M.G.; Santos, T.D.; Moraes, L.; Vaz, B.S.; Morais, E.G.; Costa, J.A.V. Exopolysaccharides from Microalgae: Production in a Biorefinery Framework and Potential Applications. *Bioresour. Technol. Rep.* **2022**, *18*, 101006. [CrossRef]
34. Poole, K. At the Nexus of Antibiotics and Metals: The Impact of Cu and Zn on Antibiotic Activity and Resistance. *Trends Microbiol.* **2017**, *25*, 820–832. [CrossRef] [PubMed]
35. Zayed, A.; Mansour, M.K.; Sedeek, M.S.; Habib, M.H.; Ulber, R.; Farag, M.A. Rediscovering Bacterial Exopolysaccharides of Terrestrial and Marine Origins: Novel Insights on Their Distribution, Biosynthesis, Biotechnological Production, and Future Perspectives. *Crit. Rev. Biotechnol.* **2022**, *42*, 597–617. [CrossRef] [PubMed]
36. Pushkar, B.; Sevak, P.; Parab, S.; Nilkanth, N. Chromium Pollution and Its Bioremediation Mechanisms in Bacteria: A Review. *J. Environ. Manag.* **2021**, *287*, 112279. [CrossRef]
37. Ward, N.P.; DeNicola, G.M. Sulfur Metabolism and Its Contribution to Malignancy. *Int. Rev. Cell Mol. Biol.* **2019**, *347*, 39–103.
38. Mukhopadhyay, R.; Bhattacharjee, H.; Rosen, B.P. Aquaglyceroporins: Generalized Metalloid Channels. *Biochim. Biophys. Acta* **2014**, *1840*, 1583–1591. [CrossRef]
39. Himeno, S.; Yanagiya, T.; Enomoto, S.; Kondo, Y.; Imura, N. Cellular Cadmium Uptake Mediated by the Transport System for Manganese. *Tohoku J. Exp. Med.* **2002**, *196*, 43–50. [CrossRef]
40. Khan, A.; Singh, P.; Srivastava, A. Synthesis, Nature and Utility of Universal Iron Chelator—Siderophore: A Review. *Microbiol. Res.* **2017**, *212–213*, 103–111. [CrossRef]
41. Gutiérrez-Corona, J.F.; Romo-Rodríguez, P.; Santos-Escobar, F.; Espino-Saldaña, A.E.; Hernández-Escoto, H. Microbial Interactions with Chromium: Basic Biological Processes and Applications in Environmental Biotechnology. *World J. Microbiol. Biotechnol.* **2016**, *32*, 191. [CrossRef]
42. Puzon, G.J.; Roberts, A.G.; Kramer, D.M.; Xun, L. Formation of Soluble Organo-Chromium(III) Complexes after Chromate Reduction in the Presence of Cellular Organics. *Environ. Sci. Technol.* **2005**, *39*, 2811–2817. [CrossRef]
43. Kumar, V.; Dwivedi, S.K. Hexavalent Chromium Stress Response, Reduction Capability and Bioremediation Potential of *Trichoderma* sp. Isolated from Electroplating Wastewater. *Ecotoxicol. Environ. Saf.* **2019**, *185*, 109734. [CrossRef] [PubMed]
44. Torelli, A.; Marieschi, M.; Castagnoli, B.; Zanni, C.; Gorbi, G.; Corradi, M.G. Identification of S2-T A63: A CDNA Fragment Corresponding to a Gene Differentially Expressed in a Cr-Tolerant Strain of the Unicellular Green Alga *Scenedesmus acutus*. *Aquat. Toxicol.* **2008**, *86*, 495–507. [CrossRef] [PubMed]
45. Aufschnaiter, A.; Büttner, S. The Vacuolar Shapes of Ageing: From Function to Morphology. *Biochim. Biophys. Acta Mol. Cell Res.* **2019**, *1866*, 957–970. [CrossRef]
46. Wang, J.; Chen, C. Biosorption of Heavy Metals by Saccharomyces Cerevisiae: A Review. *Biotechnol. Adv.* **2006**, *24*, 427–451. [CrossRef]
47. Lv, Q.; Yan, L.; Jiang, Y. The Importance of Vacuolar Ion Homeostasis and Trafficking in Hyphal Development and Virulence in Candida Albicans. *Front. Microbiol.* **2021**, *12*, 779176. [CrossRef] [PubMed]
48. Pivato, M.; Ballottari, M. *Chlamydomonas reinhardtii* Cellular Compartments and Their Contribution to Intracellular Calcium Signalling. *J. Exp. Bot.* **2021**, *72*, 5312–5335. [CrossRef]
49. Vasanthakumar, T.; Rubinstein, J.L. Structure and Roles of V-Type ATPases. *Trends Biochem. Sci.* **2020**, *45*, 295–307. [CrossRef]
50. Luxmi, R.; Blaby-Haas, C.; Kumar, D.; Rauniyar, N.; King, S.M.; Mains, R.E.; Eipper, B.A. Proteases Shape the Chlamydomonas Secretome: Comparison to Classical Neuropeptide Processing Machinery. *Proteomes* **2018**, *6*, 36. [CrossRef]
51. Parzych, K.R.; Klionsky, D.J. Vacuolar Hydrolysis and Efflux: Current Knowledge and Unanswered Questions. *Autophagy* **2019**, *15*, 212–227. [CrossRef]
52. Deshmukh, M.; Jangam, S.S.; Wankhede, S.B. Potential Use of Microalgal Metallothioneins and Phytochelatins in Bioremediation. In *Microbiome-Assisted Bioremediation*; Elsevier: Amsterdam, The Netherlands, 2024; pp. 367–380, ISBN 9780443219115.
53. Manikandan, A.; Suresh Babu, P.; Shyamalagowri, S.; Kamaraj, M.; Muthukumaran, P.; Aravind, J. Emerging Role of Microalgae in Heavy Metal Bioremediation. *J. Basic Microbiol.* **2022**, *62*, 330–347. [CrossRef]
54. Majhi, P.; Nayak, S.; Samantaray, S.M. Microalgal Bioremediation of Toxic Hexavalent Chromium: A Review. In *Environmental and Agricultural Microbiology: Applications for Sustainability*; Wiley: Hoboken, NJ, USA, 2021; pp. 25–37. [CrossRef]
55. Yang, H.-C.; Fu, H.-L.; Lin, Y.-F.; Rosen, B.P. Pathways of Arsenic Uptake and Efflux. *Curr. Top. Membr.* **2012**, *69*, 325–358. [PubMed]
56. Kumari, N.; Jagadevan, S. Genetic Identification of Arsenate Reductase and Arsenite Oxidase in Redox Transformations Carried out by Arsenic Metabolising Prokaryotes—A Comprehensive Review. *Chemosphere* **2016**, *163*, 400–412. [CrossRef] [PubMed]

57. Pawitwar, S.S.; Nadar, V.S.; Kandegedara, A.; Stemmler, T.L.; Rosen, B.P.; Yoshinaga, M. Biochemical Characterization of ArsI: A Novel C-as Lyase for Degradation of Environmental Organoarsenicals. *Environ. Sci. Technol.* **2017**, *51*, 11115–11125. [CrossRef] [PubMed]
58. Slyemi, D.; Bonnefoy, V. How Prokaryotes Deal with Arsenic(†). *Environ. Microbiol. Rep.* **2012**, *4*, 571–586. [CrossRef]
59. Zhu, Y.-G.; Xue, X.-M.; Kappler, A.; Rosen, B.P.; Meharg, A.A. Linking Genes to Microbial Biogeochemical Cycling: Lessons from Arsenic. *Environ. Sci. Technol.* **2017**, *51*, 7326–7339. [CrossRef]
60. Bhardwaj, A. Understanding the Diversified Microbial Operon Framework Coupled to Arsenic Transformation and Expulsion. *Biologia* **2022**, *77*, 3531–3544. [CrossRef]
61. Zhang, J.; Liu, J.; Zheng, F.; Yu, M.; Shabala, S.; Song, W.-Y. Comparative Analysis of Arsenic Transport and Tolerance Mechanisms: Evolution from Prokaryote to Higher Plants. *Cells* **2022**, *11*, 2741. [CrossRef]
62. Branco, R.; Morais, P.V. Two Superoxide Dismutases from TnOtchr Are Involved in Detoxification of Reactive Oxygen Species Induced by Chromate. *BMC Microbiol.* **2016**, *16*, 27. [CrossRef]
63. Deepa, A.; Mishra, B.K. Microbial Biotransformation of Hexavalent Chromium [Cr(VI)] in Tannery Wastewater. In *Microbial Bioremediation & Biodegradation*; Springer: Singapore, 2020; pp. 143–152, ISBN 9789811518119.
64. Ackerley, D.F.; Gonzalez, C.F.; Keyhan, M.; Blake, R., 2nd; Matin, A. Mechanism of Chromate Reduction by the *Escherichia coli* Protein, NfsA, and the Role of Different Chromate Reductases in Minimizing Oxidative Stress during Chromate Reduction. *Environ. Microbiol.* **2004**, *6*, 851–860. [CrossRef]
65. Ackerley, D.F.; Gonzalez, C.F.; Park, C.H.; Blake, R., 2nd; Keyhan, M.; Matin, A. Chromate-Reducing Properties of Soluble Flavoproteins from *Pseudomonas putida* and *Escherichia coli*. *Appl. Environ. Microbiol.* **2004**, *70*, 873–882. [CrossRef]
66. Han, H.; Ling, Z.; Zhou, T.; Xu, R.; He, Y.; Liu, P.; Li, X. Copper (II) Binding of NAD(P)H- Flavin Oxidoreductase (NfoR) Enhances Its Cr (VI)-Reducing Ability. *Sci. Rep.* **2017**, *7*, 15481. [CrossRef] [PubMed]
67. Shi, X.L.; Dalal, N.S. NADPH-Dependent Flavoenzymes Catalyze One Electron Reduction of Metal Ions and Molecular Oxygen and Generate Hydroxyl Radicals. *FEBS Lett.* **1990**, *276*, 189–191. [CrossRef] [PubMed]
68. Hui, C.-Y.; Ma, B.-C.; Wang, Y.-Q.; Yang, X.-Q.; Cai, J.-M. Designed Bacteria Based on Natural Pbr Operons for Detecting and Detoxifying Environmental Lead: A Mini-Review. *Ecotoxicol. Environ. Saf.* **2023**, *267*, 115662. [CrossRef] [PubMed]
69. Naik, M.M.; Dubey, S.K. Lead Resistant Bacteria: Lead Resistance Mechanisms, Their Applications in Lead Bioremediation and Biomonitoring. *Ecotoxicol. Environ. Saf.* **2013**, *98*, 1–7. [CrossRef]
70. Castañeda-Barba, S.; Top, E.M.; Stalder, T. Plasmids, a Molecular Cornerstone of Antimicrobial Resistance in the One Health Era. *Nat. Rev. Microbiol.* **2024**, *22*, 18–32. [CrossRef]
71. McEvoy, M.M.; Thompson, A.M.G. CusCFBA Copper/Silver Efflux System. In *Encyclopedia of Metalloproteins*; Springer: New York, NY, USA, 2013; pp. 744–748, ISBN 9781461415329.
72. Rensing, C.; McDevitt, S.F. The Copper Metallome in Prokaryotic Cells.Metallomics and the Cell. In *Metal Ions in Life Sciences*; Springer: Dordrecht, The Netherlands, 2013; Volume 12, pp. 417–450.
73. Bondarczuk, K.; Piotrowska-Seget, Z. Molecular Basis of Active Copper Resistance Mechanisms in Gram-Negative Bacteria. *Cell Biol. Toxicol.* **2013**, *29*, 397–405. [CrossRef]
74. Solioz, M.; Abicht, H.K.; Mermod, M.; Mancini, S. Response of Gram-Positive Bacteria to Copper Stress. *J. Biol. Inorg. Chem.* **2010**, *15*, 3–14. [CrossRef]
75. Solioz, M. Copper Homeostasis in Gram-Positive Bacteria. In *SpringerBriefs in Molecular Science*; Springer International Publishing: Cham, Swizterland, 2018; pp. 21–48, ISBN 9783319944388.
76. Achternbosch, M.; Kupsch, C.; Sardemann, G.; Bräutigam, K.-R. Cadmium Flows Caused by the Worldwide Production of Primary Zinc Metal. *J. Ind. Ecol.* **2009**, *13*, 438–454. [CrossRef]
77. Khan, Z.; Elahi, A.; Bukhari, D.A.; Rehman, A. Cadmium Sources, Toxicity, Resistance and Removal by Microorganisms-A Potential Strategy for Cadmium Eradication. *J. Saudi Chem. Soc.* **2022**, *26*, 101569. [CrossRef]
78. Alvarez-Ortega, C.; Olivares, J.; Martínez, J.L. RND Multidrug Efflux Pumps: What Are They Good For? *Front. Microbiol.* **2013**, *4*, 7. [CrossRef]
79. Liu, H.; Zhang, Y.; Wang, Y.; Xie, X.; Shi, Q. The Connection between Czc and Cad Systems Involved in Cadmium Resistance in *Pseudomonas putida*. *Int. J. Mol. Sci.* **2021**, *22*, 9697. [CrossRef] [PubMed]
80. Abbas, S.Z.; Rafatullah, M.; Hossain, K.; Ismail, N.; Tajarudin, H.A.; Abdul Khalil, H.P.S. A Review on Mechanism and Future Perspectives of Cadmium-Resistant Bacteria. *Int. J. Environ. Sci. Technol.* **2018**, *15*, 243–262. [CrossRef]
81. Hussain, S.; Khan, M.; Sheikh, T.M.M.; Mumtaz, M.Z.; Chohan, T.A.; Shamim, S.; Liu, Y. Zinc Essentiality, Toxicity, and Its Bacterial Bioremediation: A Comprehensive Insight. *Front. Microbiol.* **2022**, *13*, 900740. [CrossRef] [PubMed]
82. Kandari, D.; Joshi, H.; Bhatnagar, R. Zur: Zinc-Sensing Transcriptional Regulator in a Diverse Set of Bacterial Species. *Pathogens* **2021**, *10*, 344. [CrossRef]
83. Kambe, T. Molecular Architecture and Function of ZnT Transporters. *Curr. Top. Membr.* **2012**, *69*, 199–220.

84. Bui, H.B.; Inaba, K. Structures, Mechanisms, and Physiological Functions of Zinc Transporters in Different Biological Kingdoms. *Int. J. Mol. Sci.* **2024**, *25*, 3045. [CrossRef]
85. Dash, H.R.; Das, S. Bioremediation of Mercury and the Importance of Bacterial Mer Genes. *Int. Biodeterior. Biodegrad.* **2012**, *75*, 207–213. [CrossRef]
86. Barkay, T.; Miller, S.M.; Summers, A.O. Bacterial Mercury Resistance from Atoms to Ecosystems. *FEMS Microbiol. Rev.* **2003**, *27*, 355–384. [CrossRef]
87. Seiler, C.; Berendonk, T.U. Heavy Metal Driven Co-Selection of Antibiotic Resistance in Soil and Water Bodies Impacted by Agriculture and Aquaculture. *Front. Microbiol.* **2012**, *3*, 399. [CrossRef]
88. Adhikary, S.; Saha, J.; Dutta, P.; Pal, A. Bacterial Homeostasis and Tolerance to Potentially Toxic Metals and Metalloids through Diverse Transporters: Metal-Specific Insights. *Geomicrobiol. J.* **2024**, *41*, 496–518. [CrossRef]
89. Chandrangsu, P.; Rensing, C.; Helmann, J.D. Metal Homeostasis and Resistance in Bacteria. *Nat. Rev. Microbiol.* **2017**, *15*, 338–350. [CrossRef] [PubMed]
90. Gillieatt, B.F.; Coleman, N.V. Unravelling the Mechanisms of Antibiotic and Heavy Metal Resistance Co-Selection in Environmental Bacteria. *FEMS Microbiol. Rev.* **2024**, *48*, fuae017. [CrossRef] [PubMed]
91. Larsson, D.G.J.; Flach, C.-F. Antibiotic Resistance in the Environment. *Nat. Rev. Microbiol.* **2022**, *20*, 257–269. [CrossRef] [PubMed]
92. Maurya, A.P.; Rajkumari, J.; Bhattacharjee, A.; Pandey, P. Development, Spread and Persistence of Antibiotic Resistance Genes (ARGs) in the Soil Microbiomes through Co-Selection. *Rev. Environ. Health* **2020**, *35*, 371–378. [CrossRef]
93. Nguyen, T.H.T.; Nguyen, H.D.; Le, M.H.; Nguyen, T.T.H.; Nguyen, T.D.; Nguyen, D.L.; Nguyen, Q.H.; Nguyen, T.K.O.; Michalet, S.; Dijoux-Franca, M.-G.; et al. Efflux Pump Inhibitors in Controlling Antibiotic Resistance: Outlook under a Heavy Metal Contamination Context. *Molecules* **2023**, *28*, 2912. [CrossRef]
94. Rillig, M.C.; Li, C.; Rodríguez Del Río, Á.; Zhu, Y.-G.; Jin, L. Elevated Levels of Antibiotic Resistance Genes as a Factor of Human-Caused Global Environmental Change. *Glob. Change Biol.* **2024**, *30*, e17419. [CrossRef]
95. Carvalho, R.; Aburjaile, F.; Canario, M.; Nascimento, A.M.A.; Chartone-Souza, E.; de Jesus, L.; Zamyatnin, A.A., Jr.; Brenig, B.; Barh, D.; Ghosh, P.; et al. Genomic Characterization of Multidrug-Resistant *Escherichia coli* BH100 Sub-Strains. *Front. Microbiol.* **2020**, *11*, 549254. [CrossRef]
96. Tuffin, I.M.; de Groot, P.; Deane, S.M.; Rawlings, D.E. An Unusual Tn21-like Transposon Containing an Ars Operon Is Present in Highly Arsenic-Resistant Strains of the Biomining Bacterium Acidithiobacillus Caldus. *Microbiology* **2005**, *151*, 3027–3039. [CrossRef]
97. Mergeay, M.; Van Houdt, R. Cupriavidus Metallidurans CH34, a Historical Perspective on Its Discovery, Characterization and Metal Resistance. *FEMS Microbiol. Ecol.* **2021**, *97*, fiaa247. [CrossRef]
98. Yang, S.; Deng, W.; Liu, S.; Yu, X.; Mustafa, G.R.; Chen, S.; He, L.; Ao, X.; Yang, Y.; Zhou, K.; et al. Presence of Heavy Metal Resistance Genes in *Escherichia coli* and *Salmonella* Isolates and Analysis of Resistance Gene Structure in *E. coli* E308. *J. Glob. Antimicrob. Resist.* **2020**, *21*, 420–426. [CrossRef]
99. Urquhart, A.S.; Chong, N.F.; Yang, Y.; Idnurm, A. A Large Transposable Element Mediates Metal Resistance in the Fungus *Paecilomyces variotii*. *Curr. Biol.* **2022**, *32*, 937–950.e5. [CrossRef] [PubMed]
100. Bukowski, M.; Piwowarczyk, R.; Madry, A.; Zagorski-Przybylo, R.; Hydzik, M.; Wladyka, B. Prevalence of Antibiotic and Heavy Metal Resistance Determinants and Virulence-Related Genetic Elements in Plasmids of *Staphylococcus aureus*. *Front. Microbiol.* **2019**, *10*, 805. [CrossRef] [PubMed]
101. Galetti, R.; Penha Filho, R.A.C.; Ferreira, J.C.; Varani, A.M.; Darini, A.L.C. Antibiotic Resistance and Heavy Metal Tolerance Plasmids: The Antimicrobial Bulletproof Properties of *Escherichia fergusonii* Isolated from Poultry. *Infect. Drug Resist.* **2019**, *12*, 1029–1033. [CrossRef] [PubMed]
102. Parsons, C.; Lee, S.; Kathariou, S. Dissemination and Conservation of Cadmium and Arsenic Resistance Determinants in *Listeria* and Other Gram-Positive Bacteria. *Mol. Microbiol.* **2020**, *113*, 560–569. [CrossRef] [PubMed]
103. Souza, S.S.R.; Turcotte, M.R.; Li, J.; Zhang, X.; Wolfe, K.L.; Gao, F.; Benton, C.S.; Andam, C.P. Population Analysis of Heavy Metal and Biocide Resistance Genes in *Salmonella enterica* from Human Clinical Cases in New Hampshire, United States. *Front. Microbiol.* **2022**, *13*, 983083. [CrossRef] [PubMed]
104. Mei, L.; Song, Y.; Liu, D.; Li, Y.; Liu, L.; Yu, K.; Jiang, M.; Wang, D.; Wei, Q. Characterization of a Mobilizable Megaplasmid Carrying Multiple Resistance Genes from a Clinical Isolate of *Pseudomonas aeruginosa*. *Front. Microbiol.* **2023**, *14*, 1293443. [CrossRef]
105. Rankin, D.J.; Rocha, E.P.C.; Brown, S.P. What Traits Are Carried on Mobile Genetic Elements, and Why? *Heredity* **2011**, *106*, 1–10. [CrossRef]
106. Feng, G.; Huang, H.; Chen, Y. Effects of Emerging Pollutants on the Occurrence and Transfer of Antibiotic Resistance Genes: A Review. *J. Hazard. Mater.* **2021**, *420*, 126602. [CrossRef]
107. Bahram, M.; Hildebrand, F.; Forslund, S.K.; Anderson, J.L.; Soudzilovskaia, N.A.; Bodegom, P.M.; Bengtsson-Palme, J.; Anslan, S.; Coelho, L.P.; Harend, H.; et al. Structure and Function of the Global Topsoil Microbiome. *Nature* **2018**, *560*, 233–237. [CrossRef]

108. Apreja, M.; Sharma, A.; Balda, S.; Kataria, K.; Capalash, N.; Sharma, P. Antibiotic Residues in Environment: Antimicrobial Resistance Development, Ecological Risks, and Bioremediation. *Environ. Sci. Pollut. Res. Int.* **2022**, *29*, 3355–3371. [CrossRef]
109. Li, L.; Meng, D.; Yin, H.; Zhang, T.; Liu, Y. Genome-Resolved Metagenomics Provides Insights into the Ecological Roles of the Keystone Taxa in Heavy-Metal-Contaminated Soils. *Front. Microbiol.* **2023**, *14*, 1203164. [CrossRef] [PubMed]
110. Zhang, H.; Huang, J.; Zeng, W.; Xiao, Q.; Zhu, Y.; Kong, W.; Zou, J.; Liu, T.; Yin, H. Dissecting the Metal Resistance Genes Contributed by Virome from Mining-Affected Metal Contaminated Soils. *Front. Environ. Sci.* **2023**, *11*, 1182673. [CrossRef]
111. Khatoon, Z.; Orozco-Mosqueda, M.D.C.; Santoyo, G. Microbial Contributions to Heavy Metal Phytoremediation in Agricultural Soils: A Review. *Microorganisms* **2024**, *12*, 1945. [CrossRef] [PubMed]
112. Shoumik, B.A.A.; Khan, M.Z.; Mahmud, U.; Sultan, M.T.; Baloch, M.Y.J. Bioremediation of Heavy Metals in Soil by Rhizobacteria for Sustainable Agriculture and Food Security. In *Bio-Organic Amendments for Heavy Metal Remediation*; Elsevier: Amsterdam, The Netherlands, 2024; pp. 561–600, ISBN 9780443216107.
113. Saharan, B.S.; Chaudhary, T.; Mandal, B.S.; Kumar, D.; Kumar, R.; Sadh, P.K.; Duhan, J.S. Microbe-Plant Interactions Targeting Metal Stress: New Dimensions for Bioremediation Applications. *J. Xenobiot.* **2023**, *13*, 252–269. [CrossRef]
114. Laborda, P.; Sanz-García, F.; Ochoa-Sánchez, L.E.; Gil-Gil, T.; Hernando-Amado, S.; Martínez, J.L. Wildlife and Antibiotic Resistance. *Front. Cell. Infect. Microbiol.* **2022**, *12*, 873989. [CrossRef]
115. Sun, H.; Li, H.; Zhang, X.; Liu, Y.; Chen, H.; Zheng, L.; Zhai, Y.; Zheng, H. The Honeybee Gut Resistome and Its Role in Antibiotic Resistance Dissemination. *Integr. Zool.* **2023**, *18*, 1014–1026. [CrossRef]
116. Chen, C.; Li, J.; Chen, P.; Ding, R.; Zhang, P.; Li, X. Occurrence of Antibiotics and Antibiotic Resistances in Soils from Wastewater Irrigation Areas in Beijing and Tianjin, China. *Environ. Pollut.* **2014**, *193*, 94–101. [CrossRef]
117. Zhang, R.; Yang, S.; An, Y.; Wang, Y.; Lei, Y.; Song, L. Antibiotics and Antibiotic Resistance Genes in Landfills: A Review. *Sci. Total Environ.* **2022**, *806*, 150647. [CrossRef]
118. Sanganyado, E.; Gwenzi, W. Antibiotic Resistance in Drinking Water Systems: Occurrence, Removal, and Human Health Risks. *Sci. Total Environ.* **2019**, *669*, 785–797. [CrossRef]
119. Argudín, M.A.; Hoefer, A.; Butaye, P. Heavy Metal Resistance in Bacteria from Animals. *Res. Veterin. Sci.* **2019**, *122*, 132–147. [CrossRef]
120. Karwowska, E. Antibiotic Resistance in the Farming Environment. *Appl. Sci.* **2024**, *14*, 5776. [CrossRef]
121. Hynninen, A.; Touzé, T.; Pitkänen, L.; Mengin-Lecreulx, D.; Virta, M. An Efflux Transporter PbrA and a Phosphatase PbrB Cooperate in a Lead-Resistance Mechanism in Bacteria. *Mol. Microbiol.* **2009**, *74*, 384–394. [CrossRef]
122. van Boeckel, T.; Pires, J.; Silvester, R.; Zhao, C.; Song, J.; Criscuolo, N.; Gilbert, M.; Bonhoeffer, S.; Laxminarayan, R. Global Trends in Antimicrobial Resistance in Animals in Low- and Middle-Income Countries. *Int. J. Infect. Dis.* **2020**, *101*, 19. [CrossRef]
123. Hovorukha, V.; Moliszewska, E.; Havryliuk, O.; Bida, I.; Tashyrev, O. Metal Resistance of Microorganisms as a Crucial Factor for Their Homeostasis and Sustainable Environment. *Sustainability* **2024**, *16*, 9655. [CrossRef]
124. Sazykin, I.; Khmelevtsova, L.; Azhogina, T.; Sazykina, M. Heavy Metals Influence on the Bacterial Community of Soils: A Review. *Agriculture* **2023**, *13*, 653. [CrossRef]
125. Oyetibo, G.O.; Enahoro, J.A.; Ikwubuzo, C.A.; Ukwuoma, C.S. Microbiome of Highly Polluted Coal Mine Drainage from Onyeama, Nigeria, and Its Potential for Sequestrating Toxic Heavy Metals. *Sci. Rep.* **2021**, *11*, 17496. [CrossRef]
126. Thavamani, P.; Samkumar, R.A.; Satheesh, V.; Subashchandrabose, S.R.; Ramadass, K.; Naidu, R.; Venkateswarlu, K.; Megharaj, M. Microbes from Mined Sites: Harnessing Their Potential for Reclamation of Derelict Mine Sites. *Environ. Pollut.* **2017**, *230*, 495–505. [CrossRef]
127. Cason, C.; D'Accolti, M.; Soffritti, I.; Mazzacane, S.; Comar, M.; Caselli, E. Next-Generation Sequencing and PCR Technologies in Monitoring the Hospital Microbiome and Its Drug Resistance. *Front. Microbiol.* **2022**, *13*, 969863. [CrossRef]
128. Bazzi, W.; Abou Fayad, A.G.; Nasser, A.; Haraoui, L.-P.; Dewachi, O.; Abou-Sitta, G.; Nguyen, V.-K.; Abara, A.; Karah, N.; Landecker, H.; et al. Heavy Metal Toxicity in Armed Conflicts Potentiates AMR in A. Baumannii by Selecting for Antibiotic and Heavy Metal Co-Resistance Mechanisms. *Front. Microbiol.* **2020**, *11*, 68. [CrossRef]
129. Eggers, S.; Safdar, N.; Kates, A.; Sethi, A.K.; Peppard, P.E.; Kanarek, M.S.; Malecki, K.M.C. Urinary Lead Level and Colonization by Antibiotic Resistant Bacteria: Evidence from a Population-Based Study. *Environ. Epidemiol.* **2021**, *5*, e175. [CrossRef]

MDPI AG
Grosspeteranlage 5
4052 Basel
Switzerland
Tel.: +41 61 683 77 34

Genes Editorial Office
E-mail: genes@mdpi.com
www.mdpi.com/journal/genes

www.ingramcontent.com/pod-product-compliance
Lightning Source LLC
Chambersburg PA
CBHW061606210926
51850CB00009B/2224
* 9 7 8 3 7 2 5 8 8 0 6 7 6 *